WIRELESS COMMUNICATIONS IN THE 21ST CENTURY

WIRELESS COMMUNICATIONS IN THE 21ST CENTURY

EDITED BY

Mansoor Shafi
Telecom New Zealand, Wellington, New Zealand

Shigeaki Ogose
Kagawa University, Takamatsu, Japan

Takeshi Hattori
Sophia University, Tokyo, Japan

IEEE SERIES ON
**DIGITAL
& MOBILE
COMMUNICATION**

John B. Anderson, *Series Editor*

IEEE PRESS

A JOHN WILEY & SONS, INC. PUBLICATION

This book is printed on acid-free paper. ∞

For ordering and customer service, call 1-800-CALL WILEY.

Library of Congress Cataloging-in-Publication is available.
ISBN 0-471-15041-X

Printed in the United States of America.
10 9 8 7 6 5 4 3 2 1

CONTENTS

▆▆▆▆▆ PREFACE

Wireless communications is an area of communications that is expanding very rapidly and is one of the most promising areas of research. Third-generation wireless systems, which provide streaming video, access to the internet and much more, are now becoming commercial realities. Standardization of systems beyond third-generation is already underway. The aim of the new wireless networks is to make personal communications available anywhere, anytime. This book consists of individual chapters written by notable specialists in their fields. It has a bias toward the physical layer. The key objectives of this book are to provide the reader with a broad overview based on:

- The kinds of services can we expect from wireless networks in the 21st century,
- The standardization efforts underway to ensure that such networks will be commercially realized,
- The technology building blocks that are needed, and
- Examples of some new systems that are being deployed.

Preparing an overview text with these broad aims also poses a key difficulty, especially in a dynamic area such as wireless communications. This is because the ideas that were "new" some years ago (when work on this text began) are somewhat dated now. Furthermore, "new" areas are always emerging that may be not covered in this book at all.

This book is aimed at a wide audience consisting of researchers, practising engineers, and design engineers. With this intention, the contributing authors have prepared specialist articles that provide a comprehensive tutorial style overview of the scope of the article. The first part consists of an introductory article written by the editors. This is followed by three visionary chapters on wireless network developments. Standardization efforts are covered in Part 2 by two chapters. Part 3 consists of four chapters on propagation issues, because the vagaries on the radio channel continue to provide a continuous challenge to radio engineers. Part 4 consists of 5 chapters on key technologies that form the building blocks of the physical layer of a wireless system. Parts 5 and 6 consist of 5 chapters on the examples of the new systems being deployed.

Our last words in this Preface must be ones of thanks to our respective families, who have provided us moral support and the friendship needed during this project. We also wish to express our gratitude to our respective organizations for providing us the facilities to complete this work. We would like to acknowledge the efforts of Ms. Lisa Van Horn and her colleagues at John Wiley & Sons, Inc. in the final completion of this work.

MANSOOR SHAFI
SHIGEAKI OGOSE
TAKESHI HATTORI

January 2002

WIRELESS COMMUNICATIONS IN THE 21ST CENTURY

CHAPTER 1

Introduction

MANSOOR SHAFI, SHIGEAKI OGOSE, and KEITH BUTTERWORTH

1.1 HISTORY OF MOBILE RADIO COMMUNICATIONS

More than a century ago, in 1898, Lord Kelvin asked Guglielmo Marconi to send a message on his wireless telegraph. This was to become the world's first commercial wireless telegram. Since then, due to the efforts of many notable scientists and organizations [1], radio is an essential part of our daily lives today. Applications such as, audio and video broadcasting, fixed and mobile communication systems, radar, radio navigation systems, and so on are almost taken for granted.

This book focuses on a part of radio communications—the area of mobile communications. Mobile (wireless communications) consists of a communication system where at least one user is on the move. Mobile systems may have a terrestrial component and/or a satellite component. Our focus here is largely on the terrestrial component.

Today, wireless communications have captured the love of the media. Articles on this subject often appear in the worldwide daily newspapers. Numerous trade conferences and seminars, and so on are frequently held on this subject worldwide. Each year, the IEEE sponsors conferences for the ICC, GlobeCom, VTC, PIMRC, ICUPC which focus on mobile communications. It is almost impossible to keep track of the technical journals and magazines, symposia, and so on concerning this subject. It is clear, therefore, that wireless communications are by any measure, one of the most rapidly growing segment of the telecommunications market.

The first mobile radio systems were introduced by the military and were limited only to voice communication systems [2]. The handsets provided very poor voice quality, low talk time (typically some tens of minutes), low stand-by time (at most, a couple of hours) and were rather bulky in size. The first public cellular phone system, known as AMPS, was introduced in 1979 in the United States [3]. This was followed shortly by the introduction of the NMT systems in Scandinavia and the TACS and NAMTS systems in the UK and Japan, respectively. In Europe, there was a plethora of country-specific systems each being totally different from the other. These first-generation systems were based on analog FM.

Wireless Communications in the 21st Century, Edited by Shafi, Ogose, and Hattori.
ISBN 0-471-155041-X   © 2002 by the IEEE.

In the 1990s, second-generation mobile systems, such as GSM, PDC, IS-54 (now succeeded by IS-136), and IS-95 systems were introduced (see [4] Cox for a table of characteristics of the second-generation systems and also the dedicated chapters of [5]). All these systems are now commercially successful and deployed in many parts of the world—more than 110 countries with subscriber numbers reaching in excess of 400 million [6].

1.2   TELECOMMUNICATION NEEDS FOR THE 21ST CENTURY

Present day telecommunication services are dominated by voice. The public switched telephone network (PSTN)—almost taken for granted today—was built on the "Field of Dreams" concept; "if we build it, they will come" and they (subscribers) came, indeed. As we move to the next millennium, telecommunication needs of tomorrow are less clear. Telecommunication operators worldwide are hypothesizing and forecasting the telecommunication needs of the 21st Century. A few observations may be made [7]:

- The world is becoming a global village with the advent of satellite communications, CNN, and the Internet;
- Communications will involve the concurrent use of various modes (voice, data, video): multimedia communications;
- Information, and therefore bandwidth needs, are exponentially increasing; and
- People want to be free from tethers: physical connection to the networks.

Regardless of the difficulties in forecasting tomorrow's needs, operators are considering the introduction of networks to support the introduction of broadband services (with no limit on the numerical value defining the word "broad"). Access to the Internet and the World Wide Web is growing rapidly; this has the potential to significantly change the sociology of work and personal lives.

The demand for mobility continues to surpass all forecasts and prove them wrong. Many industry practitioners believe that by the year 2005, the number of mobile phones worldwide will exceed 1 billion (equivalent to one phone per four persons). The mobile network will undoubtedly continue to provide voice-based services but also provide a slimmed-down version of all the fixed network capability

Wireless Internet access presents formidable challenges to industry practitioners and researchers. The wireless systems of today are not really designed to provide high-speed access (and that too via an end-to-end packet network). Mobile wireless data applications may be categorized in the increasing order of complexity [8]:

Simple Messaging: Text messaging is already available today on almost all second-generation mobile systems. Data speeds of 9.6 or 14.4 kbps are required.

Basic Access to Internet: Such as, downloading weather, stocks, and news, and so on. Already these features are becoming available via Wireless Application Protocol (WAP)-capable. The i-mode service in Japan has in excess of 12 million customers and permits low-cost access to e-mail services and other Web content customized for i-mode.

Network-Enhanced Applications: These will require interaction with the network and the user. A high degree of network intelligence is required. Examples are various location-sensitive services. The data rate requirement for these services are expected to vary from modest (10 kbps) to many tens of kbps.

Secure Communications: A number of applications will require secure access to corporate LANs, electronic commerce, and so on. These applications may need high bandwidth access and may require the data be encrypted.

Advanced Access to Internet: Considerable enhanced Internet access encompassing high-speed data access—many hundred of kbps over an end-to-end packetized network. Voice over IP (VoIP) would also be available at this stage.

1.3 DATA RATE ROAD MAP TO 3G

The wireless systems are all evolving to provide broadband data rate capability besides voiced. Table 1.1 lists the maximum data rates per user that are achieved by the various technologies under ideal conditions. When user numbers increase, and if all the users share the same carrier, the data rate per user will decrease. New services that utilize the data speed capability (e.g., internet-based services, video services, location-based services) are going to provide a significantly broader and enhanced range of services in a mobile environment, besides just voice.

1.4 MOBILE NETWORKS OF TOMORROW

Architecture

In order to provide broadband multimedia communications, mobile networks are aiming to support bit rates up to 384 kbps and further up to 2 Mbps [9]. The International Telecommunications Union has just approved the detailed specifications of IMT-2000, a family of radio interfaces that will provide the high data rates [10]. Standards for systems beyond IMT-2000 are also currently being drafted by ITU-R WP 8F. Wireless ATM is aiming to provide tens of Mbps per user in limited mobility (5–10 km/h) environments [9,11]; mainly for portable computing and multimedia devices.

In order to realize the multimedia aspect of the communications in a cost-effective manner, one needs to examine if the present day networks are built around a suitable architecture. The architecture of today's mobile networks shown in Fig. 1.1 is optimized for voice and is based on the principle of circuit-switched calls with separate packet-switched data components handling data calls. This means that:

- The radio resources in the air interface are maintained throughout the call regardless of the state of activity on the call.
- Voice calls are processed by the mobile services switching center (MSC) that also performs mobility management and radio resource management functions.
- Interconnection with the PSTN is via the MSC.
- Subscriber management is done via the HLR/VLR (location registers).

Table 1.1 Network Technology Migration Paths and their Associated Data Speeds

Technology		Maximum Data Rates	End User Data Rates
European Mobile Technologies (GSM)	GSM circuit switched	9.6–14.4 kbps	10–56 kbps
	GPRS	115 kbps (8 channels)	
	WCDMA	2 Mbps stationary	50 kbps uplink
		384 kbps mobile	150–200 kbps downlink
North American Mobile Technologies (CDMA)	CDMA circuit switched	9.6–14.4 kbps	
	CDMA 1×	153 kbps	90–130 kbps (depending on the numbers of users and distance from base station)
	CDMA 1×EV DO (Data Only)	2.4 Mbps	700 kbps
	CDMA 1×EV DV (Data and Voice)	3–5 Mbps	>1 Mbps
North American Mobile Technologies (TDMA)	TDMA circuit switched	14.4 kbps	
	EDGE	64 kbps uplink (initial rollout)	Initial rollout in 2001/2002: 45–50 kbps uplink, 80–90 kbps downlink
		384 kbps	2003: 45–50 kbps uplink, 150–200 kbps downlink
CDPD		19.2 kbps	
Analog		9.6 kbps	4.8–9.6 kbps

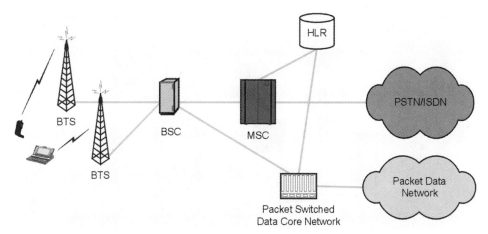

FIGURE 1.1 Architecture of today's mobile voice and data cellular networks.

- Data calls are handled by separate packet-switched data components which enable connectivity to the desired source of data—say, an ISP or connection to the packet data network.
- There is little ability for the operator to control the quality of service (QoS) for a particular application.

Telecom operators worldwide have on the average 3–4 parallel networks that are optimized for a specific application, such as the PSTN is used for voice and enhanced services, the mobile network is used for mobile services, there may be a data network, a network for video services, or a network for Internet services, and so on. All these networks have their own interfaces, protocols, management, and support systems. Such a vast overlay of parallel networks clearly results in high costs.

To support multimedia communications, a session may concurrently consist of calls involving the various independent constituent networks. Each of the calls may also have a different QoS. This could be quite cumbersome and expensive to realize on present-day networks.

There are two dominant core network footprints in the world today. These are ANSI-41 and GSM-MAP and their respective enhancements. The architecture of IMT-2000 systems proposed is based around the respective enhancements of ANSI-41 or GSM-MAP.

The IP has the advantage of seamlessly interconnecting dissimilar networks into a global integrated network by offering a common interface to higher protocol layers. Also, the Internet is now rapidly expanding its domain to include new media and networks—wireless media and wireless networks are no exception! Therefore, IP offers a low-cost way of integrating the present legacy networks, introducing new capabilities, and enabling multimedia communications. Systems beyond the present versions of IMT-2000 [10] are based on existing or evolving Internet Engineering Task Force (IETF) protocols [12]. The architecture of these systems is based on a common IP core network that is independent of access technology and will provide end-to-end IP services and will work with both legacy core networks and the PSTN.

The architecture of a future generation mobile voice and data cellular network is shown in Fig. 1.2. This network consists of a common IP core network supporting multiple IMT-2000 wireless technologies—in this case cdma2000 and Universal Mobile Telecommuni-

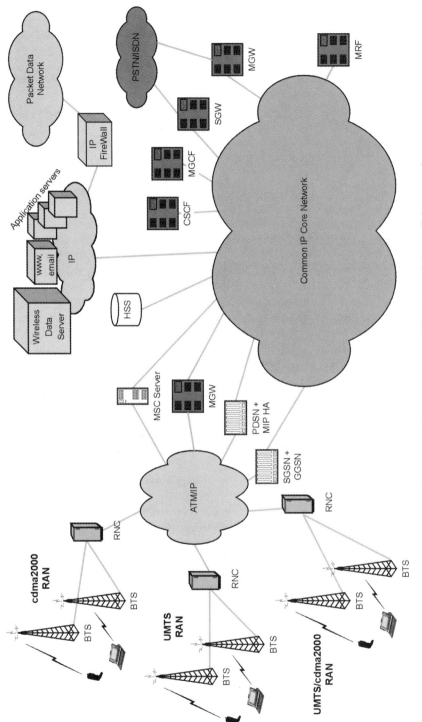

FIGURE 1.2 Architecture of third-generation mobile voice and data cellular networks.

cations System (UMTS). There are a number of new elements such as Call State Control Functions (CSCF), Media Gateway Control Functions (MGCF), Media Resource Functions (MRF). Connectoin to the PSTN is achieved via Signalling Gateways (SGW) and Media Gateways (MGW). This architecture is based on the following principles:

- A single global all-IP core network to be independent of access technology (multiple IMT-2000 wireless technologies, wireless LANs, wireline access technologies all connect to the same network. Note that Fig. 1.2 only shows IMT-2000 radio access networks connected to the common IP core network). The access network connects to the core network via an Access Gateway. This allows the access network technology to be hidden from the core network
- Core network is defined by IP-based protocols and is designed with IP-based multimedia services
- Embrace IETF protocols such as:
  - RADIUS for Authentication, Authorization, Accounting
  - SIP or H.323 for call control
- Separation of services, control, and transport
- All interfaces in the access and core networks to be made open to enable plug and play
- Scalable distributed architecture
- Quality (flexibility to apply QoS to a wide variety of services), reliability, and adoption of Internet security
- Feature servers provide the necessary intelligence to realize the particular application

Much work remains to be done to translate the above vision into robust architectures that will in turn be used in commercial hardware.

1.5   4G MOBILE SYSTEMS

There is no agreed-upon definition for 4G mobile systems. Wireless systems beyond 3G will consist of a layered combination of different access technologies:

- Cellular systems (e.g., existing 2G and 3G systems for wide area mobility)
- Wireless LANs (e.g., IEEE 802.11(a), 802.11(b), HIPERLANs for dedicated indoor applications)
- Personal LANs for short range and low mobility applications (e.g. Bluetooth, IrDA, etc.) around a room in the office or at home.

These access systems will be connected via a common IP-based core network that will also handle working between the different systems. The core network will enable inter and intra access handover. The European countries are also considering digital video, and audio broadcasting accessed via the common IP core network [13].

The peak bit rates of 3G systems are around 10 times more than 2G/2.5G systems. Fig. 1.3 shows the mobility and bit rate perspective of 4G systems [14]. The 4G systems may be expected to provide 10 times higher data speeds relative to 3G systems [14]. It is

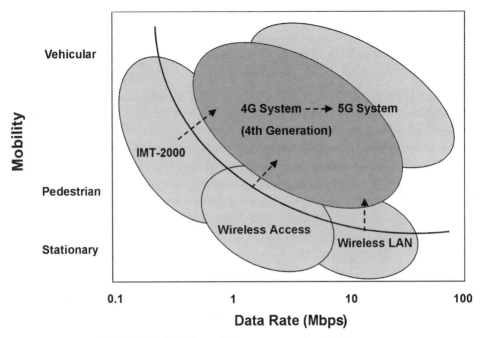

FIGURE 1.3 Mobility and bit rate perspective of 4G systems.

expected that the above layered combination will result in user bit rates of 2 Mbps for vehicular and 20 Mbps for indoor applications. 4G systems must also meet the requirements of next generation Internet through compliance with IPv6, Mobile IP, QoS control, and so on.

The 4G systems will need a fresh approach towards system design. The following key aspects of mobile systems design pose the need for breakthroughs:

Aspect	Challenge
Multiple access scheme	CDMA direct spread may require extra wideband spreading (~ 100 MHz).
	TDMA will need high-speed equalizers.
	OFDM will need linearity of amplifiers.
Multiple antennas	Receiver complexity and challenges to battery life need to be managed for MIMO systems.
	Smart antennas at the terminal will also impose constraints on handset size and battery life.
IP Header	IP header will required almost doubling of the information to be carried. IPv6 will further degrade in this situation.
	Header compression is needed. The performance of a header compression in challenging radio frequency (RF) conditions will also be an issue.
MAC layer	Resource allocation to efficiently manage varying QoS needs.
Multiple bands and systems	The ability to access various types of access systems that use different RF and basebands will need software radios for both RF and baseband

1.6 HANDSETS

Since the advent of commercial mobile telephony, handsets have undergone many significant improvements in the following areas:

- *Cost* Low range handsets now cost around US $100. This is one of the major forces driving the rapid increase of users.
- *Size* Handsets available today have a weight of less than 100 g and a volume of less than 80 cc. The lightest cellular terminal in the world (as of August 1999) was 59 g.
- *Features* Recent handsets can provide new features, such as text message displays including e-mail, phone book of numbers, automatic power on/off, alarm clock, voice recording, and so on. There are other enhanced features that can be activated by keystrokes. Dialing function based on voice recognition is already introduced in the Japanese PDC. WAP and i-mode capable phones that enable Internet access, e-mail, and various location-based services are currently being introduced to the market. The current array of cell phones is already truly bewildering. A recent issue of the *Time Magazine* [15] states that the following additional features are already available in handsets:
  - Built in FM radio
  - Combined key board and PDA functions with Web access and voice recognition
  - Support of micro browsers for the web
  - Palm pilots with wireless modem capability that enables e-mail access and access to the web, stock trades, weather information access, etc.
- *Battery life* Mobile communication terminals use batteries as the source of electrical energy. Longer battery lifetime is required for user convenience. The advent of high-energy density rechargeable batteries and low power consumption circuit components, such as LSI, and the function of intermittent activated operation all help to realize longer battery life. The three main areas of large power consumption are, lighting the display, data transfer to the LCD controller, and running a higher color depth display. The standby time is impacted by new activities such as, MP3 players, radio, and games. Nominal Ni-MH (nickel metal hydride) type batteries achieve the

FIGURE 1.4 FOMA "Introductory Service" handsets. © NTT DoCoMo Technical Journal, Vol. 3, No. 2, 2001.

capacity of 35 Wh/lb [11]. As a result, terminals with a talk time of 2–5 hours and stand-by time of up to several hundred hours are already developed. The current trend is towards lithium-ion (Li-Ion) batteries which realize twice the energy for the same mass, albeit at a higher cost. In the future Li-P (lithium polymer) technology is likely to be used with yet higher energy density.

- *Shape/style*   Variety of shapes/styles of handsets will give us the wider selection. People can choose their handsets based on their own preference. Flip-flap, hold-typed terminals and handsets with special colors/cases are already popular among youth in Japan and the United States.

It is difficult to predict the types of terminals that will support IMT-2000. Terminals for 3G mobile communication systems will have the capability of handling motion pictures as well as text messages. Various DoCoMo terminals are shown in Fig 1.4. Indications are that the future handsets may be of four different types:

- Basic: voice, messaging and data, WAP
- Standard: voice, 2-way video, Bluetooth, (and IrDA) with browser, WAP
- High-end: high-speed data, Bluetooth, voice capable
  - Industrial and portable
  - wearable and body friendly
  - PC card
- Data appliances: high-end, niche, customization, including PC card, camera, video, storage capability

Besides these capabilities, phones are expected to be multimode (support of new and legacy modes), multiband (support of various frequency bands), and multienvironment (support of various environments, e.g., satellite/terrestrial, etc.).

Software defined radio (SDR) technology is one of the promising methods to realize multimode and multiband terminals [17]. SDR is the technology that makes use of the digital signal processing method. Digital signal processor (DSP) chips operate according to the software that is downloaded through the air interface from the base station. The major subjects to realize SDR-based handsets are adoption of low power consumption high-speed/high-resolution analog to digital converters (ADC) and DSP chips [18]. ADCs with the resolution of 14–18 bits under the sampling rate of 100 MSPS (million samples per second) to 10 GSPS (giga samples per second) and DSP chips with the processing speed of several thousand MIPS (million instructions per second) are necessary. Efficient software download method is another important subject. Adoption of an improved Java networking programming method could be one of the solutions for this purpose. Improvement is necessary because the Java method requires larger memory size and lower speed of processing for completion.

1.7   MOBILITY MANAGEMENT IN AN IP WORLD

As noted above wireless networks may be used to support IP-based services. How does one manage mobility in an IP-centric world? In a traditional mobile network, mobility management is done via home and visitor location registers. In the wireline world, mobility management is done via home agents and foreign agents [16,17,19]. In an IP-centric mobile network, there may be a need to provide backward compatibility to legacy systems. Therefore, present-day devices may continue to be managed via location registers. If voice continues to be circuit-switched, then two forms of mobility management may be needed. Voice calls may be managed in the traditional manner (i.e., mobility management for radio handoff, radio resource control, etc.) and may be done by the location registers; mobility management of IP packets may be done via home and foreign agents. The synchronization between these aspects of mobility managements needs to be studied.

1.8 MOBILE IP

Mobile IP [16,17,19] will enable subscribers [mobile nodes (MNs)] to communicate using the same IP address at all times regardless of their point of access to the Internet. If this were not the case, the active transmission control protocol (TCP) sessions would be broken off each time MNs move from one IP subnet to another; and it would not be possible to guarantee service continuity and ensure that movement is completely transparent to the applications.

Mobile IP allows an MN to have effectively two addresses: one is a long-term address on its home network used for identification and a care of address (CoA) used for routing packets to the MNs current point of attachment. The home address remains associated with a care of address for a specified lifetime via a mobility binding.

Mobile IP defines new operations for location and handoff management:

- Discovery (how an MN finds a new attachment point)
- Registration (how an MN registers with its HA)
- Routing and tunnelling (how an MN receives data grams when it is away from home).

In simple words, mobile nodes register with either a home agent (HA) or a foreign agent (FA). The HA receives all the IP data grams destined for a particular mobile and delivers these to the MN via a path known as a tunnel.

While handoff is understood in traditional mobility call models, it needs special attention in the IP world. When changing their points of attachments, MNs must ensure smooth handoff to meet quality of service guarantees. A mobile may have also changed a foreign agent. IPv4 allows a previous FA to maintain a binding for its former visitors. As a consequence, IP packets at the old CoA are still received by the MN. If the binding lifetime with the previous FA has expired, the IP packets at the old CoA are re-routed to the HA that in turn sends them to a new CoA [21].

Upon introduction of IPv6, the only substantial difference between the solutions proposed for IPv4 and for IPv6 consists in the fact that in IPv4 traffic forwarding to the mobile terminal is almost always managed through a foreign agent. Whereas, in IPv6 the foreign agent no longer exists and it is assumed that the mobile terminal is always able to acquire a co-located CoA belonging to the visited subnet. The foreign agent, in fact, was conceived expressly to reduce the demand for IP addresses by sharing the same CoA among several mobile terminals. The foreign agent thus made it possible to avoid aggravating the problem of limited IPv4 addressing space; but it is no longer needed with IPv6, which has a virtually unlimited addressing space and efficient auto-configuration mechanisms which the mobile terminal can use to acquire a valid address in the visited subnet.

Applying IP mobility support protocols in the Internet depends critically on security management.

First of all, the home agent must be able to authenticate messages it receives from the mobile terminal in order to ensure that a false registration cannot cause all of the traffic intended for the mobile terminal to be redirected to an IP subnet other than that effectively visited.

Moreover, further complications emerge when the route optimization mechanism is used, given that in this case each correspondent node must be able to authenticate the Binding Update messages received from the mobile terminal (IPv6) or from its home agent (IPv4), respectively. In fact, while we can readily accept that the mobile terminal and its home agent, which are normally stations belonging to the same organization, can be configured manually with a shared secret key used for the authentication algorithms, it is much harder to imagine a similar scenario between the mobile terminal and the correspondent, or between the home agent and the correspondent node, given that the latter may be any Internet station. For this purpose, a mechanism with an appropriate level of security must be developed which enables two stations to agree dynamically on the secret key used. A mechanism of this kind has not yet been fully specified by the IETF, though the attention given to this problem by the "ipsec" workgroup is considerable.

1.9 SPECTRUM FOR MOBILE SYSTEMS

Mobile systems currently operate in the following bands:

824-849 MHz–869-894 MHz
890-915 MHz–935-960 MHz
1710-1785 MHz–1805-1880 MHz
1880-1900 MHz
1900-1920 MHz
1850-1910 MHz–1930-1990 MHz
1920-1980 MHz–2110-2170 MHz

WRC2000 allocated additional spectrum to IMT-2000 [22]. This consists of refarmed spectrum in the range 806–960 MHz, 1710–1785 MHz paired with 1805–1880 MHz and new spectrum in 2500–2690 MHz. ITU-R WP 8F is currently preparing band plans for additional spectrum. As noted previously, 4G systems are expected to operate in bands above 3 GHz. Some Wireless LANs now already operate in the 5 GHz band.

There are various regional variations in the use of the above spectrum. Unfortunately at this moment, there is no global spectrum that is universally available over the entire world for mobile use. To further complicate, duplex directions of mobile and base transmit also do not match worldwide. Software radios, including the radio frontend, are a promising option to solve the problems of global roaming.

1.9.1   Spectrum for Future Mobile Systems

The present spectrum below 3 GHz is heavily encumbered and is unlikely to be made available to mobile beyond the allocations of WRC2000. Therefore, new frequency bands above 3 GHz must be considered. The higher frequency bands result in increasing RF propagation loss and RF circuitry losses. In turn, these losses impose restrictions on the link budget and, hence, cell sizes but are also beneficial in reducing system interference. The increased data speeds may also pose further reductions on the link budget. The combined effects of increased bit rates and higher frequencies will require transmissions at higher output power that will impose challenges in maintaining long battery life that present-day mobiles have achieved.

1.10   ORGANIZATION OF THIS BOOK

In order to address the many aspects of wireless communications discussed in this book, the following structure is adopted:

Part 1: Visions of Wireless Communications Applications in the 21st Century

This part has four chapters that discuss the visions from the different regions of the world about the role of wireless communications, needs of different markets, and rates of wireless network deployment.

Chapter 2 by Raymond Steele presents a very ambitious and a rather unconventional view of the role of wireless communications in the 21st century. The chapter describes the necessity of realization of global mobility functions supported by software-defined radio (SDR) because of the existence of different systems.

ATM is the promising method to provide the advanced services that support both circuit-switched and connectionless services including the use of IP. The attributes of third- and fourth-generation (4G) systems are presented. 4G system networks have powerful attributes, flexibility, and adaptability. With little solar-powered wireless nodes (made from biodegradable material) embedded in our clothes and designed to read the sensors monitoring our body functions, we will be able to instantly communicate with super-computers via a global family of public and private wireless networks. We will become superhuman beings. Eventually wireless communication will enhance our body functions enabling each of us to become the realization of the "Six Million Dollar Man" and the "Bionic Woman" at a throw-away price!

Chapter 3 by Chung Liu and Wayne Strom takes a relatively more pragmatic and near-term view than Steele and discusses how today's wireless networks that are largely deployed to provide narrowband circuit-switched services will migrate to packet-based networks providing voice over IP and multimedia communications. The architectural aspects of wireless radio and packet core networks are discussed.

In Chapter 4 Hatori presents a view from Japan. The number of mobile telephone terminals in Japan reached 69 million (57% penetration) by the end of June 2001. Based on the increase in subscribers, the views from Japan to 3G and 4G systems and their related aspects are provided after describing the features of current mobile communication systems such as PDC and Personal Handyphone System (PHS).

The Japanese standardization body, the Association of Radio Industries and Businesses (ARIB), pioneered the development of the wideband CDMA (W-CDMA) system (now known as the DS CDMA mode) that is now one of the IMT-2000 approved radio interface family [10]. The key features of W-CDMA, including transmission technology such as multiple spreading factor, two-layer spreading code allocation, are given. The 4G systems should accommodate increased data/multimedia traffic in 2010. The requirements for 4G systems and technical approaches to realize them are mentioned. This chapter also covers other future systems such as multimedia mobile access communication (MMAC) systems, wireless home link, and intelligent transportation system (ITS).

Part 2: Developments in International Standards

This part has two chapters and they focus on the development of standards in wireless communications networks. The chapters discuss radio-specific standards, network standards, and the relevant IP standards.

The development of international standards is critical to the globalization of wireless communications. The deployment of technologies that are compliant with agreed standards will result in significant economies of scale. Chapter 5 by Jane Brownley, Fran O'Brien, Maria Palamara, Derek Richards, and Lynne Sinclair presents a broad overview of the standardization activities of the ITU, special-interest groups (third-generation partnership projects), IETF, and also gives a summary of several other related activities such as Bluetooth and WAP.

Chapter 6 by Umehira is about the standardization of wireless ATM and IP. The latest activities in WATM-WG of ATM Forum, IEEE802.11 of the United States, ETSI-BRAN of Europe, and MMAC of Japan are introduced focusing on radio access layer and networking protocols. Harmonization on radio physical layer specification among IEEE802.11, ETSI-BRAN and MMAC groups is the most notified result. They achieved the 48 subcarrier orthogonal frequency divisioin multiplexing (OFDM) with channel spacing of 20 MHz. Frequency spectra for broadband wireless systems are in the 5 GHz band because of the requirement of higher bit rate transmission and spectrum congestion in bands below about 3 GHz. IP mobility support options such as gateway architecture, overlay network architecture, and integrated network architecture are also given to interconnect the PCs over heterogeneous networks.

Part 3: Propagation Issues

The radio channel presents unique problems and challenges to wireless communications. It is one of the principal contributors to the limitations that beset the capacity and performance of wireless systems. A major characteristic of the radio channel is multipath propagation. It is due to the diffraction and scattering of radio waves due to the terrain and man-made objects. Numerous studies (see [23] and [24] for a comprehensive bibliography) have appeared involving the modeling and characterisation of the radio channel so that an accurate prediction of system performance can be made and an evaluation of methods to mitigate the deleterious effects can be assessed.

Chapter 7 by Bertoni gives a comprehensive review of the radio channel characteristics that have been observed by various research groups under different conditions for frequencies below about 3 GHz. The chapter describes how multipath rays are produced at the base station and the subscriber terminal and describes models to characterize the fading resulting from multipath. The chapter also gives models for signal variation resulting from shadowing by buildings and range dependence.

Chapter 8 by Hashemi describes the modeling of the indoor radio channel. The channel is modeled as a linear time-varying filter in three-dimensional space, and properties of the filter's impulse response are described. Theoretical distributions of the sequences of arrival time, amplitudes, and phases are described. Other relevant concepts such as special and temporal variation of the channel, large-scale path-loss, and delay spread are also mentioned.

Chapter 9 by Hata gives an overview of the propagation loss models commonly used in mobile communications design and discusses their applications in various mobile propagation environments, for example, macrocells, microcells, and the like. The frequency range of the models described is up to 2 GHz. It is important to select the prediction model appropriate for the intended cell size. Ray tracing methods and the Walfish–Bertoni model are useful for microcells. The Okumura–Hata model with Akeyama correction and terrain correction is effective for macrocell with cell radii larger than 1 km. For future wideband transmission on 2 to 5 GHz band, DOA (direction of arrival) and TOA (time of arrival) of each path under microcell or picocell environment should be considered.

In Chapter 10 Har and Xia gives a comprehensive review of propagation measurements for indoor microcell to outdoor macrocell environments for the 800 MHz and 2 GHz bands. Measurements reviewed were classified into two categories. These are:

- Path-loss measurements done by the use of narrowband continuous wave (CW) signal
- Multipath characteristic measurements employing wideband pulse signal.

Comparison of propagation models is also given for the case of microcell models such as the COST231 Walfish–Ikegami model and the Har–Xia–Bertoni models.

Part 4: Technologies

This section focuses on the major technological areas that distinguish wireless systems and presents chapters that describe a new range of capabilities that future systems must possess.

Chapter 11 by Biglieri, Caire, and Taricco describes coding and modulation considerations for wireless channels. It gives an overview of the issues involved in the selection of coding and modulation schemes for the wireless channel. It rightfully points out that our perspective of coding/modulation schemes for the wireless channel is influenced by the Gaussian channel. The later is a poor model for many of the wireless channels. Using more appropriate channel models, the chapter derives code selection, power allocation, and information-theoretic bounds for the channel capacity of the radio channel.

Chapter 12 by Sampei discusses modulation/demodulation techniques and derives expressions for their BER for both Rayleigh and additive white Gaussian noise (AWGN) channels. It also presents the promising methods for next-generation high-bit-rate transmission system such as OFDM and adaptive modulation schemes. For the latter method, estimation for modulation parameter and instantaneous delay profile is provided. The adaptive modulation scheme enables the optimization of modulation parameters (modulation level, symbol rate, coding rate) during different propagation conditions or traffic load variations.

In Chapter 13 Adachi describes fundamentals of multiple access schemes used in wireless systems. It gives capacity estimates for demand assign based multiple access schemes [frequency division multiple acces (FDMA), time division multiple access (TDMA), and code division multiple access (CDMA)]. Packet-based communication systems use a random multiple access method such as ALOHA, carrier sense multiple access, and the like. The throughput for these random multiple access schemes is also given.

Smart antennas are a very powerful tool to improve the spectral efficiency and performance of existing and future wireless communication systems. The benefits achieved from smart antennas have not been fully realized in present-day commercial wireless systems. Spatial and temporal combining methods described by Falconer in Chapter 14 are effective to eliminate the intersymbol interference arising from multipath and co-channel/adjacent channel interference. This chapter describes the configuration and principle of space–time processing, channel models for multiple element arrays, receiver/transmitter space–time processing, recent space-time wireless communication architectures, and adaptation issues. It also describes that spatial-temporal signal processing method is applicable to smart antenna for mobile radio systems.

In a companion chapter (Chapter 15), Kohno, describes the principles of multiuser detection and gives an overview of various receiver structures to combat multiuser interference. An optimum receiver structure and various practical applications and their performance are discussed. The applications of spatial filtering to combat multiuser interference are also discussed.

Part 5: Wireless Systems and Applications

The evolution of TDMA-based systems to provide 3G services is based on the enhanced data rates for global evolution (EDGE) concept. Both IS136 and GSM systems may be evolved to provide 3G capabilities. EDGE can be introduced in a smooth way in existing frequency plans of already deployed networks. The physical layer of EDGE is based on selecting a combination of modulation levels, forward error correction (FEC) rates, and combining appropriate slots to achieve the required data rate. Chapter 16 by Javerbring describes the physical layer, link layer, and EGPRS (packet-switched services) performance of EDGE.

In Chapter 17 Viterbi discusses the origins of CDMA, arising from military applications. During the last decade, CDMA has developed into a widespread commercial technology. Evolutionary improvements and enhancements are underway that will increase voice capacity and provide medium-speed data services. High-speed Internet access can be achieved as well, even without requiring wide bandwidth allocation. Viterbi shows that CDMA, arising from military application, has developed into a widespread commercial technology only in the last decade. Evolutionary improvements and enhancements are underway that will increase voice capacity and provide medium-speed data services. High-speed Internet access can be achieved as well, even without requiring wide bandwidth allocation. This chapter shows that the average data rate can be more than tripled, with peak rates exceeding 2 Mbps, by employing variable latency packet allocation systems utilizing currently operational base stations and corresponding radio frequency (RF) carriers. This chapter pioneers future enhancements to IMT-2000.

Chapter 18 by Dahlman et al. describes the wideband CDMA (W-CDMA) radio access technology standardized on a global basis as the radio-access technology for 3G mobile communication. The chapter gives the background for W-CDMA and an in-depth description of this technology, including channel structures, radio protocols operation, resource management, and performance enhancing technologies such as transmitter diversity and adaptive antenna.

Satellite systems are also expected to play a major part in future personal communication systems. Various satellite-based systems are being considered. These systems use satellite constellations in low earth orbits (LEO), intermediate circular orbits (ICO), and will provide global mobile satellite coverage to complement terrestrial mobile systems. New systems to provide data and multimedia services with up/down link capacities of 2 to 20 Mbps for fixed applications are also being considered. These issues are discussed in Chapter 19 by Evans.

Part 6: Wireless ATM Networks

Chapter 20 by Raychaudori et al. represents an overview of wireless ATM (WATM) network technologies being developed as a potential solution for delivery of broadband services to portable computing devices. The WATM reference architecture, wireless access protocols for WATM, including medium access control (MAC) layer and data link connection (DLC) layer, are described. Mobile ATM network infrastructure concept and selected system-level issues are considered including methods for QoS control and alternative approaches for support of IP-over-WATM.

1.11 SUMMARY

Wireless networks have come a long way since their introduction. In the early days of mobile communication, no one would ever forecast the penetration of wireless services that exist today and the impressive range of services/capabilities expected by tomorrow's networks. The wireless channel continues to provide an extensive range of stimulating and challenging research opportunities.

The chapters in this book are only a snapshot of the current worldwide efforts in enabling to provide a rich range of capabilities on tomorrow's wireless networks. The field of mobile communications is very dynamic and already the contents of this book appear to be somewhat dated as new ideas/technologies are coming to the horizon. The editors have made an effort to cover a large scope—ranging from future visions, standards, physical layer aspects, systems, and network aspects. We are unable to cover many new and challenging ideas that are the subject of wide interest among the research community. In particular, we would have liked to have chapters on software radio [17], space–time codes, multiple input, multiple output antenna systems, turbo coding [25–28] techniques to achieve fast packet access, and the like. Many good chapters could not be printed here due to limitations of space. The interested readers are referred to excellent bibliographies contained in the individual chapters for further reading. We hope that this book will serve as a good reference for researchers and practicing engineers alike.

Acknowledgments The editors would like to take this opportunity to express their appreciation to the various authors from the four corners of the world. Without their contributions this project would not have gotten off the ground. We are also grateful to our respective employers for providing the facilities to complete this work. The author (Mansoor Shafi) would also like to express his appreciation to NTT Japan, because it was during a sabbatical there, that this work began.

ACRONYMS

ADC	analog-to-digital converter
AMPS	Advanced Mobile Phone System
ANSI	American National Standards Institute
ARIB	Association of Radio Industries and Businesses
ATM	asynchronous transfer mode
AWGN	additive white Gaussian noise
BRAN	Broadband Radio Access Networks
BER	bit error rate/bit error ratio
CDMA	code division multiple access
CNN	Cable News Network
CoA	care of address
CSCF	Call State Control Function
DLC	data link connection
DOA	direction of arrival
DSP	digital signal processor
EDGE	enhanced date rates for global evolution
EGPRS	Enhanced GPRS (general packet radio service)
ETSI	European Telecommunications Standards Institute
FA	foreign agent
FDMA	frequency division multiple access
FEC	forward error correction
FOMA	Freedom of Mobile multimedia Access
GlobeCom	Global Communications Conference
GPRS	general packet radio service
GSM	Global Systems for Mobile Communication
HA	home agent
HLR	home location register
ICC	International Conference on Communications
ICUPC	International Conference on Universal Personal Communications
IETF	Internet Engineering Task Force
IMT-2000	International Mobile Telecommunications 2000
IP	internet protocol
ISP	Internet service provider
ITS	intelligent transport system
MAC	medium access control
MAP	mobile application part
MGCF	Media Gateway Control Function
MGW	Media Gateway
MIMO	Multiple Input Multiple Output

MMAC	multimedia mobile access communication systems
MN	mobile node
MRF	Media Resource Function
MSC	mobile services switching center
NAMTS	North America Mobile Telephone System
NMT	Nordic Mobile Telephone
OFDM	orthogonal frequency division multiplexing
PDA	personal digital assistant
PDC	personal digital cellular
PHS	Personal Handyphone System
PIMRC	Personal Indoors and Mobile Radio Conference
PSTN	Public Switched Telephone Network
QoS	quality of service
RF	radio frequency
SIP	session initiation protocol
SDR	software defined radio
SGW	Signalling Gateway
TACS	Total Access Communication System
TDMA	time division multiple access
TOA	time of arrival
VLR	visitor location register
VTC	Vehicular Technology Conference
WAP	wireless application protocol
WATM	wireless ATM
W-CDMA	wideband CDMA
WIN	Wireless Intelligent Network
WRC	World Radiocommunication Conference

REFERENCES

1. Mobile Radio Centennial, *IEEE Proceedings Special Issue*, July 1998.

2. L. Hanzo, "Bandwidth Efficient Wireless Multimedia Communications," *Proc. IEEE*, Vol. 86, No. 6, pp.1342–1382, June 1998.

3. W. R. Young, "Advanced Mobile Phone Services—Introduction, Background and Objectives," *Bell Sys. Tech. J.*, Vol. 58, pp. 1–14, 1979.

4. D. C. Cox, "Wireless Personal Communications: What Is It?" *IEEE Personal Commun.*, Vol. 2, No. 2, pp.20–35, Apr. 1995.

5. J. D. Gibson, *The Mobile Communications Handbook*, IEEE Press/CRC Press, New York, 1996.

6. J. Uddenfelt, "Digital Cellular—Its Roots and Its Future," *Proc. IEEE*, Vol. 86, No. 7, pp. 1319–1324, July 1998.

7. R. W. Lucky, "New Communications Services—What Does Society Want?" *Proc. IEEE*, Vol. 85, No. 10, pp. 1536–1543, Oct. 1997.

8. IP Broadband Wireless Access, Doc. 8F/128, ITU-R WP 8F meeting, Geneva, Oct. 2000.

9. M. Shafi, A. Hashimoto, M. Umehira, S. Ogose, T. Murase, "Wireless Communications in the Twenty-First Century: A Perspective," *Proc. IEEE*, Vol. 85, No. 10, pp. 1622–1639, Oct. 1997.

10. Doc. 8-1/TEMP/275, IMT-RSPC's: Detailed Specifications of The Radio Interfaces of IMT-2000, 18th Meeting of ITU-R TG 8/1, Helsinki Finland, Nov. 1999.

11. D. Raychaudhuri and N. Wilson, "Multimedia Personal Communication Networks: System Design Issues," Third WINLAB Workshop on Third-Generation Wireless Information Networks, pp. 259–288, Apr.1992 (also in J. M. Holtzman and D. J. Goodman, Eds., *Wireless Communications*, Kluwer, 289–304, 1993).

12. http://www.mwif.org.

13. R. Keller, T. Lohmar, R. Tönjes, and J. Thielecke, "Convergence of Cellular and Broadcast Networks from a Multi-Radio Perspective," *IEEE Pers. Commun.*, Vol. 8, No. 2, p. 51, April 2001.

14. N. Nakajima, "Future Mobile Communications Systems in Japan," *Wireless Personal Communications*, Vol. 17, No. 2–3, pp. 209–223, June 2001.

15. *Time Magazine*, May 29, 2000.

16. P. Lettieri and M. B. Srivastava, "Advances in Wireless Terminals," *IEEE Pers. Commun.*, Vol. 6, No. 1, pp. 6–19, Feb. 1999.

17. J. Mitola, "The Software Radio Architecture," *IEEE Commun.*, Vol. 33, No. 5, pp. 26–38, May 1995.

18. A. K. Salkintzis et al., "ADC and DSP Challenges in the Development of Software Radio Base Stations," *IEEE Personal Commun.*, Vol. 6, No. 4, pp. 47–55, Aug. 1999.

19. D. Raychaudhuri and N. D.Wilson, "ATM Based Transport Architecture for Multiservices Wireless Personal Communication Networks," *IEEE J. Selected Areas in Commun.*, Vol. 12, No. 8, pp. 1401–1414, Oct. 1994.

20. 3GPP, "Combined GSM and Mobile IP Mobility Handling in UMTS IP CN" 3R TR 23.923 V.3.0.0, pp. 69–73, May 2000.

21. I. F. Akyildiz, J. McNair, J. S. Ho, H. Uzunalioglu, and W. Wang, "Mobility Management in Next Generation Wireless Systems," *Proc. IEEE*, Vol. 87, No. 8, pp. 1347–1384, Aug. 1999.

22. CPM Report on Technical, Operational and Regulatory/Procedural Matters to be considered by the 2000 World Radio Conference, Doc. CPM 99/1, 7, June 1999.

23. H. Bertoni, "UHF Propagation Prediction for Wireless Personal Communications," *Proc. IEEE*, Vol. 82, pp. 1333–1359, Sept. 1994.

24. H. Hashemi, "The Indoor Radio Propagation Channel," *Proc. IEEE*, Vol. 81, pp. 943–967, July 1993.

25. S. L. Ariyavisitakul, "Turbo Space-Time Processing to Improve Wireless Channel Capacity," *IEEE Trans. Commun.*, Vol. 48, No. 8, pp. 1347–1359, Aug. 2000.

26. G. J. Foschini, "On Limits of Wireless Communications in a Fading Environment When Using Multiple Antennas," *Wireless Personal Communications*, Vol. 6, No. 3, pp. 311–335, March 1998.

27. S. Benedetto and G. Montorsi, "Unveiling Turbo Codes: Some Results on Parallel Concatenated Coding Schemes," *IEEE Trans. Information Theory*, Vol. 42, No. 2, pp. 409–428, March 1996.

28. V. Tarokh, N. Seshadri, and A. R. Calderbank, "Space-Time Codes for High Data Rate Wireless Communications: Performance Analysis and Code Construction," *IEEE Trans. Information Theory*, Vol. 44, No. 2, pp. 744–765, March 1998.

PART 1

VISIONS OF WIRELESS COMMUNICATIONS APPLICATIONS IN THE 21ST CENTURY

CHAPTER 2

Vision of Wireless Communications in the 21st Century

RAYMOND STEELE

The current scene in wireless communications is described followed by a discussion of what third-generation (3G) networks are likely to be during the first decade of the 21st century. Armed with this knowledge, we speculate on post-3G systems using a mixture of prediction and preconceptions based on how technology and society are evolving. We arrive at the global family of ad hoc networks, such as wearable networks, home networks, networks using aerial platforms, and maritime networks. Then we peer through the future haze and consider the key role of software agents and wonder if our global information network will become a global brain.

2.1 INTRODUCTION

Predicting the future in wireless communications is perilous because of the reliance on sequential developments in the industry being reasonably correlated. Near-term improvements can often be foretold, but long-term events are difficult to predict. For example, the invention of fiber communications, with its truly mega capacity, microelectronics providing vast processing power and memory storage, and even the explosive growth of the Internet as a social phenomenon, came as a great surprise to most engineers. We do not, however, need to rely exclusively on correlation to predict the future as we may also employ preconception. A person at the end of the 19th century could have preconceived the transmission of visual scenes. The telephone had already been invented enabling people to speak to one another over long distances, so one might have anticipated that eventually people would be able to see and talk to each other when they were well beyond the line-of-sight. So although a visionary could predict video communications he or she would be totally stumped on how to realize them.

This then will be our approach: a mixture of prediction based on a reasonably high correlation between previous and current significant events in wireless communications;

Wireless Communications in the 21st Century, Edited by Shafi, Ogose, and Hattori.
ISBN 0-471-155041-X © 2002 by the IEEE.

and preconceptions based on how technology and society are broadly evolving. Therefore, before we begin our predictions of a vision of wireless communications in the 21st century, we need to briefly describe the current situation, followed by the third-generation systems planned for the first decade, and then use this information in the prediction process for events later in the century. After that we will use preconception based on guesses on how society is likely to evolve.

2.2 CURRENT SCENE IN WIRELESS COMMUNICATIONS

The first commercial cellular radio service was the Nordic Mobile Telephone (NMT) system, introduced in 1981. In 1983 the United States deployed the Advanced Mobile Phone System (AMPS), and this was followed by a number of systems in different countries. All these networks are called the first-generation (1G) networks, and they are said to be analog networks because the speech signals are not digitized. However, all control and supervision of the networks are digital [1–4].

These 1G networks are still in use today, but they are being phased out by the second-generation (2G) systems, which are all digital, although they remain basically telephone networks. However, low-speed data is supported. There are three prominent 2G systems: the European global system of mobile (GSM) communications, and the American cdmaOne and IS-136 systems [5]. GSM is the market leader and has become a defacto world standard [6]. cdmaOne, previously called IS-95, has become the second world standard [7, 8]. IS-136 is also deployed in various parts of the world [9]. GSM and IS-136 are based on time division multiple access (TDMA), and GSM has a comprehensive back-haul network. cdmaOne uses a code division multiple access (CDMA) radio interface that is, perhaps, more accurately described as a 2.5G interface in that it represents a bridge to (3G) networks [7, 8].

We now have mobile satellite systems (MSSs) that provide mobile services in most parts of the world, and users only need a small hand-held terminal if the satellites are in low earth orbits (LEOs). MSSs are not new, but until recently the satellites have been in higher geostationary orbits requiring greater transmission powers from the mobile stations (MSs), and incurring much higher round-trip signal delays ($\simeq 250$ ms) compared to LEO-based MSSs. Even LEO MSSs using multiple-beam antennas form terrestrial cells that at their smallest are tens of kilometers in size, and this means that the spectral efficiency of MSSs are low. MSSs are basically deployed to support low bit rate services on a global basis [10], but it is the terrestrial-based networks that provide high capacity where it is needed.

In the so-called developed countries, cellular networks provide ubiquitous coverage by an arrangement of different cells that vary from large cells of tens of kilometers formed by antennas on towers to multilayering of cells in cities where minicells of <2 km may oversail street microcells whose shape depends on the local buildings, vegetation, terrain, rivers, and the like and may be <300 m to indoor cells, often called picocells that may be <10 m [11, 12]. Allocating channels in multilayered cells is often done by partitioning them so that each layer has its own channel set. For example, GSM1800 may be used for indoor picocells, while the frequency band of GSM900 may be divided between the street microcells and the minicells. This example requires the use of dual-band 900/1800 MHz MSs.

There are private mobile radio (PMR) and special mobile radio (SMR) networks that provide wide area coverage of both low bit rate data and voice to their users. These networks were initially simplistic with a dispatcher communicating with a set of MSs over a wide area and often with only one channel. There has been a historic development where two generically different wireless systems, namely PMR and cellular, have been integrated, as exemplified by TETRA and iDEN networks, and further they are also able to operate over a wide range of cell sizes.

While 2G cellular systems were being conceived, analog cordless communications [cordless telephones (CTs)] were already available, and their digital versions, namely CT2, followed by DECT, followed by PHS, then PACS were introduced [13, 14]. These CTs were basically designed for in-building use (CT2, DECT), but their deployment on the city streets was anticipated (particularly by WACS, a forerunner of PACS) and implemented with PHS. In the exclusive in-building and adjacent environment role, CTs are essentially radio tails of the fixed network, and their use imposes no additional financial penalty to the user. But in the mobility stakes, cellular is king, and CTs will be a blip in history as cellular operators place small base stations within buildings and offer competing tariffs to the users.

The smaller the cell, the higher the spectral efficiency, and the easier it is to support high bit rate services, provided the regulatory authority has allocated sufficient radio spectrum. The smallest cells are to be found within buildings, and it is here that CTs and cellular networks may be required to compete with wireless local area networks (LANs) and wireless ad hoc networks that use higher propagation frequencies, for example, 60 GHz and infrared.

GSM evolution is a continuous process. With the phenomenal growth of the Internet has come the expectation that the users will want to explore and retract data from Web sites. Wide area protocol (WAP) services are now being offered on GSM networks enabling the user to get limited data services, for example, the latest price of stocks and shares. Slot aggregation will be used to provide high-speed circuit switched data (HSCSD), and the general packet radio service (GPRS) will offer packet transmissions in excess of 150 kb/s. Multilevel modulation will increase the bit rates of both HSCSD and GPRS. Called the enhanced data rates for GSM evolution (EDGE), this GSM Phase 2+ development will support many new services, ensuring that GSM evolves from a 2G to a 3G system.

2.3   3G SCENE FOR THE BEGINNING OF THE 21ST CENTURY

In 1988 the Europeans launched their RACE 1043 project with the aim of identifying services and technologies for an advanced (3G) system to be deployed by the year 2000. The European research programs started with RACE I (which included the 1043 project), then RACE II was followed by ACTS (advanced communications, technologies, and systems), which led to their universal mobile telecommunication system (UMTS) [7, 15]. It was paralleled by work undertaken by the International Telecommunications Union (ITU) Committee TG8/1. The aspiration was that UMTS would be a global system offering a wide range of multimedia services operating up to 2 Mb/s, and further, it would integrate the numerous types of wireless networks, such as MSSs, W-LANs, PMRs, and cellular. As the countdown for the 3G deployment approached and decisions had to be made, the ITU renamed the title of its 3G system as the international mobile telecommunications of the year 2000 (IMT-2000), and on receiving 16 proposals for IMT-2000 it was realized that no single proposal would be unanimously adopted, and instead a family of IMT-2000 systems that were mutually compatible was the only practical decision. So the dream of IMT-2000 as a single global network embracing *all* types of genetic wireless networks will not be realized within the next decade.

The radio interface proposed by the Japanese Association of Radio Industries and Businesses (ARIB) is very similar to the European Universal Terrestrial Radio Access (UTRA) interface, and certain companies in Japan and Europe have worked closely together to bring this about. Although the back-haul network will be based on the GSM one, the multiple-access method for the radio interface is wideband code division multiple access (CDMA) (W-CDMA). This means that the radio interfaces of GSM and UTRA are very different, although some similarities have been retained. In the United States a considerable portion of the IMT-2000 spectrum has already been auctioned for personal communication systems (PCS), and this has persuaded the Americans to evolve their cdmaOne and IS-136 networks toward 3G ones called cdma2000 [16] and UWC-136, respectively. Because IS-136 is TDMA, many of the enhancements to GSM, known as GSM Phase2+ will be part of UWC-136, and indeed it appears that UWC-136 might evolve to UMTS in the same way that GSM networks will evolve to UMTS networks. cdmaOne is essentially a subset of cdma2000, with less backward compatibility problems compared to the UTRA system.

2.3.1 MSS Component of IMT-2000

Regarding MSSs, ICO and Globalstar (along with the apparently defunct Iridium) were IMT-2000 proposals. They provide mobility and support hand-held terminals, but in their initial versions these global mobile personal communications by satellite (GMPCS) systems will only operate at 9.6 kb/s. There are broadband MSSs, for example, Teledesic, Skybridge, Spaceway, Astrolink, that can deliver 2 Mb/s but require relatively large terminals and are not conceived to support mobility. Imarsat Horizon is an IMT-2000 proposal that lies between the LEO and geostsationary earth orbit (GEO) systems. If all theses MSSs are to be components of IMT-2000, then once again we will end up with a family of IMT-2000 MSSs with different air interfaces that will have to interface with the family of five terrestrial IMT-2000 systems. We note that the GMPCS systems have accepted the use of subscriber identity module (SIM) cards to support global roaming.

2.3.2 Global Mobility

Global mobility is defined as the ability of a user to originate and receive calls anywhere in the world [17]. If a user would like to do this by having a single 3G mobile station (MS), then this MS must be able to interface with any of the IMT-2000 family of networks, as well as the prevailing 2G networks when no 3G networks exist in that region, assuming that mobile satellite systems (MSSs) do not have sufficient capacity. If the 3G base stations (BSs) have adaptive radio interfaces, they might be able to emulate the interface of a visiting 3G MS. Another approach is for the MS to accept a software down load over its radio interface so that it can operate as the host network requires. Alternatively, the MS could be multimode so that it can accommodate all 3G, and leading 2G, variant radio interfaces. To realize this type of MS, a primitive form of software radio (SR) is required, having analog front ends, perhaps multiple radio frequency (RF) transceivers, and a combination of dedicated hardware and digital signal processing (DSP) chips to provide the SR element. A specified common radio access is necessary so that a MS can make the initial approach to a network.

If the radio interface between a MS and BS is agreed, the next step is for the host network to identify the MS's home network. If IMT-2000 standardizes on the use of SIMs, then there is no problem, as the GSM network has a well-established procedure for authenticating users, as distinct from their MS. For example, the designers of Globalstar, whose radio links are based on IS-95, have designed an air interface that is able to interface with both the GSM core network and the IS-41 core network, thanks to a mapping technology between the two protocols, and because Globalstar has agreed to use SIM cards. However, if IS-41C continues where the MS and the user are not independent, then validating a visiting European 3G user with his SIM card will be a problem. The same applies for an American subscriber visiting Europe. If the internationally agreed access procedure specifies that each user has a unique user number, and given the low cost of signaling, it should be possible to verify the authenticity of a user by his own network, whether a SIM is used or not.

2.3.3   Global Transportation of 3G Services

Cellular operators who provide 3G services lease capacity from the large international photonic networks owned by the telephone companies (telcos). When a user requests a 3G service from a local cellular network, the quality of service (QoS) required is associated with an acceptable delay, throughput, and perceptual parameters. If the network agrees to provide the service, the source material will be digitized and packetized by the MS and its BS(s) during a call. There is widespread use of Internet protocol (IP) in desk-top and local area network (LAN) environments and the deployment of wireless access protocol (WAP) mobile services. Currently IP does not explicitly reserve resources for specific data transmissions, although standardization processes to do this are in progress. Asynchronous transfer mode (ATM) overcomes these problems, with QoS guaranteed bandwidth allocated on demand, and both circuit-switched and connectionless services handled. ATM can also be the transport mechanism for the wireless links. Transmission from an originating MS to a receiving station can be conveyed over the optical fiber networks using just IP, just ATM, or IP over ATM [18].

2.3.4   Provision of Global Coverage for 3G Services

MSSs will provide global coverage but not high capacity, which is required in urban and suburban areas. Global mobility requires global radio coverage. High capacity requires small cells ranging from a few meters in offices to outdoor cells of a few hundred meters. The BSs will be smaller than a coffee mug and generally connected by fiber. The innate cost of these BSs is low, as is the cost of fiber. Although the site rental charges are low or nonexistent in private campus-type environments, they may be large in public areas. Microcell sites could be established universally by placing radio tails on the telco network outlets, for example, on phone sockets.

2.3.5   Beginnings of Reconfigurable Systems and Networks

Throughout the first decade of the 21st century we will see the introduction of basic reconfigurable radio interfaces and networks [19, 20]. The radio interface consists of the hand-held terminal and the base station and all the subsystems and software that enables them to interact. Reconfigurable radio interfaces are software radios, which will initially have analog RF front-end transceivers, but digital, zero intermediate frequency circuits to provide the baseband signals. Because of the amount of signal processing required, particularly for down conversion and forward error correction (FEC) decoding, the software radio operating only at baseband will have a mixture of dedicated hardware and digital signal processor (DSP) chips.

The user will have a software agent (SA), but its role will be limited to basic tasks [21, 22]. The network will begin to become reconfigurable according to traffic load, offered services, and the like. The hand-held mobile station will disappear and be replaced by novel forms of information presentation. Basic wearable computers and communication systems will become available, and voice recognition systems will begin to replace the keyboard and the light-pen for keying in information.

2.4 POST-3G SYSTEMS

Extrapolating to 2010 is not too difficult, although surprising innovations will occur whose consequences we cannot guess. In the previous section we mentioned briefly the 3G scene and what the mobile communications scene may be like by the year 2010. To be specific the IMT-2000 family of networks will offer seamless global communications for mobile users, user-definable services, support for a huge increase in the number of subscribers, separation of service provision from network operation, unique subscriber number, low-cost terminals and services, SIM roaming, multimode terminals, provision of services without significant additional network investment from service providers, high level of communication security, wide area services up to 384 kb/s, small area services up to 2 Mb/s, and circuit- and packet-switched transmissions. These are some of the key provisions we may expect.

While 3G systems are technologically more advanced and will offer a much wider range of services compared to 2G systems, they are not particularly advanced in terms of their radio interfaces, nor in their maximum bit rates they can support. Specifying the radio interface is still required; up-grading specifications will be slow and expensive as they must be processed by the standardization committees. Later generations of mobile networks will not suffer from these deficiencies, and will have new and powerful attributes. So let us now look a little further into the 21st century and the post-3G systems.

2.4.1 Flexibility and Adaptability

The next generation of mobile communication systems should have a distinguishing feature, and we suggest that it will be "flexibility and adaptability" of the network, and this includes the radio interface. A user will have an intelligent multimode terminal (IMT) [23], namely a fully implemented software radio (SR) [24] that interfaces into a global network that is becoming increasingly soft in that it will be reconfigurable to software commands [25]. The block diagram of an ideal IMT is shown in Figure 2.1. We observe that the ADC and DAC are located just after the LNA and just before the PA, respectively, that is, essentially at the antenna system (AS). The received signal following amplification is therefore converted into a digital signal and applied to the DSP. The DSP has a number of roles to play. It provides down conversion, selecting the wanted signal, followed by

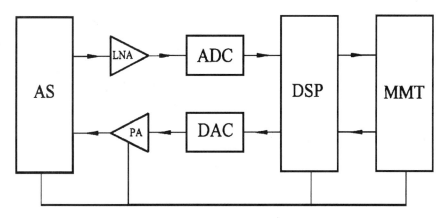

FIGURE 2.1   Ideal IMT where AS, LNA, PA, ADC, DAC, DSP, and MMT are the antenna system, low noise amplifier, power amplifier, analog-to-digital converter, digital-to-analog converter, digital signal processor, and multimedia transducers, respectively.

establishing the bit pipe, which may include the demodulator, RAKE receiver, FEC decoder and deinterleaver, multiple-access interference (MAI) canceller, and so forth, in other words, an appropriate receiver configuration for the type of transmission received. It will also form the transmitter, namely the source codec, channel codec and interleaver, multiple-access procedure, modulation, and the like to produce a signal that will be DAC, power amplified, and transmitted. The DSP is also involved with signaling communications with the BS and will be able to reconfigure the radio pipe of the IMT within a packet period if required. For example, it could rapidly change from a CDMA receiver to a frequency division multiple access (FDMA) one. Part of the service creation environment used by the intelligent network (IN) will reside in the DSP, and the DSP will also cater to the user's personal software, such as the users software agent (SA). There will be an assembly of multimedia transducers that input and output the source signals, for example, audio, video, and data.

Body LANs (B-LAN) embedded in a user's clothing will be increasingly popular. The LAN could be connected via either fiber or micropower radio using a descendent of Bluetooth and will monitor bodily functions as well as supporting the IMT, whose components will be conveniently distributed around the LAN as required.

The IMT, and the SR in the BS, will cause the demise of most of the standardization processes. As the IMT is able to reconfigure rapidly, increasing the power of the FEC codec, adding another stage in a parallel multiple-access interference (MAI) canceller, and the like, there will be too many configurations to standardize. The MS will work in harmony with the network, but it will be the network that has the final say as to the IMT's configuration. However, it is desirable that the radio access methods, as well as internetworking procedures be internationally specified.

A cellular BS will be able to do more than the IMT can achieve. It may choose to down load software to the IMT should this be necessary, and it will interact with other fixed network nodes and local system control to provide the agreed QoSs for the network users. The IMT would not be confined to the cellular networks. It could interact with a W-LAN to obtain very high bit services in offices, or it would act as a PMR transceiver or an MSS one. The important point is its ability to be what a network requires it to be. For a user to access a wide range of types of networks, new methods of authentication and billing will be employed.

2.5 GLOBAL FAMILY OF AD HOC NETWORKS

As the 21st century progresses, our environment will change. To the ubiquitous embedded computers in virtually every artefact will be added a wireless communicator. People will have their own body wireless network (BWN) with which they will communicate for services. If these cannot be satisfied by their BWN network, communications will be made with the local area network around them, for example, with the intelligent home network (IHN). Should the person be in the street, then communications will be with the network that resides there. Wherever the user is located his BWN will be able to communicate by wireless to a network, which in a wilderness will be via a MSS. We will now discuss the BWN, a network that is mobile, followed by the other networks that will be components of the global network.

2.5.1 Wearable Wireless Networks

It has been apparent for some time that the role of the computer could be extended to also be an adaptive communication terminal, an IMT, as well as a home entertainment center supporting television, and with other features, such as the ability to transport the user by means of virtual reality into an environment and situation of their choice [23]. For laypeople a computer is characterized by a computer screen, a keyboard, and the need for disk and CD-ROMs. They are often only vaguely aware of the key component: the DSP chip. To the initiated, computers are ubiquitous, residing in wrist watches, domestic appliances, cars, mobile phones, airplanes, indeed, almost in every technological piece of equipment and appliance. The keyboard and the screen are usually not present because they are not required.

This combination of computer, IMT, entertainment center, and general wireless interface to a whole plethora of technological entities and systems, we will call *technology fusion* (TF). Here we are interested in the wireless aspects of TF, that is, the TF that is a vital component of the BWN. A person in the middle of the 21st century having a BWN would be unaware of its presence as it would be of near-zero weight and not visible. Only when the person was required to interact with the BWN would the person be conscious of its presence. The BWN would be a distributed micropower wireless ad hoc network that connects the nodes and sensors residing in the persons clothing, or on the persons body for certain monitoring functions. For example, the microphone and loudspeaker system could reside in a shirt collar or resemble jewelry, and they could be linked to processing nodes in the clothing via wireless. There would be no need for a keyboard. Instead there would be voice recognition and voice synthesizer systems with which to talk with the user's personal software agent (SA). The user would be wearing an enormous amount of processing power as well as a vast amount of memory containing most of the files the SA will need when dealing with his user's requests. Sometimes the user will request a task or service that requires new code to be written. The SA will need to specify exactly what the user wants and employ software programmers (SPs) residing within the BWN to write the code required. We make the important point that codes will be written by other autonomous codes, namely SPs. These SPs will be housed in tiny chips at nodes that may also act as buttons on a garment. If the SPs have difficulty in writing the required code, the SA of the SPs will use the radio interface node of the BWN to communicate via radio to the huge computing systems within other networks for assistance.

The user will sometimes need to "see" the fruits of his request. For example, a person may be at an airport waiting for a plane and would like to watch a movie that is stored on one of his BWN memory banks. The person makes himself comfortable and then tells his SA the movie he requires and where the display will be shown. The user may wish to beam the display from his wrist watch onto a nearby wall to create a large image, or he may want to view it in an arbitrary space in front of him in a holographic form generated from sensors within his clothing.

When the user hangs his clothes up at night, the hanging structure would provide the power to charge up the very low power devices in the wearable BWN. Similarly, power sources within the heels of shoes could be recharged while stored overnight. Solar panels and energy generated from body movement would provide additional power for the wearable BWN during the day. The microelectronics of the BWN needs to be of near-zero cost, a throw-away item that is biodegradable. This means the use of organic microelectronics that utilize semiconducting polymers. Mechanical microcomponents based on semiconductor polymers will be used to provide the electronic power from body movements.

The wearable BWN replaces the IMT used at the beginning of the 20th century. In addition to being a complete, albeit physically small, network, it will also interact with BWNs worn by other people to provide a repeater function and thereby support ad hoc BWNs. The relaying of signals via BWNs will be relatively easy because of the large memory capacity and the wide choice of propagation frequencies and multiple-access methods available to BWNs. When idle, BWNs will determine who are the other BWNs within their vicinity, pass this information on to these BWNs and receive similar lists from them. In this way ad hoc BWNs can be set up so that all BWNs will know how to be connected to other BWNs and hence to those nodes with connections to the back-haul fixed network with all its facilities.

2.5.2 Home Wireless Network

The future networks will consist of fixed networks and mobile networks, that is, networks that move as exemplified by BWN, and networks within aircraft, trains, cars, and ships. The fixed networks would be of vastly different sizes. One of the smallest will be the network deployed in the intelligent home. Processors with their microtransceivers will abound. They will be in every household appliance, such as refrigerators, washing machines, lamps, fans, and kettles. While we may anticipate a fiber backbone network connecting some appliances, others whose movement may change will be linked by wireless. Some home networks will be completely wireless, some will have a central control processor, while others will form autonomous ad hoc networks (AAHN). Fuel tanks will report on the level of fuel they contain, ditto for other containers such as milk and orange juice cartons, house plants containers will report the degree of moisture they have.

The wearable BWN will enable an occupier to move about his house interacting as required with the home wireless network (HWN), making tea, closing windows, and the like. The HWN will communicate via wireless to the urban wireless network (UWN), which will be connected to the wider area networks, and so on until, if necessary, the global networks. The result is that communications with the HWN can be made irrespective of distance. This means that appliances in the home can be monitored and often serviced remotely, fuel consumption monitored and charged, the condition of the home observed by experts at a distance, and so on.

2.5.3 Urban Wireless Network

Similar to the HWNs will be the office wireless networks (OWNs). These will communicate with office workers via their wearable BWN. In large office complexes the OWNs will resemble the integration of many HWNs, as each office area will in effect have an HWN. Outside the offices, the streets, urban open spaces such as parks and stadia, will have radio coverage from outdoor radio nodes (and from some indoor nodes). It may be that high-capacity optical fibers will form packet highways that link the OWNs, HWNs, and the street wireless networks (SWNs). While loop networks may be the preferred basic network configuration, there will be many of them, interconnected in a mesh arrangement. The hybrid nodes, that is, fiber on one side, wireless on the other, will be connected to the high-capacity optical fiber network that also conveys traffic that originated from hardwire terminals (that are treated from a protocol point of view as stationary mobile terminals). The ubiquitous UWN will include the BWN, HWN, OWN, and SWH, which facilitates wireless communications in the urban environment. A simple pictorial representation of this situation is illustrated in Figure 2.2.

We may summarize our view of an urban communication system as having fixed nodes with input and output optical trunks that employ optical switching using wavelength division multiplexing (WDM) under complex software control. On the periphery of the network will be hybrid nodes that provide wireless coverage, as well as small autonomous networks such as HWNs. The BWN may be fiber or a low-power radio, but its interaction with other networks will be via radio. The BWN is essentially a very advanced distributed IMT.

Autonomous ad hoc Urban Network The QoS and capacity of an UWN is dependent on both radio and network factors. The network will know the location of all mobiles, here BWNs, those which are currently using high bit rate services, or have negotiated a high QoS agreement, or have other stringent requirements. In addition, the network will be briefed on the resources available at the different wireless nodes. Armed with this information, the network will reconfigure both the wireless links of the users, reassign the available radio resources and alter the virtual paths within the packet networks to improve the overall network performance. For this to be accomplished, the urban network needs to make many rapid decisions, often local ones, and this suggests a flat management structure rather than a hierarchical one. An ad hoc network has the required flat structure, which needs to be both autonomous and adaptive. Within the physical network we propose a virtual overlay facilitator network (FN) that provides those nodes participating in decision processes with the latest information. Each node has an SA and these SAs meet with the SA of the FN. Generally there will be a cookbook of responses as most problems will have been experienced before. Should an unexperienced event occur, it will be up to theses SAs to decide on network strategy. Notice that while the BWN, HWN, and the like may inform the network of their difficulties, it is the network's SAs that decide as an executive committee on the response to be made to any situation.

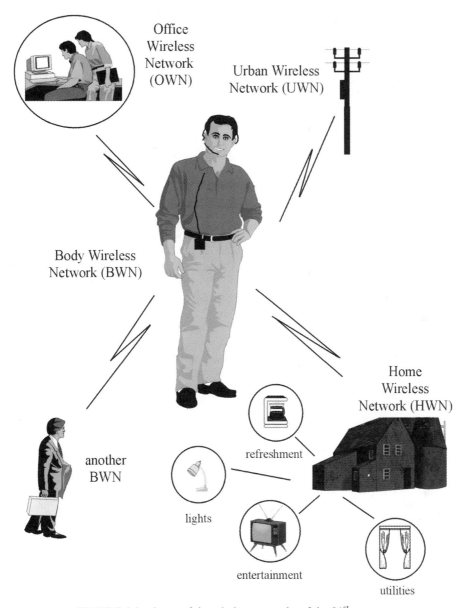

FIGURE 2.2 Some of the wireless networks of the 21st century.

As to the size of the ad hoc network, this is dependent on the collection of nodes whose SAs must be involved in the decision process. These are essentially the local fixed, hybrid, and wireless nodes. If the communications stretch over large distances, then the SAs of main trunk nodes and destination local nodes must be brought into the negotiations. The trunk nodes could be national or internationally located. Interworking SAs may become involved. These latter agents could be crucial if the QoS involves a delay limit on communications when, for example, MSS should be avoided.

2.5.4 Sky Communications

Sky communications enable capacity to be instantaneously available as and where required in outdoor environments by creating *softcells*. These softcells can be formulated, moved and changed in size in a fraction of a second. They are created from aerial platforms (APs) in the sky, and as a consequence terrestrial BSs can be avoided with their site rental charges and back-haul infrastructure [26–28].

An aerial network is composed of APs that are located in the stratosphere or lower and facilitate sky communications. The APs are at their lowest on tall man-made structures, for example, on the top of the Sears Tower in Chicago or on the Eiffel Tower in Paris. More typically they will be higher, and a popular height is 21 km (70,000 ft) where the stratospheric winds are minimal and tend to be planar. The first generation of APs are likely to be high-altitude aircraft that fly in tight circles on-station, that is, at the location and height required, or unmanned airships that are untethered and can also be maintained at a fixed location in the sky. For example, by the first decade of the 21st century the high-altitude long-endurance (HALE) AP will be launched. Its length will be 600 ft with a diameter of 160 ft. There will be 80,000 ft^2 of solar cells, a 98 kW motor, a 30-ft propeller, with an ability to handle 100,000 telephone calls simultaneously over a 400-mile radius.

The softcells on the earth's surface are formed by beams from an antenna array mounted below an AP. Initially these beams will be from antennas with multiple feeds where each antenna will create a cluster of softcells on the earth. The cluster will be contiguous over the region serviced by the AP. Later switched beams will be used, then phased arrays that will be able to move the softcells and change their size under system control. Smart antennas will then be introduced to create both the softcells to accommodate communications between MSs and the AP via an antenna beam, while at the same time nulling-out interferers from adjacent softcells.

An AP will have beams that also connect it to other APs. By this means beams from the APs will provide coverage over large expanses of the world. The APs may be viewed as network nodes, and communication from mobile terrestrial users in different parts of the world will be linked via this aerial network. The APs will also have a terrestrial beam(s) to a ground station (GS) that interfaces with other networks. The aerial network will be ad hoc with distributed control. Figure 2.3 shows the concept where the APs are represented by airships.

Each cell from an AP is defined as the half-power beamwidth on the earth's surface, and the cells are formed from a phased array on the AP. Adjacent cells therefore overlap at lower power levels creating intercellular interference. For TDMA a multiple cell cluster will be required, while CDMA can operate with one cell per cluster. As an example, if CDMA is used with a chip rate of 4.4 Mcps/s at 2.2 GHz and if the minimum E_b/I_0 for acceptable operation is 3 dB, then the number of users per cell for service rates of 8 and 32 kbps is 85 and 22, respectively. For a 1-W transmitter power the radius of the total area covered by the AP is 70 km. This assumes the receiver temperature is 300 K, and the link margin is 15.4 dB. To achieve a microcell of 100 m radius the side of the square array is 12 m [28].

If the APs are at some 21 km and thereby above the commercial aircraft lanes, and if they are untethered, then they offer no obstruction to aircraft flights. Being close to the earth they may form soft microcells with relatively small antenna structures. The delays from the MSs to the APs are small ($\simeq 70$ μs) compared to those experienced in MSSs whose satellites are in space. Because the APs are stationary, they do not contribute to

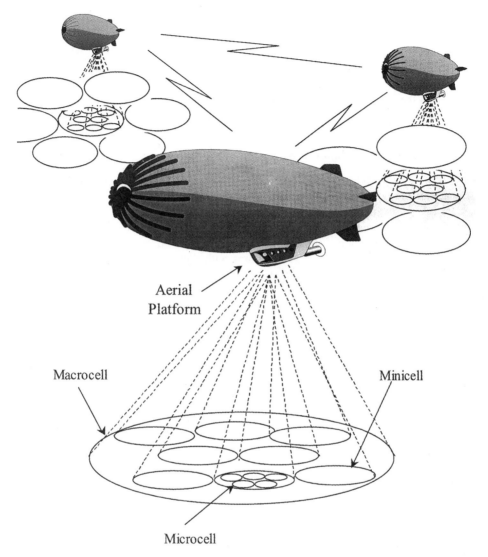

FIGURE 2.3 Aerial network having multiple aerial platforms (shown as airships) and creating a contiguous coverage by softcells whose size and position can be changed instantly as required.

Doppler effects. By creating soft microcells the aerial networks have high capacity. The propagation environment is relatively good, having approximately an inverse square law path loss until the earth is reached when Rician rather than Rayleigh fading tends to occur. The APs may also have antenna structures operating at different frequencies so that APs at high frequencies, for example 50 GHz, may deliver wireless local loop (which is sometimes called wireless fixed access) services, while lower frequencies may be employed for cellular networks, and so forth. Further the APs are inexpensive to make, launch, maintain on-station, and bring back to earth for maintenance. Beams from APs are able to track "solitons" of teletraffic as exemplified by teletraffic, from high-speed trains. AP networks will cause the demise of the MSSs that will reach their peak use in the early part of the 21st century.

As the 21st century progresses, we may expect smaller hovering APs that are unmanned, or APs that are tethered with fiber embedded in the tethering cable for back-haul communications, or remain as currently proposed, namely untethered [26]. We may also expect them to be at any height and not restricted to be above the aircraft lanes. This will be achieved by constraining the movement of aeroplanes.

2.5.5 Space Communications

At the end of the 20th century the first components of a large space station were sent into orbit around the earth, and within the first decade of the 21st century we may anticipate the space station will be fully operational. As the century proceeds, more space stations will be introduced into a range of orbits from the GEO of 36,000 km to the LEOs of some 7000 km. The space station system (SSS) will replace the mobile satellite systems (MSSs).

The space stations will be large structures with dimensions of hundreds of meters, supporting extended antenna arrays that focus many beams simultaneously on the earth's surface. The GEO space stations require a larger array than the LEO space stations for a given beam diameter on the earth's surface. As an example, the size of the central cell (defined as the broadside 3-dB beamwidth on the earth's surface) directly below a space station as a function of the length of the side of a square antenna array carried by a GEO and LEO space station is shown in Figure 2.4. The propagation frequency is 10 GHz. Observe that for a GEO the array side size needs to be 500 m to create a cell of 900 m diameter, whereas an LEO station can form a 100-m radius cell with a much smaller array of 100 m. The 10-GHz propagation frequency means that line-of-sight links between the space stations and the earth terminal must be used. The LEO and GEO SSS are able to use such large antenna structures that the antenna gains are sufficiently high to support small hand-held mobile terminals. Decreasing the carrier frequency f_0 increase the radius of the terrestrial cells [proportional to $\tan(1/f_0)$] and increases the array size (proportional to $1/f_0$).

The LEO SSS will be a moving mesh of space stations, as exemplified by the Iridium MSS concept with its moving configuration of satellites. Although the SSS can provide global coverage and with much smaller cells than MSSs, they will be at a disadvantage compared to APs, which are much closer to the earth's surface and can therefore use much smaller arrays to achieve the same cell size of a SSS but at the lower frequencies and deceased transmit powers that are the same as those used by the terrestrial-based cellular systems.

Different operational scenarios can be envisaged regarding stationary ad hoc AP systems that are close to the earth, stationary GEO SSS that are very far from the earth, and LEO SSS, and perhaps intermediate earth orbit SSS, that are dynamic in that all the space stations are moving relative to the earth, but that they are well below the GEO SSS and well above the AP network. Each system could be operated independently but interworked via sky gateways. An operator may own a GEO SSS, an LEO SSS, and an AP network, in addition to a terrestrial base network. The later may be used for confined spaces (e.g., inside of buildings) that are shielded from sky communications, while the AP network provides outdoor microcells, minicells, and macrocells, leaving the LEO SSS to cover sparsely populated parts of the world as well as providing isolated spot beams of high capacity as and when required. For example, the LEO SSS may use a spot beam for a

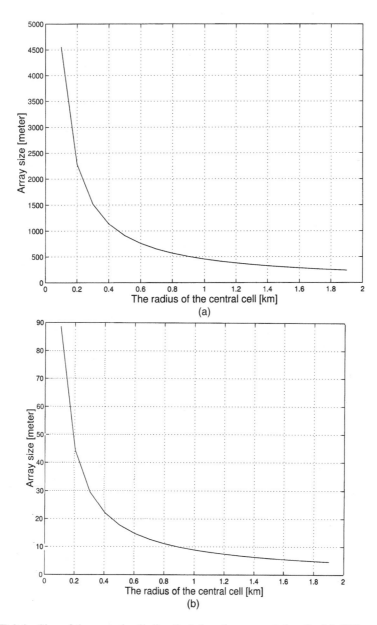

FIGURE 2.4 Size of the central cell directly below the space station for (a) GEO and (b) LEO space stations as the length of the side of a square antenna array. The propagation frequency is 10 GHz.

sports stadium or an area where a traffic accident has occurred. These beams will not be subject to cochannel interference and hence the capacity of an SSS will be very high. The GEO SSS would handle high signal capacity asymmetric mobile traffic where the longer signal delay is acceptable, as well as providing broadcasting services. There will be an FN between the different AP networks, SSS and terrestrial networks; and SAs will be used in the FN as discussed earlier.

2.5.6 Maritime Wireless Networks

Probably the first mobile station was housed on a ship. Wireless telegraphy using Morse code was adopted during the first decade of the 20th century for ship-to-ship and ship-to-shore links. These communications were via the long-wave band, and later during the century the high-frequency band was used utilizing the ionosphere as an electromagnetic mirror, while in recent times satellite links have found favor. During the first decade of the 21st century, shipping will increasingly rely on the MMS component of IMT-2000 for their communications.

The last two decades of the 20th century witnessed the laying of many submarine cables around the world, linking the continents and islands with a massive communications channel capacity. The delay of the submarine links is significantly lower than those associated with GEO MSS links, in spite of the propagation delay in fiber being greater than in free space. As the 21st century progresses, a series of unmanned surface vessels will connect into the submarine fibers laying beneath them. These vessels will take energy from the sea to assist them to stay on-station.

Each surface vessel will house a sea platform (SP) whose communication equipment is a hybrid network node, with photonic communications to the submarine cable to which it is tethered, and wireless communications over the surface of the sea and into the sky above it. The SPs form sea cells of different sizes such that a pattern of slightly overlapping cells is established over the entire sea or ocean, as shown in Figure 2.5. In addition the adaptive antenna arrays on the SPs will provide spot beams to individual ships or airplanes that

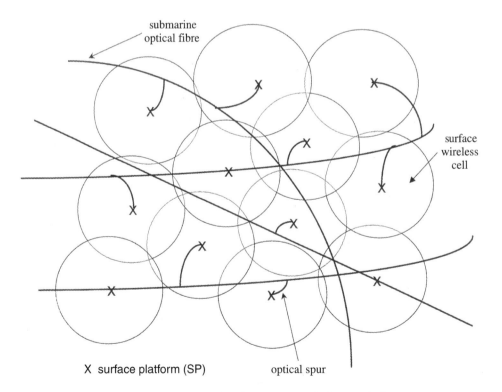

FIGURE 2.5 Creation of surface wireless cells by sea platforms.

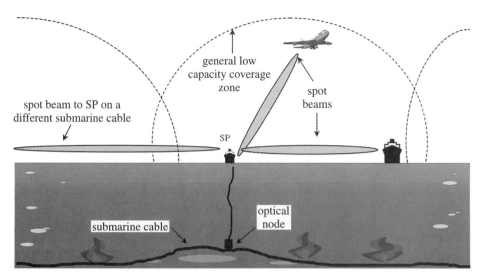

FIGURE 2.6   High capacity spot beams from an SP.

require a high capacity. SPs will also connect via line-of-sight spot beams to nearby SPs that are not tethered to the same submarine cable. The arrangement is illustrated in Figure 2.6. The submarine cables and the wireless network they support enable the fibers to be interconnected via radio spot beams. Ships will carry maritime platforms (MPs), communication equipment that can interact with the maritime wireless networks (MWN) illustrated by Figures 2.5 and 2.6, or with the SSS. They will have a repeating capability enabling maritime ad hoc networks to be established. They will also extend the range of MWN when there are no submarine cables within a region.

2.6 PEERING THROUGH THE FUTURE HAZE

The global network will become exceedingly dense with a profusion of nodes that will collect very low power wireless signals before conveying them to their destination via the complex fiber and wireless networks. The processing power within the global network will be enormous, as the network will be ad hoc and soft, making billions of decisions per second. The myriad of SAs will have created a virtual world that mirrors our own [29]. Their huge number and their diversity will mean that they may start to display human characteristics. They will be able to solve complex organizational problems, and be innovative. SAs will be the *oil* of the 21st century global network, the equivalent of people in a human society. We may expect them to genetically develop, initially according to our designs, but later by themselves. This could have severe consequences for mankind; a global network out of our control, behaving as it thinks best. SAs will develop with a wide range of characteristics, from the docile to the energetic, the unintelligent to the super-intelligent, the well-behaved to the rogue, and so forth. They will have the power to do much good and great harm, and containing their behaviour will be a key societal task. SAs have the potential to transform the global information network to a *global brain*. By global brain we do not mean a single co-ordinated intelligence, but rather heterogeneous distributed layers of intelligence with behavioral abilities. The alarming factor is the speed of innovation in electronics and software that will result in increasingly complex SAs with their ability to write complex code, to be cloned, to be genetically modified, to move at vast speeds, and to interconnect with other SAs in numerous ways. It may be that when the development of SAs reaches a certain threshold the acceleration toward a global brain will be very fast. And, what will be the mind of this global brain? [30]

By contrast to SAs, our biological development is exceedingly slow. We evolved to survive, and although we have developed physically in recent years, the enhancement of our intelligence is essentially static. By acquiring vast amounts of knowledge that is stored,

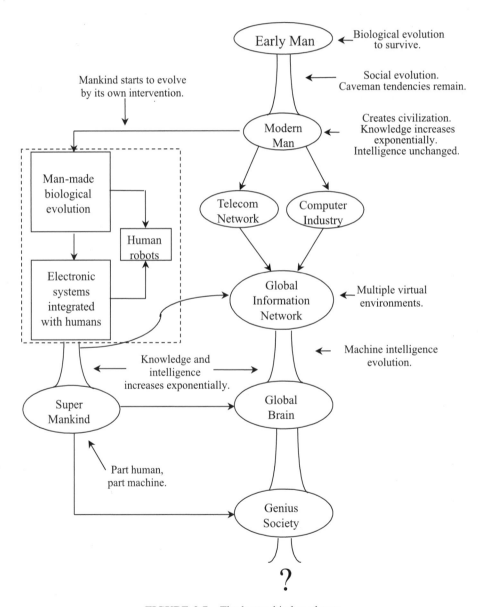

FIGURE 2.7   The humankind road map.

not biologically, but in books and electronic databases, humankind has been able to build complex machinery, computers, telecommunication networks, and SAs.

Human intelligence may not be increasing but it is high, whereas we have created machines whose intelligence is minimal but with the potential to accelerate it at a prodigious rate. A response to this situation is to understand the mechanics of our own intelligence and then biologically evolve it. Indeed, the inclusion of technological enhancements to our bodies has been in progress for a number of decades during the late 20th century, as exemplified by embedding pacemakers into peoples' chests to regulate their heart rate. While this process of improving our health by inserting electronics within our bodies will continue during the 21st century, we may expect the biological manipulation to enhance our life expectancy and mental prowess, and technology incursions within our bodies to occur to improve our abilities, for example, strengthen our limbs (as in the TV series, "The Six Million Dollar Man"). But, we may also expect the embedding of computers to allow us to calculate at machine speed, to work out complex solutions to life-threatening situations in a fraction of a second, and so on. Part, or all, of the BWN will move inside our bodies and be powered by muscular movement. The internal BWN will communicate by very low power radio to a repeater housed in a person's wearable BWN, which will then in turn relay messages to the nearest network as required.

Homo sapiens survival route would therefore appear to transform ourselves to a hybrid creature—part-human, part-machine—if we are to avoid being overtaken by the very machines we initiated. Figure 2.7 is a proposed humankind road map where super-humankind keeps the machines at bay and creates a genius society. The reader should not presume that the writer advocates this road map, but merely points out that Pandora's box is open and that humankind should be aware that it has initiated technological changes that may result in fundamental changes to ourselves.

REFERENCES

1. Advanced Mobile Phone Service. Special Issue, *Bell Sys. Tech. J.*, Vol. 58, Jan. 1983.

2. W. C. Jakes, Jr., *Microwave Mobile Communications*, Wiley, 1974.

3. V. H. MacDonald, "The Cellular Concept," *Bell Sys. Tech. J.*, Vol. 58, No. 1, pp. 15–41, Jan. 1979.

4. W. Y. C. Lee, *Mobile Cellular Telecommunications*, 2nd ed., McGraw-Hill, New York, 1995.

5. D. J. Goodman, *Wireless Personal Communications Systems*, Addison-Wesley, Reading, MA, 1997.

6. M. Mouly and M. B. Paultet, *The GSM System for Mobile Communications*, 1992.

7. R. Steele, C. C. Lee, and P. Gould, *GSM cdmaOne and 3G Systems*, Wiley, New York, 2001.

8. R. Steele and L. Hanzo, *Mobile Radio Communications*, 2nd ed., Wiley, New York, 1999.

9. EIA/TIA Interim Standard, "Cellular System Dual Mode Mobile Station—Land Station Compatibility Specifications," IS-54, Electronics Industries Association, May 1990.

10. G. Maral, J.-J. De Ridder, B. G. Evans, and M. Richharia, "Low Earth Orbit Satellite Systems for Communications," *Int. J. Satellite Commun.*, Vol. 9, 209–225, 1991.

11. S. Dheghan and R. Steele, "Small Cell City," *IEEE Commun. Mag.*, Vol. 35, No. 8, pp. 52–59, Aug. 1997.

12. R. Steele, "Speech Codecs for Personal Communications," *IEEE Commun. Mag.,* Vol. 31, No. 11, pp. 76–83, Nov. 1993.

13. W. H. W. Tuttlebee, (Ed.), *Cordless Telecommunications Worlwide*, Springer, 1996.

14. J. E. Padgett, C. Günther, and T. Hattori, "Overview of Wireless Personal Communications," *IEEE Commun. Mag.*, pp. 28–41, Jan. 1995.

15. Special Issue on the Universal Mobile Telecommunication System (UMTS), *IEEE Trans. Veh. Tech.*, Vol. 47, No. 4, Nov. 1998.

16. D. N. Knisely, Q. Li, and N. S. Ramesh, "cdma2000: A Third-Generation Radio Transmission Technology," *Bell Labs Tech. J.*, Vol. 3, No. 3, pp. 63–78, July–Sept. 1998.

17. R. Steele, "Global Mobility," IST'98, Vienna, 30 November–2 December 1998, pp. 203–204.

18. E. Guarene, P. Fasano, and V. Vercellone, "IP and ATM Integration Perspectives," *IEEE Commun. Mag.*, Vol. 36, No. 1, pp. 74–80, Jan. 1998.

19. Special Issue: Active and Programmable Networks, *IEEE Network*, Vol. 12, No. 3, May/June 1998.

20. Software Radios, *IEEE JSAC*, Vol. 17, No. 4, April 1999.

21. M. Breugst and T. Magedanz, "Mobile Agents—Enabling Technology for Active Intelligent Network Implementation," *IEEE Network*, Vol. 12, No. 3, pp. 53–60, May/June 1998.

22. A. Pham and A. Karmouch, "Mobile Software Agents: An overview," *IEEE Commun. Mag.*, Vol. 36, No. 37, July 1998.

23. R. Steele and J. E. B. Williams, "Third Generation PCN and the Intelligent Multimode Mobile Portable," *IEE Electronics and Commun. Engineering J.*, Vol. 5, No. 3, pp. 147–156, June 1993.

24. Proceedings of Software Radio Workshop, organized by the European Commission DGXIII-B Brussels, 29 May 1997.

25. G. Fettweis, P. Charas, and R. Steele, "Mobile Software Telecommunications," EMPC'97, Bonn, Germany, September 30–October 2, 1997, pp. 321–325.

26. R. Steele, "Mobile Communications in the 21st Century," In D. E. N. Davies, C. Hilsum, and A. W. Rudge, (eds.), *Communications after AD2000*, Chapman and Hall for the Royal Society, 1993, pp. 135–147.

27. R. Steele, "An Update on Personal Communications," Guest Editorial, *IEEE Commun. Mag.*, pp. 30–31, Dec. 1992.

28. B. El-Jabu and R. Steele. "Aerial Platforms: A Promising Means of 3G Communications," IEEE VTC'99, Houston, May 1999.

29. R. Steele, "Communications ++: Do We Know What We Are Creating," EPMC'97, Bonn, Germany, September 30–October 2, 1997, pp. 19–23.

30. R. Steele, "Beyond 3G," 2000 Int. Zurich Seminar on Broadband Communications, Zurich, Feb 15–17, 2000, pp. 1–7.

CHAPTER 3

Wireless Migration to Packet Network: U.S. Viewpoint

CHUNG LIU and WAYNE STROM

3.1 FUTURE WIRELESS NETWORK VISION

3.1.1 Future Network Vision

The explosive demand for [Internet protocol (IP) based] Internet access and packet data applications is changing the nature of future communications from circuit-switching-based architecture to a packet-data focused architecture. Concurrent voice, data, and multimedia applications also blur the lines between telecommunications and data services and therefore drive future architectures to multiservice platforms. In addition, some telecommunications service providers want to utilize common platforms to support multiple services and access technologies (e.g., wireline and wireless). To meet these demands, future networks envision low-cost and flexible architectures for multiservice applications and environments. As shown in Figure 3.1, this vision includes common core networks and application servers that can be shared by multiple-access technologies (e.g., wireline, wireless, and cable). Core networks are packet based on IP or asynchronous transfer mode (ATM) protocols and transport both voice and data traffic.

Application servers support feature control, resource allocation, databases, OA&M as well as other Internet functionality. Shared core network and application servers not only mean lower infrastructure deployment and operational costs but also allow common features to be shared across multiple-access networks (fixed or mobile, wired or wireless). This chapter will focus on how wireless networks will evolve to support this future network vision. Wireless-specific architecture enhancements and issues will also be identified and discussed.

Wireless Communications in the 21st Century, Edited by Shafi, Ogose, and Hattori.
ISBN 0-471-155041-X © 2002 by the IEEE.

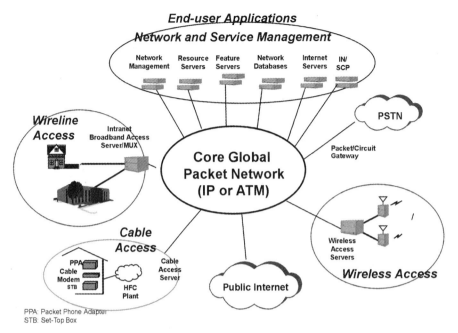

FIGURE 3.1 Future network vision.

3.1.2 Future Wireless Network Vision

Future wireless networks will allow wireless radio access networks to share common packet core networks and services with other access technologies. The radio access network could be a second-generation digital technology [code division multiple access (CDMA) or time division multiple access (TDMA)] or third-generation radio technology. To support higher speed data and more advanced services, third-generation (3G) wireless access technology using wideband CDMA [1] or EDGE [2] (enhanced data for global evolution) technology will be required. Wireless access networks will ultimately use packet data radio technologies so that end-to-end packet services (including voice, data, and multimedia) can be supported from end-user terminals. Open interfaces will be required between the wireless access network and core networks to leverage the economies of scale of the multipurpose core networks. Wireless gateways will be provided to terminate wireless-specific protocols and to provide a standard interface into the core networks.

The wireless network vision also supports direct IP access for voice, data, and multimedia services. Next-generation wireless network elements will be integrated with Internet-derived client–server platforms and packet-switching routers and servers and will share platforms and services with wired access systems. Enhanced Internet Engineering Task Force (IETF) standards for open interfaces (e.g., Mobile-IP [3], Servers Firewall, Private Network) will be used to manage services and provide interoperability across networks with multiple vendors and between networks operated by different carriers. Centralized feature and intelligent network (IN) servers will also be shared for wireline and wireless applications to ensure that features behave the same in all environments. The Inernational Telecommunications Union (ITU) is defining the virtual home environment

(VHE) that will leverage IN techniques to allow a subscriber features to follow them across networks and access technologies. Other servers such as common OA&M, resource, and Internet-related servers will be shared across voice and data as well as wired and wireless networks and thus provide greater economies of scale in manufacturing and efficiencies in operations. All of this provides a convergence path between wireline and wireless services (e.g., voice over IP) and among wireless data systems (cdmaOne, cdma2000, TDMA, EDGE, GSM, GPRS, and UMTS).

3.2 FUTURE WIRELESS NETWORK ARCHITECTURE

3.2.1 New Services and Functions

If wireless users are to take full advantage of these emerging networks and the packet data services they promise (e.g., Internet access, FTP file transfer, and real-time wireless voice over IP and multimedia), the following functions need to be enhanced in the radio access network and packet core network architecture:

- *High-Speed Packet Data Radio Access* Currently most wireless systems only support voice and low-speed (e.g., 9.6 or 14.4 kbps) data over the radio Interface. Second-generation wireless radio technologies are being enhanced to allow higher speed data (64 kbps and higher) services while third-generation systems are being designed to meet even higher IMT2000 bandwidth requirements for different environments: 144 kbps for high mobility, 384 kbps for low (pedestrian) mobility, and 2 Mbps for fixed service.
- *IP over the Radio Interface* This capability allows the end users to access IP-based services (e.g., Internet, intranet, packet voice, wireless multimedia) using special wireless packet terminals or terminals similar to those used in the wireline environment. Some improvements in the air interface are needed in order to dynamically manage multiple services with different quality requirements while maintaining radio spectrum efficiency.
- *IP-Based Mobility* Mobile IP provides packet data mobility across wired and wireless environments. The mobile IP foreign agent located in a foreign network currently serving a mobile subscriber registers with the home agent in the subscriber's home network. The mobile IP home agent will keep track of the current location of the foreign agent and thus of the mobile user. The home agent also acts as an anchor for the incoming traffic, keeping the mobility aspects of the user transparent to regular packet data IP network. GPRS also provides similar mobility functionality via serving GPRS system node (SGSN) and global GPRS system node (GGSN). Some enhancement are needed for real-time applications; for example, handoff from one packet zone to another packet zone.
- *End-to-End Performance and quality of service (QoS)* In order to ensure that the quality of new packet voice service is as good as that provided by the circuit network, the packet network needs to be enhanced with IP-based QoS to support multiple service classes (e.g., real-time conversational, interactive, and non-real-time background).

- *Multiservice Platform*   Multiple services (e.g., voice, data, video, and multimedia) are supported in one platform that can be accessed by multiple access networks.
- Virtual Home Environment (VHE)   VHE provides the same services in a visited environment as the subscriber receives in his home environment.

3.2.2 Architect Principles

From a network deployment and migration viewpoint, continuous cost reduction and flexible architecture are always major goals of network evolution. To reach this goal, the following major key architecture principles should be considered:

- *Separate the Radio Access Network from the Core Network*   This will allow the different types of access (e.g., wireline or wireless) networks to be connected to a common packet core network. This common core network can be used not only for traffic transport (e.g., IP traffic) but also for control signaling (e.g., mobile IP or SIP [4]/H.323 [5]). For example, existing wireless-specified signaling (e.g., ANSI41 or GSM MAP) could be supported over the IP network instead of the SS7 signaling network by replacing lower layer protocols.
- *Maximize the Advantages of Packet Data Networking*   In general, packet routing supports higher efficiency with greater flexibility compared with circuit switching. packet based core networks allow wireless systems to leverage lower cost packet transmission and switching infrastructure (e.g., commercial routers) and to provide compatibility with emerging IP-based service nodes. Wireless architectures should be compatible with both IP and ATM core networks. Network operators will then be free to choose either IP or ATM depending on their needs for backward compatibility, migration plan and configuration requirements (e.g., reliability and performance).
- *Separate Control from Transport*   Since the characteristics of control and transport planes are different, the separation allows control and transport platforms to be optimized independently. For example, a transport platform needs high throughput and real-time processing capabilities. Control platforms generally require more program flexibility and more storage but have less stringent timing constraints and lower input/output volumes.
- Share Common Application and OA&M Servers   Common IP service platforms (e.g., feature servers) support feature transparency between wireline and wireless environment. Centralized service and control logic also supports the same services in a mobile environment. It also lowers end-to-end life-cycle costs by using a single OA&M platform.
- Backward Compatibility   For most service providers, it is very important to evolve to the new architecture while leveraging their existing infrastructure. For example, the circuit-based infrastructure can co-exist with packet data/voice services until a complete IP-packet-based system is deployed. This allows service providers to protect their existing investment while maintaining a competitive status with advanced technology.

The remaining sections of this chapter will discuss wireless architecture alternatives in more detail.

3.3 WIRELESS PACKET NETWORK EVOLUTION

One potential wireless packet data network migration roadmap is shown in Figure 3.2 and outline below:

- *Now*   Currently, circuit voice is widely deployed with some circuit data capabilities using existing circuit radio access networks and the circuit-based public switched telephone network (PSTN). Low-speed data services with packet radio and/or circuit radio with access to packet networks are being introduced into some wireless networks. Circuit and packet radio access networks have independent interfaces to the PSTN and the packet core network.

- *Phase 1*   This phase allows existing wireless systems to utilize packet core networks for voice as well as data. This architecture is also referred as "trunking replacement" or "virtual switching network." Existing circuit voice mobiles and radio interfaces (e.g., IS95, IS136, etc.) are used and the existing MSC continues to support basic call and supplemental services. This allows existing wireless users to use the existing mobile terminal and have the same features with feature transparency. Once wireless VoIP access is introduced in phase 1, the packet data core network can begin to supplement or replace the PSTN. New or additional traffic is redirected to the packet data network (instead of existing PSTN) to take advantage of packet routing efficiency and lower cost infrastructure.

   The advantage of phase 1 is to reduce long-distance network access fees, leverage capabilities/cost structures of newer high-bandwidth core network providers, and also support existing base of mobiles and features.

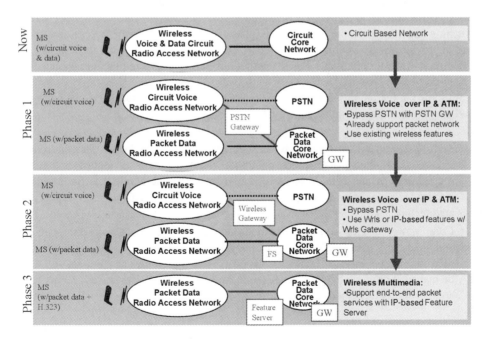

FIGURE 3.2   Wireless packet network migration.

- *Phase 2*   This phase allows existing wireless terminals to obtain features from the packet network and ultimately eliminates the need for a circuit switch. Existing wireless features and additional new IP-based features could be supported depending on the application and evolution needs of each service provider and market segment. For service providers who need the backward compatibility, existing MSC features can be implemented in the packet mode by deploying a new feature server or by evolving the MSC into a feature server. The same circuit voice terminal and radio interface is used with either feature server. If additional new IP-based features are required, IP-based (e.g., SIP) feature servers could be supported with existing MSC servers.

  The advantage of phase 2 has the benefits of phase 1 plus the opportunity to migrate to lower cost packet technology. It results from eliminating circuit switching from the network and optimizing network topology.

- *Phase 3*   Once the wireless packet data radio interface and terminal interfaces are enhanced to support real-time voice applications (e.g., SIP or VoIP and multimedia), an end-to-end packet data architecture can be introduced. This means that wireless terminals running SIP applications can communicate with the same network-based feature servers used in the wireline environment. Access to these applications therefore provides feature transparency between wireless, wireline, and other access system. This is target architecture for service providers to build packet networks with advanced multimedia services.

  The advantage of phase 3 has consolidated voice/data or wireline/wireless network by leveraging creativity of Internet for new advanced services of all technologies. It allows rapid feature customization using centralized feature server.

In general, the smooth migration starts with the replacement of the PSTN with a packet data core network in phase 1. Then feature servers are introduced to eliminate circuit switching in phase 2. In the target phase (phase 3), packet data radio access network is used to provide an end-to-end packet solution. Some service providers might skip phase 2 with circuit voice terminal and directly migrate to phase 3 with multimedia applications over packet-based terminal.

As shown in the migration roadmap, wireless packet networks consist of two parts that can be evolved to packet independently:

- *Radio Access Network*   This wireless-related network includes interconnections among base stations, base station controllers, MSCs, mobile IP routers, and GPRS routers. It is assumed that this network is managed by the wireless carrier to ensure the needed capacity and QoS. See Section 3.4 for a more detailed description of this architecture.

- *Core Network*   This common backbone network interconnects wireless, wireline, and other types of access networks and may also provide network-level services. This network may be operated by the wireless carrier or by a third party but also needs to be managed properly to ensure the desired capacity and QoS objectives are met. This architecture is described in more detail in Section 3.5.

While both networks can be evolved independently, they can share significant portions of the packet data infrastructure if the same technology (e.g., IP or ATM) is used. For

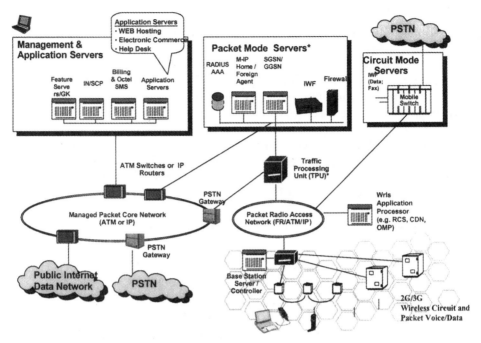

FIGURE 3.3 Wireless packet data network architecture.

example, a packet network might first be introduced to provide packet data services and Internet access. Voice services could then be migrated to the packet core network while still using the circuit voice radio access network. At some point in time, the radio access network can be converted to packet using some of the same packet network resources. For example, ATM switches used in the core network might also be used in the radio access network for interconnection of base stations with MSCs and wireless gateways.

One potential wireless architecture is shown in Figure 3.3. The left side of the figure shows packet core network with common servers (e.g., Internet, feature server, SCP/IN). The right side of Figure 3.3 shows the packet radio access network consisting of base stations and traffic processing units managed by control servers. Both networks could share the same packet data infrastructure if the same technology (e.g., IP or ATM) is used. The existing PSTN is included for backward compatibility purpose.

3.4 MIGRATION OF WIRELESS RADIO ACCESS TO PACKET DATA NETWORK (INCLUDING M-IP AND GPRS)

The existing radio access network is optimized for circuit-based services (e.g., voice and circuit data). Future radio access networks will need to support advanced packet data services (e.g., data transfer, Internet/intranet access) with higher data rates. North American standards are continuing to evolve second-generation systems (ANSI-95 and IS-136) to higher data rates and packet data services. The ANSI-95 radio interface has been enhanced from the original circuit voice and data service to support higher speed packet data service. The evolution to 3G cdma2000 will introduce a more flexible protocol structure [e.g., additional media access control (MAC) and link access control (LAC) layers] to support concurrent circuit and packet voice/data services. cdma2000 will also support higher speed data rates. Mobile IP as defined by the IETF is also being enhanced to support packet data mobility and other functions (e.g., authentication, dynamic IP addressing, QoS) required for wireless applications. These enhanced functions are described in more detail in the TIA TR45.6 standard [6].

For TDMA (IS-136) the second-generation data rates are enhanced by 136+. Third-generation services are being defined around the EDGE [7] air interface derived from GSM to support higher speed data services. 136+ and EDGE leverage network elements (e.g., SGSN and GGSN) based on ESTI defined GPRS (general packet radio service). These elements allow the 136+/EDGE data services to be overlaid on existing TDMA systems with minimal spectrum as they provide access to packet networks independent of the existing circuit access. GPRS also relies on a separate GSM-based HLR for user authentication and mobility management for packet services.

Industry technologies such as advanced computing platforms and client–server architecture also offer new opportunities for the evolution of wireless networks. Figure 3.4 shows how such a client–server architecture provides distributed wireless control and packet data network functionality within the radio access network. At the same time the packet-based (e.g., ATM) radio access network supports higher capacity base stations (e.g., 3G) with higher throughput and more efficient facilities utilization. A flexible network such as this promises to support new packet data services while reducing the infrastructure cost.

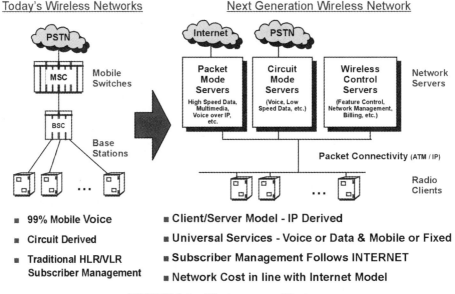

FIGURE 3.4   Client–server architecture.

3.5 WIRELESS PACKET NETWORK

3.5.1 Wireless VoIP architecture with PSTN Gateway (Phase 1)

Functional Architecture In the first phase of VoIP, packet-based IP or ATM core networks can begin to replace the circuit-based PSTN. Existing wireless circuit networks are connected directly to the PSTN using traditional trunks. This is accomplished using a PSTN gateway to interwork between the wireless circuit network and the packet core network (see Fig. 3.5). This gateway performs two major functions: signaling interworking and voice traffic interworking. Existing ISUP signaling from the wireless MSC will terminate on the PSTN gateway (instead of the regular PSTN SS7 switches), which will provide interworking with IP-based control signaling (e.g., SIP or H.323). Voice PCM traffic from the MSC will be converted by the voice gateway to IP or ATM data frames and sent over the core network.

Architecture Impacts and Considerations

Call termination Within the packet core network, calls are routed by a network feature server function that translates the dialed number into the address of the destination gateway. A mobile customer, however, might be anywhere in the network or even in another network. Ideally, calls made to mobile subscribers via the core network should be terminated to the gateway where the mobile is currently located. In this case, the gatekeeper in the core network should query the subscriber's SCP/HLR to determine the subscriber's current location and translate that to the nearest gateway.

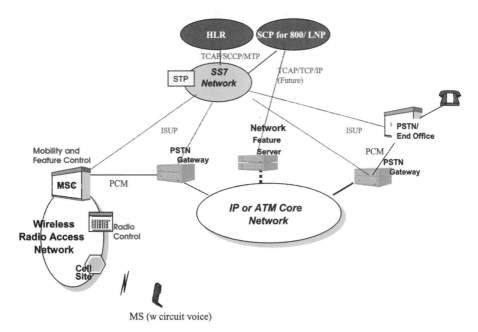

FIGURE 3.5 Wireless VoIP architecture with PSTN gateway.

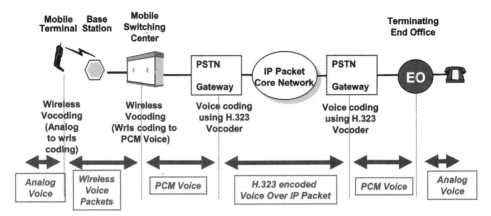

FIGURE 3.6   Wireless VoIP traffic path.

End-to-End Delay   When wireless propagation delays are added to packet network delays, the result approaches the limits for satisfactory two way conversations. In Figure 3.6, if we assume that there are 90- to 100-ms delay in the radio access network (mobile and MSC), 70- to 100-ms delay in gateway and core network, and 10 ms in the PSTN network, then the total one-way delay could be 170 to 210 ms. The delay in the core network depends on the number of routers or switches transversed, the type of packet network (IP or ATM), buffering size, and voice coding techniques use. The total network delay could be reduced by using higher rate, lower delay, or H.323 vocoding. This would improve voice quality at the expense of using additional bandwidth in the core network.

Bypassing wireless vocoding in the MSC and handling wireless vocoding in the far end gateway could also reduce the total delay. This would require, however, that all gateways in the core network be equipped with wireless speech coders that are part of the H.323 standards. In addition, it would require that speech coding be negotiated at call setup and for some scenarios (e.g., conferencing and handoff to analog systems), coding would need to be renegotiated during the call.

3.5.2   Wireless VoIP Architecture Using an IP-Based Feature Server (Phase 2)

Functional Architecture   As for the phase 1, the phase 2 architecture uses existing mobile phones and the circuit voice radio interface. The major difference is that features are performed through direct interworking with the IP/ATM core network. This eliminates the need for the circuit-switching facility in the radio access network by wireless access gateway. Wireless feature servers or MSC servers are used to support existing wireless features for existing mobile that is transparent to end users.

As shown in Figure 3.7, wireless feature server (e.g., new feature server or one based on an existing MSC) would provide the same features as phase 1 using packet techniques instead of circuit techniques. This feature server would use a wireless call model but use IP-based resource servers and gateway to perform services and gateways for direct interfacing with the IP/ATM network without converting to a circuit interface. This provides the feature transparency that allows the wireless subscriber to receive the same feature treatment as they move between circuit- and packet-based network.

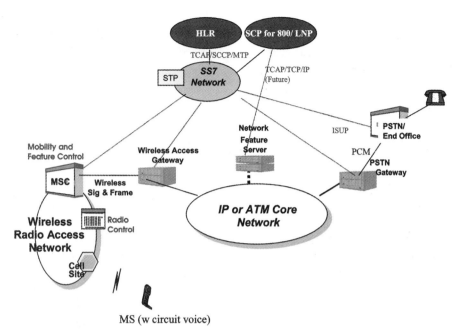

FIGURE 3.7   Wireless VoIP with wireless feature server (phase 2).

Architecture Considerations   Since wireless features are implemented as part of the MSC, the wireless access gateway (WAG) can provide the signaling and voice interfaces directly to the IP/ATM core network without ISUP signaling or conversion to PCM. In general, WAG integrates MSC and gateway functions with a packet-based platform.

3.5.3   Wireless Multimedia Architecture (Phase 3)

Functional Architecture   This target architecture supports an end-to-end packet data solution including wireless voice, data, and multimedia services. In accordance with the wireless network vision, while the access network can share infrastructure with the core network, it is separated logically from it. Application servers can be distributed across the core network and could be centrally managed and shared with other access technologies as desired. Multimedia client applications (e.g., SIP) should be treated the same as other data applications over the wireless packet data network (e.g., using mobile IP or GPRS for access and mobility). Figure 3.8 shows wireless multimedia support for a cdma2000 network using mobile IP.

For a TDMA/EDGE network using GPRS for mobility, a similar architecture is applied with feature servers accessed via the GGSN (Fig. 3.9). The embedded packet network provides the lower layer transport from the mobile terminal to another mobile terminal (end-to-end packet call) or to a packet gateway (packet to PSTN call). The client application will directly access the feature server to support voice or multimedia applications. The feature server also supports basic and supplementary services (e.g., call forwarding, call waiting, and multiple way conferencing). The resource server provides the bearer channel services (e.g., tone, announcement, and conferencing) requested by the feature server. Both resource servers and gateway are basically the same as those in phase

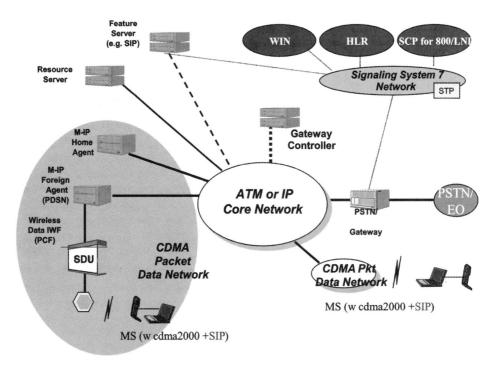

FIGURE 3.8 cdma2000 wireless multimedia architecture (phase 3).

1 or 2. Only additional IP-based feature servers are required. Legacy wireless terminals can continue while new terminals capable of running multimedia (e.g., SIP) applications and working with the enhanced air interface operate in the phase 3 mode and receive network-based multimedia services.

Architecture Considerations

Radio Resource Efficiency IP headers needed for IP-based applications transmitted over the radio interface will use more radio bandwidth because of the additional overhead. Since radio spectrum is limited in capacity and expensive, some mechanisms such as header compression (e.g., compression for RTP/UDP/IP headers) or/and header stripping (e.g., shorter temporary connection identification during data transfer) must be considered to optimize the radio resource usage.

Lower rate coding algorithms (e.g., evolution and possibly convergence of wireless and H.323 coding) should be used over radio interface to maximize radio efficiency.

End-to-End Quality of Service Different QoS criteria (e.g., delay, error rate, and jitter) are required for different types of application (e.g., real-time conversational, interactive, and non-real-time background applications). For example, lower error rate and shorter delay are required for real-time voice applications. It is easier to use the same layer (application layer is preferred) of QoS mechanism across all network segments (including air interface, radio access network, and core network). The same QoS mechanism used for wireline applications (e.g., Diff-Serv) could be used within the core network as well as the radio access network. The wireless access network also could use a different QoS approach (e.g., RSVP), however, mapping would then be required between the radio access and core

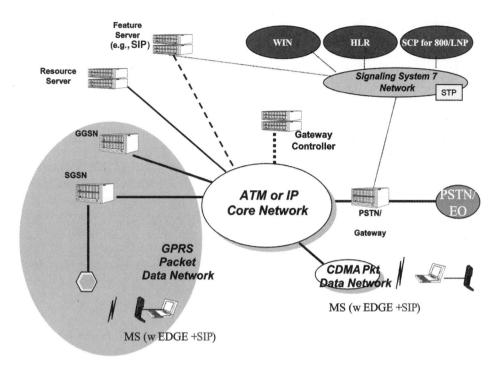

FIGURE 3.9   EDGE/GPRS wireless multimedia architecture (phase 3).

networks. The mobile terminal and the base station also need to manage the radio resource (e.g., traffic channel allocation) based on QoS parameters needed for the application and core network (e.g., multiple RLP queues).

For real-time applications, some cdma2000 or GPRS packet data protocols (e.g., media access control layer) need to be enhanced to ensure proper quality. For example, supporting nonacknowledged CDMA RLP or TDMA RLC reduces the delay over radio interface. Minimization of reactivation time from dormant states and optimization of algorithms for sharing applications with different QoS requirements across higher speed data channels are also needed to manage quality and capacity across the air interface.

Mobility Support for IP Servers and Network   As wireless IP servers (e.g., feature and resource servers) are introduced into the packet-based network, the additional impact of mobility must be considered. For example, which feature and/or resource server should be used in the home or visited system? What are the additional architecture considerations of handoff?

During registration period of time, additional IP-based discovery procedures could be used to select the feature server in additional to the provisioning approach. When a mobile roams into a visited network and originates a call, it could use an IP based feature server in the visited system, one in its home system, or a standalone server, based on subscriber or system provisioning and engineering. The home wireless service provider or a different service provider could own this server. For call termination, a home feature server needs to locate the desirable feature server and gateway before passing control message or traffic to visiting system. Handoff (e.g., soft and hard handoff, inter-base-station handoff) could be supported within a wireless access network or across base stations covered by the same wireless gateway. The ability to extend handoff across the packet core network depends on the delay, anchoring techniques, and re-routing of packet control signaling and transmission support.

M-IP (for CDMA) or GPRS (for TDMA) provide IP-based packet data mobility. The quality impact for handoff between foreign agents or SGSN reselection should be considered for real-time applications such as voice. Since feature and resource servers could be located in central locations within the IP network (e.g., home or standalone), the same servers should be accessible without service interrupt during handoff.

Quality/Performance Phase 3 offers a number of unique challenges and opportunities. One opportunity is to use the same speech coder type on both ends of the call. This requires convergence of wireless vocoders and Internet vocoders. This appears to be a reasonable objective since both deal with the same issues of delays, limited bandwidth, errors, and so forth. Of critical importance to wireless VoIP quality is the management of real-time data services over the air. Air interface standards must advance beyond those being developed for initial 3G offerings to support voice services using data techniques while retaining channel efficiencies.

3.6 SUMMARY

Wireless networks based on North American standards will evolve to take advantage of emerging low-cost packet technology and to meet the needs of third-generation wireless services. This evolution will include both the core network and the radio access network, which can be introduced independently. Migration to a packet core network can be accomplished in three phases that provide: (1) packet transport of voice and data, (2) implementation of features in a packet mode eliminating the need for circuit switches, and (3) an end-to-end packet solution. The object of the final phase is to allow wireless terminals running multimedia applications to access the same feature servers and thus the same features used by wireline applications—including telephony and Internet-enabled features. Where backward compatibility with existing wireless features is called for, today's MSC might be converted into a packet feature server. The final phase is dependent on advances in wireless standards for the air interfaces, terminal/application interfaces, as well as network interfaces and network mobility. While there are significant architectural issues to be solved to ensure performance and quality of service, with proper care, the resulting networks will extend the increasing variety of voice services of the packet world to the mobile user.

REFERENCES

1. TIA/EIA/IS-2000-1 cdma2000 Standards for Spread Spectrum Systems, TIA.
2. ITU–3rd Generation Partnership Project; Technical Specification Group Series and System Aspects: General Packet Radio Services (GPRS) (3G TS23.060.3.2.1).
3. C. Perkins, "IP Mobility Support," RFC 2002, Oct. 1996.
4. M. Handley et al., "SIP: Session Initiation Protocol," RFC 2543, March 1999.
5. ITU–Version 2 of H.323—Packet-Based Multimedia Communications Systems, January 1998.
6. TIA/EIA/IS-835 Wireless IP Network Standard, TIA.

CHAPTER 4

Vision of Wireless Communications Applications in the 21st Century: A View from Japan

MITSUTOSHI HATORI

4.1 INTRODUCTION

This chapter describes the situation in Japan regarding future mobile communications systems paying close attention to third-generation, fourth-generation, and other radio communication systems distinctive to Japan. It will be useful to review first the history of wireless communications and survey the current systems.

In 1979, the first cellular system began service in Japan. The start of this service began a succession of such services—in Sweden in 1981 and in the United States in 1982. These early systems, called the first-generation systems, adopted separate analog methods. Each country had its own system such as the Nippon Telegraph and Telephone (NTT) system, the high-capacity system of Japan, the Advanced Mobile Phone Service (AMPS) in the United States, the Total Access Communication System (TACS) in Great Britain, and the Nordic Mobile Telephone (NMT) system used in four Scandinavian countries.

The rapid development of LSI technology in the 1980s propelled the digitization of mobile communications throughout the world. The attempts at improving quality, expanding capacity, and economization, accompanied with digitization, have led to today's explosive spread of mobile communications. The four digital cellular systems or the second-generation systems, namely Personal Digital Cellular (PDC) in Japan standardized by the Research and Development Center for Radio Systems (RCR currently known as ARIB, the Association of Radio Industrial and Businesses), the Interim Standard 54 (IS-54) and IS-95 standardized by Telecommunications Industry Association (TIA) in the United States, and the Global System for Mobile Communications (GSM) standardized by the European Telecommunications Standards Institute (ETSI) are recognized as world standards by the Comité Consultatif International des Radiotélécommunications (CCIR

Wireless Communications in the 21st Century, Edited by Shafi, Ogose, and Hattori.
ISBN 0-471-155041-X © 2002 by the IEEE.

currently known as the ITU-R, the International Telecommunication Union Radiocommunication Sector).

New mobile communication systems appeared in the 1990s. In Japan, PHS came into practical use in 1995. Along with the propagation of data communications over fixed networks, the support of data communications in mobile communications has become an important issue. Dedicated mobile data communication systems such as Teleterminal and Mobitex were introduced and initially used. In Japan the Personal Digital Cellular–Packet (PDC-P) system based on PDC was developed and introduced in 1997. Among various pager systems that have been developed and introduced, FLEX-TD (aiming at increasing the speed and improving the spectrum utilization efficiency) has been developed and introduced as the second-generation pager system in Japan.

Diverse mobile communications systems have been introduced and operated worldwide, responding to the needs and the purpose of each user. Due to rapid internationalization, a more unified system has come into demand. The ITU initiated international discussion over ways to proceed with IMT-2000 (formerly FPLMTS, Future Public Land Mobile Telecommunication Systems), the standard for systems integrating cellular, cordless telephone, pager, and mobile data communications in order to establish a global and more unified system.

Mobile communications also include maritime and aeronautical mobile communications. This chapter does not deal with mobile satellite communications.

In this chapter, current major mobile communications systems are outlined. Next, the general concept of the IMT-2000, the third-generation system, is presented. Then the fourth-generation system, which is expected to come into practical use around the year 2010, is described. Finally, the situation in Japan regarding the development of wireless local area network (LAN)/wireless access, wireless home link, and information and telecommunication system (ITS) as future radio communication systems other than cellular is presented.

4.2 CURRENT WIRELESS COMMUNICATIONS SYSTEMS

4.2.1 Cellular Systems

Two digital cellular systems, PDC and IS-95, were introduced in Japan. PDC was introduced in 1993, and IS-95 followed in 1998. At the end of July 1999, PDC attained 95% of the market.

PDC is a digital mobile communications system developed and standardized in Japan. In 1991 on the initiative of ARIB, this air interface standard was established by ratification of RCR-STD-27 [1]. The standard for the Mobile Application Part (MAP) Signaling System, JJ-70.10, was drafted by the Telecommunication Technology Committee (TTC) in 1993. This standard is full-rate-3-channel and half-rate-6-channel time division multiple access, and the frequency bands are 900 MHz and 1.5 GHz. By adopting the $\pi/4$ shift quadrature phase shift keying (QPSK) modulation method, diversity technology, and low bit-rate voice CODEC [1], the spectrum utilization efficiency was improved and microportable terminals were developed. The half-rate pitch synchronous innovation code excited linear prediction (PSICELP) voice CODEC [1] in particular is the most efficient in transmission at 5.6 kbps. After establishing the standard, a series of improvements were made, and data communications at the maximum of 9.6 kbps and packet communications at the maximum of 28.8 kbps have been in force since 1995 and 1997, respectively.

PDC-P is a packet communications system based on PDC that was introduced in March 1997. Multislot transmission is used to establish data communications at 28.8 kbps, which is three times faster than the conventional circuit-switching data transmission at 9.6 kbps. Adaptive forward error correction (FEC) control and the automatic repeat request (ARQ) protocol were introduced, and the throughput was improved. These technologies made possible services that integrate voice and packet communications, and they incorporate a flexible air interface that does not rely on the user protocol.

4.2.2 PHS

PHS is a commercial digital cordless phone system developed in Japan, which began service in July 1995. PHS is a personal communications technology that is supported by digital cordless and intelligent network technologies. Its concept is to provide multimedia bidirectional communications, such as voice and data transmissions, while ensuring terminal mobility. PHS is considered a tool for computerizing personal daily life.

The network structure comprises base stations, which are connected to a switchboard through the existing integrated services digital network (ISDN) and PHS terminals. The air interface between the base stations and terminals, that is, the radio access transmission link, is the technical standard RCR-STD-28 ratified by ARIB [2]. This standard allocates 23 MHz in the 1.9-GHz band for its radio frequency, of which 12 MHz (40 channels) are for public use. Its access method is TDMA-TDD, and four slots each for transmission and reception are multiplexed at the same frequency. The multiplexed bit rate is 384 kbps, and the user rate for each slot is 32 kbps. Although the 32-kbps ADPCM full rate CODEC [3] is used for voice transmissions, it is capable of adopting a half-rate CODEC in the future. The modulation method is $\pi/4$ shift QPSK the same as in PDC.

4.2.3 Other Current Systems

Pager The pager is the easiest and the most compact one-way mobile message communications system. The pager system has multiple base stations that cover an extensive service area, and paging signals are transmitted from various base stations at the same radio frequency. As a result, signals from multiple stations arrive at the receiving point at the same time, which significantly improves reception reliability. Multicast services can be easily established using the pager.

The transmission bit rate of the pager system has been improved coinciding with advances in phase synchronization technology. The mainstream of the conventional system was 1200 bps, however, the Europe Radio Message System (ERMES), FLEX, and FLEX-TD established increased speeds of up to 6400 bps by employing high-accuracy time synchronous technology using global positioning system (GPS).

FLEX-TD, which is standardized in Japan, is a sophisticated paging system that integrates the signal bit rate and signaling method of FLEX with the technologies of variable multiple transmission and time diversity reception developed by NTT DoCoMo. With these technologies, in addition to the advantages of FLEX, reliability transmitting long messages has been improved, as well as minimizing the number of additionally required high-speed stations. Since FLEX-TD is compatible with FLEX, it is expected that it will be used globally as the second-generation paging system.

MCA Multichannel access (MCA) is a communications system that uses frequencies more efficiently than systems that occupy a certain frequency for each user. This is accomplished by sharing multiple frequencies by many users. Its basic system comprises an MCA controller station, which is shared between users and command and mobile stations, which are operated by users. Users communicate through the MCA controller station. This system employs a digital method that enables service provisioning using the TDM-TDMA communications system and 16-QAM modulation.

Maritime Mobile Communication This service has the longest history among mobile communications. This wireless telephone service was put into service in 1953 by Nippon Telegraph and Telephone Public Corporation. Initially, this manual system using the 150-MHz band was introduced in 1964, and then an automatic system was introduced using the 250-MHz band in 1979. Furthermore, by downsizing and adopting weight-saving technology cultivated through the development of mobile communications and wireless circuit control technology, which excels in its efficient use of frequency, a new maritime mobile communication system was developed in 1988. This maritime system comprising approximately 130 base stations covers an area stretching 50 to 100 km offshore nationwide, and attempts are being made to unify maritime mobile communications facilities with cellular systems as much as possible.

In March 1996, the N-STAR satellite was introduced to maritime communications. It employs the 2.6/2.5 GHz bands to meet the demand for diversification, enhancement, and service area expansion. By introducing this satellite service, problems with communications at sea, such as in the area of islands where conventional service was unavailable, were solved as well as the expansion of the service area. Moreover, while assuming the conventional maritime communication services, services exclusive to satellite communications such as multicasting service (broadcast) were started. This satellite service will sequentially replace conventional service.

Aeronautical Mobile Communication This service was started on domestic flights as an in-flight telephone service in March 1986. With six base stations throughout the country, its service area covers an altitude of more than 5000 miles nationwide. In June 1993, public switched telephone service started in response to requests from users of private airplanes. In aeronautical mobile communication systems, attempts are being made to share facilities with cellular telephones as much as possible. This would realize practical use of frequency taking into account traffic as well.

4.3 THIRD-GENERATION SYSTEMS

As mentioned previously, up to now various wireless communications systems have been developed and utilized together depending on the needs and purpose of each user. The international mobile telecommunications 2000 (IMT-2000) is one solution that integrates these various systems. A high degree of commonality of design worldwide, and the ability to provide new services and capabilities, such as multimedia services, are also main targets of the IMT-2000. Supporting high-speed data services, improving service quality, and efficiently using radio resources are paramount subjects for wireless systems. The IMT-2000 aims to utilize frequencies more efficiently, establish higher quality services than current systems, and support multirate services up to 2 Mbps. In Japan, ARIB proposed wideband code division multiple access (W-CDMA) to the ITU-R as a candidate radio transmission technology for the IMT-2000 in June 1998 [4].

Prior to this selection, ARIB investigated a number of candidates for wireless transmission technology for the third-generation mobile communication system that could satisfy the requirements and objectives of the IMT-2000 system and its services. After 2 years of intensive study with hardware demonstrations and field trials, W-CDMA was selected in January 1997 as the most suitable and the most verified wireless transmission technology for the third-generation mobile communication system. After selecting this concept, ARIB started refinement activities for the W-CDMA system proposal to make a better technical solution and, at the same time, to seek the possibility to coordinate with proposals from other regions.

Ten terrestrial radio transmission technologies (RTTs) including cdma2000 (TIA), UTRA (ETSI), and W-CDMA (ARIB) were submitted to the ITU-R. The 3G Partnership Project (3GPP) and 3GPP2 initiated coordination efforts aimed at a global IMT-2000 system based on the best technologies. 3GPP promoted W-CDMA and UTRA, while 3GPP2 promoted cdma2000, IXEV, IXEVDV, etc. Individual Japanese members took part in both projects and contributed to establishing a global standard. The Telecommunications Technology Council in Japan reported that direct spread CDMA (DS-CDMA), which corresponds to W-CDMA, and MC-CDMA (multicarrier CDMA), which corresponds to cdma2000, were as adequate as wireless transmission technologies of the third-generation mobile communication system.

Today, the W-CDMA network is operated by NTTDoCoMo in Japan. J-Phone expects to launch W-CDMA in the near future.

This section describes the key features and main radio transmission technologies of W-CDMA proposed by ARIB. It will clarify the ability and the limit of applications on the W-CDMA air interface.

4.3.1 System Design Concept

The following are the main features to be supported by the third-generation mobile communication system defined by ARIB [4]:

- High degree of design commonality worldwide
- Compatibility of services among different third-generation mobile communication systems and with fixed networks
- High service quality
- High-speed data services
- Worldwide use of small pocket terminals

The radio transmission technology of the third-generation mobile communication system should support multimedia and personal communications as well as intelligent functions. The system should meet advanced multirate services up to 2 Mbps and a quality comparable to that of fixed communications networks. At the same time, the system should aim to achieve a simple cell structure, easy channel management, high subscriber capacity, and low transmit power.

In W-CDMA, both a frequency division duplex (FDD) mode and a time division duplex (TDD) mode are utilized for the duplexing scheme. The combination of FDD and TDD makes it possible to use the allocated frequency efficiently according to the frequency conditions of each region, thus increases flexibility. Most of the key parameters including the chip rate, frame length, and modulation/demodulation schemes are made common to both modes. Only a certain part of layer 1 of the radio interface, for example, transmitter power control and diversity schemes, needs to be operated differently for each of these two modes.

Another point of difference is the synchronization requirement between base stations. The FDD mode can be operated both in synchronous and asynchronous modes, although the TDD mode requires synchronous operation based on the level of accuracy of the guard time.

4.3.2 W-CDMA Key Features

Generally, CDMA has the following features.

Highly Efficient Spectrum Utilization Spectrum utilization efficiency is easily improved in CDMA by employing transmitter power control [5] and voice operated transmission (VOX) technology, where signals are transmitted only when there is speech. The effect of these technologies is equivalent to sophisticated technologies such as dynamic channel assignment in other radio access schemes. This means that a system with a high level of spectrum efficiency can be established easily in CDMA.

Release from Frequency Management A frequency assignment plan is no longer necessary since CDMA allows the use of the same frequency in adjacent cells. With FDMA and TDMA systems, however, frequency assignment is a necessity. In an actual environment where base stations are deployed, it is quite difficult to assign frequencies while giving consideration to irregular propagation patterns and the impact of geographic topology. In addition, while imperfect frequency assignment design could lead to lowering of the spectrum utilization efficiency, CDMA can eliminate such concerns.

Low Mobile Station Transmit Power CDMA can improve the reception performance by utilizing technologies such as RAKE receiving [6]. The required transmit power at the mobile station can, thus, be reduced in CDMA systems compared to TDMA systems. Since intermittent transmission is employed in TDMA, the peak transmit power will increase in proportion to the number of timeslots. And, the same level of peak power for mobile terminals is required for even voice service. On the other hand, the peak power for CDMA can be kept low since continuous transmission is adopted. This is also advantageous in terms of keeping the impact in an electromagnetic environment to a minimum.

Use of Independent Resources for the Uplink/Downlink In CDMA, an asymmetrical uplink/downlink structure can be easily supported. In other access methods such as TDMA, it is difficult to assign separately the number of timeslots to the uplink and downlink for one user. It is also difficult to realize this with FDMA because the uplink/downlink carrier bandwidth must be changed. On the contrary, the uplink and downlink rates can be established independently in CDMA systems by using independent spreading factors for the uplink and downlink for each user. Consequently, radio resources can be used efficiently also in asymmetric traffic communications such as Internet access.

Furthermore, the wideband nature of W-CDMA can offer improved efficiency for the following points, as well.

Wide Variety of Data Rates   The use of a wider band carrier makes it possible to provide higher transmission rates. It also allows efficient provisioning of services, even in a situation where low rate services and high rate services coexist. This enables not only speech/fax services, but also high-resolution video services to be supported in the same band. Access to the Internet can be established at the same rate as from a fixed network.

Improvement of Multipath Resolution   RAKE diversity receiving technology improves the reception performance by combining the individual paths after they were received separately among multiple paths. The use of a wider band carrier can improve the capability to separate these multiple paths, which consequently reduces the required transmit power. This makes it possible to lower the transmit power of mobile stations, and at the same time decreases the interference power, which, in turn, leads to further improvement in the spectrum utilization efficiency.

Statistical Multiplexing Effect   The use of a wider band carrier increases the number of channels accommodated in one carrier. Further improvement in the use of frequency can thus be expected, thanks to the statistical multiplexing effect. In particular, when relatively fast data communication is performed, the efficiency drops in a narrowband CDMA system because the number of channels accommodated in one carrier is limited. On the other hand, the statistical multiplexing effect derived from a use of a wider band carrier can greatly enhance the efficiency.

Reduction of Intermittent Reception Ratio   The adoption of a wider band carrier can enhance the transmission rate of the control channels. When a mobile station is in a standby mode, it receives only a part of the control channel to save battery consumption, which results in a reduction of the intermittent reception ratio. This contributes to achieving longer standby time of the mobile station.

4.3.3   Wireless Transmission Technologies [6–8]

The following describes the key features of W-CDMA technology that outline the physical layer of the wireless interface.

- To provide operation flexibility, besides the 5-MHz bandwidth (3.84 Mcps), expandability to 10 MHz (7.86 Mcps) and 20 MHz (15.36 Mcps) is also established. Figure 4.1 provides a rough image of the interrelation between the chip rate and data rate.
- Carrier raster enables carrier spacing to be flexible and detailed and improves spectrum utilization efficiency.
- Both FDD and TDD modes are applied as W-CDMA duplexing schemes. The combination of FDD and TDD enables highly efficient use of the spectrum depending on the conditions in each region, and at the same time offers a high degree of flexibility. Most of the key parameters including the chip rate, the frame length, and the modulation/demodulation schemes are common to both modes.

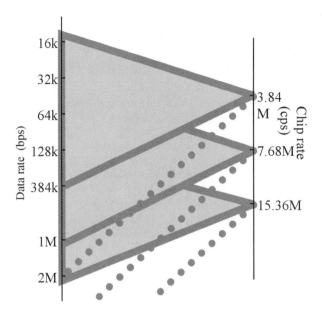

FIGURE 4.1   Interrelationship between data rate and chip rate.

- Intercell asynchronous operation, which does not require precise synchronization between base stations, was adopted for W-CDMA to ensure the freedom to deploy base stations indoors and outdoors. However, W-CDMA can also adopt intercell synchronous operation to provide operation flexibility.

- As a solution to implement multirate transmission, a variable spreading factor is used for the uplink, and an orthogonal variable spreading factor is used for the downlink, while the hardware implementation is simplified. It is possible to use multicode transmission for high-speed transmission, for example.

- Convolutional codes are employed for channel coding. Turbo codes, which offer better error correction performance, are applied to high-speed data transmission.

- Spreading codes have a two-layered structure comprising channelization codes and scrambling codes [9]. In the downlink, scrambling codes are assigned specifically to each cell (see Fig. 4.2), while they are assigned specifically to each user in the uplink. Since there is a plethora of channelization codes, the codes can be assigned to each cell without any constraints. Channelization codes are orthogonal, and all codes are used commonly for all cells, minimizing the interference between users within the cell.

- Coherent detection using pilot symbols is used as a detection method [10]. The pilot symbols on the downlink are common (CDM: code division multiplex) and dedicated (TDM: time division multiplex). Utilizing a common pilot symbol establishes the accuracy of channel estimation because the pilot power can be increased. On the other hand, TDM minimizes the control delay of the transmitter power control. The pilot symbols on the uplink are I-Q multiplexed together with data, after they are spread with different spreading codes than those for the data. This enables continuous

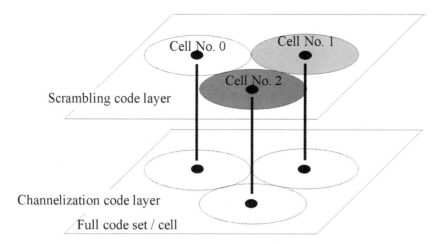

FIGURE 4.2 Two-layer spreading code allocation.

transmission when variable-rate transmission is carried out, and can minimize the peak factor of the transmit waveform. This solution is effective in reducing the impact on an electromagnetic environment and in alleviating the requirements on the mobile station transmit amplifier. However, in the TDD mode, the uplink pilots are time multiplexed similarly to the downlink. This is because there are no advantages to I-Q multiplex pilot symbols in a discontinuous TDD mode.

- A transmission scheme based on dedicated pilot symbols is adopted as the transmission method of pilot symbols. This scheme is beneficial for fast closed-loop transmitter power control on the downlink. It is also possible to apply a common pilot scheme, where the pilot symbols of common control channels are used by each traffic channel. This makes it possible to further improve the performance. The dedicated pilot scheme is an extremely effective solution in terms of securing expandability to adopt adaptive antennas [11], interference canceller [12], and other technologies in the future.

- Packet data transmission is performed to support asymmetric uplink/downlink transmission and a wide range of data transmissions from low to high data rates. For this packet transmission, adaptive channels depending on the traffic characteristics are used. For example, when the traffic is light, common physical channels are used. On the other hand, dedicated physical channels are used when the traffic is heavy.

Regarding multimedia services, the important factors are quality and quantity of contents and the way of providing these services. We have Internet access services through current cellular systems such as i-mode service in Japan. Current i-mode services are provided through the PDC-P network; therefore, the information is mainly presented as text. On the other hand, since 2001, the IMT-2000 was launched and transmission speeds gained a significant boost. Around that time, it will be possible for i-mode users to experience multimedia-rich contents through their terminals. By rich contents we mean

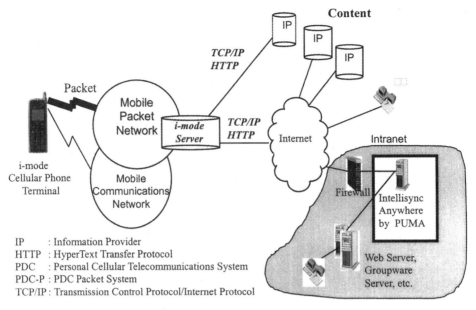

FIGURE 4.3 System configuration of i-mode service.

moving color images or high-quality music. Figure 4.3 shows the system configuration of i-mode services. Current i-mode contents will become more attractive when the technology supporting rich contents becomes available (Fig. 4.4).

4.4 FOURTH-GENERATION SYSTEMS

First-generation (1G) and second-generation (2G) cellular systems have been used mainly for voice applications and supporting switched-circuit-type services. The transition in

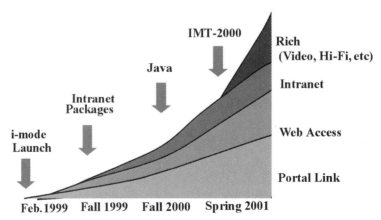

FIGURE 4.4 Evolution of i-mode service.

information types from voice (telephone) to data and multimedia will accelerate in the 21st Century, thus the third-generation (3G) systems have been designed to cover both telephone-type and data/multimedia-type communications. In 2010, this trend will be even more remarkable. The fourth-generation (4G) mobile communication system should accommodate increased data/multimedia traffic in the 2010s.

Mobile access to both the Internet and intranets will become increasingly popular and essential. Data size is increasing year by year and higher speed mobile communication systems will be required to increase user satisfaction. Mobile communication devices will be not only handheld phones or personal digital assistants (PDAS) carried by people but also installed in all kinds of goods such as luggage, wallets, or leashes for pets. After several years when penetration into the mobile phone user market becomes saturated, the number of mobile devices will still increase and reach a level several times that of the number users in the 2010s. Various new applications will explore mobile communications markets that are not apparent at present. See Figure 4.5.

4.4.1 Requirements

The 4G system should achieve the following:

High-Speed Transmission The 3G system covers up to 2 Mbps for indoor environments and 144 kbps for vehicular environments. The 5-GHz band wireless LAN and wireless broadband access systems being developed in Japan (MMAC), Europe (Hiperlan2), and United States (IEEE 802.11) have an approximate 20-Mbps transmission speed. The target speed of 4G will be 10 to 20 Mbps for quasi-static environments and at least 2 Mbps for moving vehicles.

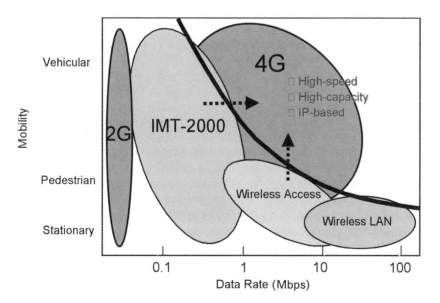

FIGURE 4.5 Position of 4G system.

Higher Capacity   The 3G system capacity will not be sufficient to handle the explosively growing data/multimedia traffic in the 2010s. Spectrum efficiency for 4G systems should be 5 to 10 times higher than that of the 3G system. To accommodate the enormous amount of traffic in dense urban areas, a spatial frequency reuse strategy is the key. At this point, seamless geographical coverage with a microcellular structure is better than a "hot-spot" coverage strategy because the former can avoid concentrating geographical traffic. Thus, constructing microcellular networks at a reasonable cost is an important issue.

Good Area Coverage with Variable-Speed Transmission   As transmission speeds increase, the required receive signal level will increase. In the existing cellular systems, the transmission speed is relatively low (up to 64 kbps), thus they can provide good coverage with medium-size cells. Since the target speed of the 4G system is more than 100 times higher than the existing systems, the cell radius will be decreased and in-building coverage will be severely degraded. Wide-range variable-speed transmission is necessary to extend coverage to indoor areas where there are neither 4G base stations nor other wireless networks connected to the 4G system.

Wireless QoS Resource Control   For Internet services, the best effort service is very attractive because it has the potential to lower service cost. However, wireless systems use limited radio resources (frequency bandwidth and transmitting power) and suffer from congestion. Therefore, wireless quality of service (QoS) resource control is necessary to maintain the service quality and to support various applications and service classes.

Seamless Service with 3G, Wireless LAN (W-LAN), and Fixed Networks   By constructing networks based on IP (Internet protocol) technology, seamless connection between 4G, 3G, wireless LANs (W-LANs), and fixed networks will be possible. As a result, users will be able to choose the network that is best suited to their situations (place, time, cost, etc.).

Reduced Network Cost   The bit cost in the 2010s should be decreased, perhaps to the level of one tenth that at the beginning of the 2000s.

4.4.2   Techniques to Establish High-Speed and High-Capacity Communications

High-speed mobile transmission suffers from frequency selective fading. Robust modulation/demodulation schemes should be studied to find a way to withstand frequency selective fading. Multiple carrier modulation and single carrier modulation with adaptive equalizers are candidates.

The other key demand for high speed is an extremely low required E_b/N_0 value. Since the noise bandwidth at the receiver is wide in a high-speed system, low E_b/N_0 values are required to achieve reasonable area coverage. High-speed transmitter power control to mitigate Rayleigh fading and a pilot-added fast-tracking coherent demodulator are effective ways to achieve this goal. Frequency domain antifading measures such as RAKE combining spread spectrum receivers or frequency hoping techniques are also necessary.

Interference canceling with an adaptive array antenna and an interference canceling equalizer (ICE) is a promising candidate to increase capacity. One problem in making this candidate feasible is in implementing a fading channel estimation circuit. Since the channel estimation requires many calculations, the estimation algorithm and processor configuration are very important. A systolic array RLS processor is a candidate to achieve this function.

4.4.3 Techniques to Handle Multimedia Traffic

Intelligent wireless resource management is a key technique in handling multimedia traffic. For high-speed mobile communications, not only the spectrum resource but also the transmission power available in the base station and mobile station restricts the user's transmission speed. A wireless resource manager should check the available resources, the quality of the forward and reverse links, the application type, and user class in the QoS services and then assign the appropriate forward and reverse link speeds and transmitting power to the user.

The IP-technology-based network structure can handle IP packet traffic efficiently and at a low cost. It can also easily provide the broadcast and multicast functions essential for push-type information services. The key issue is a routing/hand-over scheme and an authentication strategy that does not affect mobility or throughput.

4.5 OTHER FUTURE SYSTEMS

4.5.1 Wireless LAN/Wireless Access

Multimedia mobile access communication (MMAC) systems aim to provide broadband wireless access for mobile/nomadic multimedia services. In 1995, the Japanese Ministry of Post and Telecommunications (MPT) formed the MMAC study committee to conduct a study on the next generation of broadband mobile communication systems. The committee presented its report in May 1996. It suggested that MMAC should enable seamless wireless access to broadband backbone networks, for example, ATM networks, and it should be deployed early in the 2000s. The MMAC report identified two system concepts, the first of which is mobility oriented "high-speed wireless access" for broadband mobile/ubiquitous multimedia services up to 30 Mbps using 3 to 60 GHz for both indoor and outdoor applications in both private and public environments. Since it will provide more than 20 Mbps, 10 times higher than IMT-2000, its mobility is limited to pedestrian speeds. This limitation is mainly due to its relatively small cell radius and the relatively high frequency to be used. The other MMAC system is a quality oriented "ultra-high-speed wireless LAN" up to 156 Mbps using 30 to 300 GHz, for corporate LAN use.

Following the recommendations in the MMAC report, the MMAC Promotion Council was established in December 1996 to promote the development of MMAC systems in cooperation with ARIB [13]. MMAC-PC comprises a technical committee and a promotion committee. The MMAC technical committee comprises the High-Speed Wireless Access Subcommittee, Ultra-High-Speed Wireless LAN Subcommittee, and two ad hoc committees, that is, the 5-GHz Band Mobile Access Ad-hoc Committee and the Wireless Home Link Ad-hoc Committee, which were established in December 1998. The 5-GHz Band Mobile Access Ad-hoc Committee has two working groups: the first is the ATM-WG working with the High Speed Wireless Access Subcommittee and the other is the Ethernet-WG. The Wireless Home Link Ad-hoc Committee is developing specifications for IEEE Standard 1394, which is based on wireless LANs. The targets of the above-mentioned MMAC systems are summarized in Table 4.1. Those MMAC family systems will enable wireless access in various environments, that is, home (indoor), office (indoor), premises, and public spaces (outdoors and indoors).

The MMAC-PC has a close liaison relationship with other standardization bodies. The MMAC High-Speed Wireless Access Subcommittee and ATM-WG has a liaison relationship with ETSI-BRAN and ATM Forum WATM-WG, and MMAC Ethernet-WG has a similar relationship with IEEE 802.11, to coordinate the specifications. Most notably, these three groups (MMAC, ETSI-BRAN, and IEEE 802.11) have selected coded OFDM as the modulation scheme for wireless access systems using the 5-GHz band because of its robustness against frequency selective fading, large coding gain, large diversity gain, and scalability in its transmission speed. Coded OFDM for these systems has the same basic parameters, that is, 48 subcarriers plus 4 pilots based on 64-point FFT/IFFT, 20-MHz channel separation, 4 μs per OFDM symbol, and $R = \frac{1}{2}$, $K = 7$ convolutional coding with interleaving within the OFDM symbol. This means that the OFDM modem LSI can be applied to wireless access devices according to the above-mentioned standards although

TABLE 4.1 Target of MMAC Systems

	High-Speed Wireless Access	Ultra-High-Speed Wireless LAN	5-GHZ Mobile Access (Wireless ATM & Wireless LAN)	Wireless Home Link
Service area	Public: outdoor, indoor	Private: indoor	Public: outdoor, indoor	Private: indoor
	Private: indoor, premises		Private: indoor, premises	
Network and interface	Public: ATM, IP, etc.	Private: ATM	Public: ATM, IP, etc.	IEEE 1394, etc.
	Private: ATM, IP, Ethernet, etc.		Private: ATM, IP, Ethernet, etc.	
Information rate	30 Mbps	156 Mbps	20 to 25 Mbps	30 to 100 Mbps
Terminal equipment	Notebook-type PCs, etc.	Desktop PCs and WSs, etc.	Notebook-type PCs, handy terminals, etc.	PCs and audio visual equipment etc.
Mobility	Stationary or pedestrian (with handover)	Stationary (with handover)	Stationary or pedestrian (with handover)	Stationary or pedestrian (with handover)
Radio frequency	25/40/60 GHz	60 GHz	5 GHz	5/25/40/60 GHz
Bandwidth	500 to 1000 MHz	1 to 2 GHz	Greater than 100 MHz	Greater than 100 MHz
Bit error rate	Around 10^{-6}	Equivalent to wired neworks (around 10^{-8} to 10^{-10})	Around 10^{-6}	Equivalent to wired networks (around 10^{-8} to 10^{-10})

there are small differences among them, for example, preamble patterns. This will be helpful in reducing the wireless terminal cost. Draft MMAC specifications will be released in early 2000s.

4.5.2 Wireless Home Link

There are currently many electric devices in the house and will continue to be so in the future. One group of these devices is household appliances such as air-conditioners, refrigerators, microwave ovens, and lights. Another is multimedia devices such as desktop PCs, printers, TVs, set-top boxes, audio sets, and digital video cameras.

In the MMAC-PC, an air interface protocol is being developed for the wireless home link that connects these electric devices to one another and with telecommunication networks and cable TV networks. The assumed frequency bands are the 5-GHz and millimeter-wave, and they are compatible with IEEE 1394. The 5-GHz band link has the advantage that it can be used throughout the household. However, its transmission speed is limited to around 30 Mbps, which is suitable to transmit digital video camera signals, but lower than the original IEEE 1394 speeds (100 Mbps and above) and requires a bridge device to adapt the speed. The other millimeter-wave link supports the full 100-Mbps transmission speed [13].

4.5.3 ITS

ITS, the information and telecommunication system, in Japan is outlined as follows [14,15]:

1. *First Phase* "Dawn of ITS" (to be established around the early 2000s): High-performance ITS systems are achieved by improving and adding applications to the various stand-alone systems that are already in use today or soon to be introduced, such as car navigation systems and electronic road toll systems.

2. *Second Phase* "Advancement of ITS" (to be established around 2005): ITS becomes more advanced thanks to improved network facilities and multiple communication methods, varying from simple voice and data to multimedia communications including real-time video transmission. Terminal equipment will feature high-performance functionality, multifunctionality, and an interface that is user friendly and safe to use while driving.

3. *Third Phase* "Completion of ITS" (to be established around 2010): ITS becomes sufficiently advanced and automated and integrated functions become available such that fully automated driving becomes a reality. Fully automated driving makes vehicles safer and more pleasant to ride in.

Among the wide range of ITS applications, some of the most interesting areas from the end users' perspective are as follows

- Road and traffic information provisioning and car multimedia
- Dedicated short range communication (DSRC) including electronic toll collection (ETC)

- Provisioning of public transport information and efficient commercial vehicle operation
- Advanced driving-assist systems and automated driving

Many key technologies are required to develop and complete ITS. The following are the major technologies:

- Wireless agents and highly reliable distributed control technologies that can handle incomplete or ambiguous user input
- Optical and radio conversion devices that achieve high-speed and broadband applications in the millimeter wave band
- Road-to-vehicle and vehicle-to-vehicle communications that provide continuous communication between vehicle and the road for automatic driving
- Multicast routing that uses various advanced telecommunications and broadcasting systems
- Advanced human–machine interfaces that enable automatic driving
- Multimode terminal equipment that achieves flexible changes in communication methods with software technologies

4.6 SUMMARY

Today, we have various mobile communications systems such as cellular systems, cordless systems, paging systems, MCA, maritime and aeronautical mobile communications systems, mobile satellite communications systems, to name a few. IMT-2000 is one solution that can integrate these systems. It achieves a worldwide design and will provide new services particularly in the field of multimedia with high-speed data services. ARIB has selected W-CDMA as the most suitable and the most verified radio transmission technology for IMT-2000. Higher speed transmission, higher capacity, and good area coverage with variable-speed transmission will be pursued in 4G systems. Other than cellular, MMAC, wireless home link, and ITS are future radio communication systems that are currently being studied in Japan.

REFERENCES

1. RCR STD-27, "Personal Digital Cellular Telecommunication System RCR Standard," Feb. 1999.
2. RCR STD-28, "Personal Handy Phone System RCR Standard," Feb. 1999.
3. ITU-T Recommendation G.726, "40, 32, 24, 16 kbps Adaptive Differential Pulse Code Modulation (ADPCM)," Dec. 1990.
4. ARIB IMT-2000 Study Committee, "Japan's Revised Proposal for Candidate Radio Transmission Technology on IMT-2000: W-CDMA," Association of Radio Industries and Businesses (ARIB), Sept. 1999.
5. T. Dohi et al., "Performance of SIR Based Power Control in the Presence of Non-Uniform Traffic Distribution," IEEE ICUPC'95, pp. 334–338, 1995.
6. S. Onoe et al., "Wideband-CDMA Radio Control Techniques for Third Generation Mobile Communication Systems," IEEE VTC'97, pp. 835–844, 1997.

7. F. Adachi et al., "Coherent Multi-rate Wideband DS-CDMA for Next Generation Mobile Radio Access: Link Design and Performance," APCC'97, pp. 1479–1483, 1997.

8. M. Sawahashi, et al., "Wideband CDMA Mobile Radio Access for IMT-2000," 2nd CIC, pp. 571–576.

9. K. Higuchi et al., "Fast Cell Search Algorithm in DS-CDMA Mobile Radio Using Long Spreading Codes," IEEE VTC'97, pp. 1430–1434, May 1997.

10. H. Andoh et al., "Channel Estimation Using Time Multiplexed Pilot Symbols for Coherent Rake Combining for DS-CDMA Mobile Radio," IEEE PIMRC'97, pp. 954–958, 1997.

11. S. Tanaka et al., "Pilot Symbol Assisted Decision Directed Coherent Adaptive Antenna Array Diversity for DS-CDMA Mobile Radio Reverse Link," *IEICE Trans. Fundamentals*, Vol. E80-A, pp. 2445–2454, Dec. 1997.

12. M. Sawahashi et al., "Pilot Symbol Assisted Coherent Multistage Interference Canceller Using Recursive Channel Estimation for DS-CDMA Mobile Radio," *IEICE Trans. Commun.*, Vol. E79-B, pp. 1262–1270, Sept. 1996.

13. Multimedia Mobile Access Communication Systems Home Page, *http://www.arib.or.jp/mmac/e/index.htm.*

14. ITS Telecommunications Business, "Report of Telecommunications Technology Council in Ministry of Posts and Telecommunications," Great Cruise Co. Ltd., April 1999.

15. *ITS Handbook in Japan*, Handbook Supervised by Ministry of Construction, HIDO (Highway Industry Development Organization), Oct. 1998.

DEVELOPMENTS IN INTERNATIONAL STANDARDS

■■■■■■ **CHAPTER 5**

Developments in International Standards

JANE BROWNLEY, FRAN O'BRIEN, MARIA PALAMARA, DEREK RICHARDS,
and LYNNE SINCLAIR

5.1 OVERVIEW

5.1.1 Applications Driving the Need for New Standards in the 21st Century

Thanks to the rapid growth of the Internet and the wide variety of information and applications that it offers to users, data traffic has overtaken voice traffic on wireline networks in many areas of the world. Users have quickly become dependent on rapid access to the multimedia information afforded by the Internet and the convenience of doing business from the home regardless of the hour of the day. The ability of wireless technology to satisfy user needs for access to the Internet while mobile has not kept up with user interest.

With the emergence of third-generation (3G) technologies, data rates available on wireless are just now beginning to meet user requirements for their applications. Global standardization efforts have made substantial impact on wireless data ease of use by increasing data rates, improving Internet connectivity through enhanced packet data standards that solve security issues associated with private network access, and addressing wireless network interoperability issues to permit users to roam freely.

5.1.2 ITU's Vision for IMT-2000 and Today's "Family of Systems" Approach

The International Telecommuncations Union (ITU) has been coordinating global efforts to address these standards objectives in its initiative known as IMT-2000.* IMT-2000 was initially envisioned to be a single, worldwide standard offering high-speed, multimedia

---

*International Mobile Telephony-2000, formerly known as FPLMTS (Future Public Land Mobile Telephony System).

Wireless Communications in the 21st Century, Edited by Shafi, Ogose, and Hattori.
ISBN 0-471-155041-X © 2002 by the IEEE.

service with global roaming capabilities. Target data rates included 144 kbps for high-speed mobile service, 384 kbps for pedestrian-level mobility, and 2 Mbps for fixed service. It was targeted for a single-spectrum band in the range of 1.8 to 2.2 GHz, and countries were requested to clear this band for third-generation wireless service.

Today, with the vast presence of second-generation (2G) digital wireless networks representing billions of dollars of investment, operators would like to leverage their investment to ease the economic burden of deploying third-generation networks. Also, the United States has allocated a large portion of the designated frequency band for personal communication systems (PCS) service, with multiple competing 2G technologies already deployed in that band. These key economic drivers have forced the ITU to reconsider its original vision, resulting in a "family of systems" concept that allows operators to evolve their 2G networks to third generation and focuses on the interoperability issues associated with networking the resulting 3G systems to allow for global roaming.

5.1.3 3G and Related Standardization Efforts Around the World

The standardization process has been changing over recent years to reflect the current trend toward global standards and the need for rapid standards development. More often than not, special-interest groups (SIGs) are forming that are chartered to advance the specifications for certain technologies that are then adopted by formal standards development organizations (SDOs). The primary organizations that have emerged in setting third-generation specifications are the Third-Generation Partnership Projects, 3GPP and 3GPP2. Several other groups are also developing proposals for inclusion in 3G standards through these partnership projects. These SIGs and their efforts will be described in the following pages. While IMT-2000 and the future of wireless also include a significant presence for satellite-based systems, this chapter will concentrate on terrestrial-based networks, where most of the current interest in standards has been focused.

5.2 ITU'S IMT-2000 STANDARDIZATON EFFORTS

IMT-2000 is defined by a set of interdependent ITU recommendations. This work has been focused in two main groups, Task Group 8/1 of the ITU-R and Study Group 11 of the ITU-T. TG8/1 has been focusing on radio-related issues, while SG11 has been addressing the network-related issues related to IMT-2000.

5.2.1 ITU-R Standardization Efforts

Task Group 8/1 Charter   ITU-R Task Group 8/1 was the key organization involved in defining the radio aspects of IMT-2000. It was a design objective of IMT-2000 that the number of radio interfaces should be minimal and, if more than one interface is required, that there should be a high degree of commonality among them.*

---

*These radio interfaces will serve a variety of radio operating environments, as indicated in ITU-R, Rec. M.1034 on the requirements for the radio interface(s) for future public land mobile telecommunication system.

RTT Submissions and Evaluation Efforts   Sixteen candidate radio transmission technologies (RTTs) for IMT-2000 were submitted to the ITU:* 10 for the terrestrial component and 6 for satellite. The 10 terrestrial candidate RTTs included:

- cdma2000 proposed by the Telecommunications Industry Association (TIA) TR45.5
- DECT proposed by the European Telecommunications Standards Institute (ETSI) Project DECT
- Global code division multiple access (CDMA) I and II proposed by TTA of Korea
- TD-SCDMA proposed by China Academy of Telecommunications Technology
- UTRA proposed by ETSI SMG
- UWC-136 proposed by TIA TR45.3
- W-CDMA proposed by the Association of Radio Industries and Businessess (ARIB) of Japan
- WCDMA/NA proposed by ATIS T1P1
- WIMS W-CDMA proposed by TIA TR46

These RTTs were evaluated under a procedure given as guidelines in recommendation ITU-R M.1225 to determine their conformity with the minimum performance capabilities, such as 144 kbps for vehicular environments, 384 kbps for pedestrian environments, and 2 Mbps for indoor environments. Fourteen evaluation groups submitted evaluation reports on one or more of these RTTs to the ITU. With these evaluation reports, ITU-R TG8/1 concluded that all terrestrial and satellite RTT proposals met the minimum performance capability requirements and published these specifications as IMT.RSPC. TG8/1 has since been disbanded, and continued maintenance of the radio specification criteria and post-IMT-2000 issues such as intelligent antenna support and impact of all-IP (Internet protocol) core networks are currently managed under ITU-R Working Party 8F.

Global Harmonization Efforts   Several global wireless operators have collaborated to determine how the various CDMA proposals for the IMT-2000 system could be harmonized. An Operators Harmonization Group (OHG) was established to discuss key parameters—chip rate, pilot structure, and synchronization method. As a result of these efforts, the OHG has now agreed to a harmonized global third-generation (G3G) CDMA standards framework. They recommend:

- A harmonized standard for direct sequence (DS) mode based on the UTRA W-CDMA proposal with an additional common CDM (code division multiplexed) pilot and a chip rate of 3.84 Mcps
- A harmonized standard for multicarrier (MC) mode based on the cdma2000 proposal with a chip rate of 1.2288 and 3.6864 Mcps
- A chip rate of 3.84 Mcps for TDD mode

*In accordance with the procedures specified in the ITU Circular Letter 8/LCCE/47.

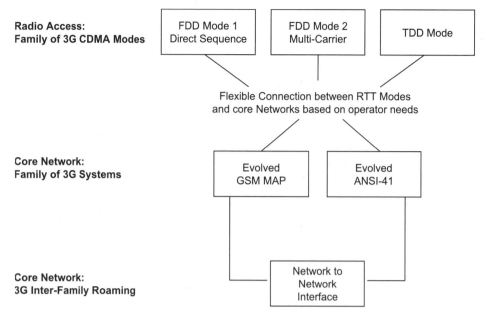

FIGURE 5.1   Modular 3G harmonization proposal.

The OHG further proposed a modular approach (see Fig. 5.1) by which ANSI-41 and GSM-MAP-based services will be fully supported in the radio access network associated with all three 3G CDMA modes.

In addition to harmonization efforts for CDMA, the time division multiple access (TDMA) community—advocates of the UWC-136 RTT—are actively working to converge the ANSI-136 TDMA standard's evolution with that of GSM radio technology's evolution, with a common migration path toward an enhanced data rates for global evolution (EDGE) radio interface using general packet radio service (GPRS) networking for packet data. They are also actively involved in specifying methods for roaming between ANSI-136 networks and GSM networks through the GAIT (GSM-ANSI-136 Interoperability Team) initiative. They are working with ETSI and 3GPP to assure that the resulting standards for EDGE radio characteristics, GPRS networking characteristics, and GSM/ANSI-41 internetworking do not diverge.

5.2.2   ITU-T

Relevant Study Groups   In ITU-T Study Group 11 (SG11) is the lead study group for IMT-2000-related work. SG11's IMT-2000-related work has focused on capabilities that include broadband data, multimedia, Internet access, and global roaming. In addition to SG11, there is work proceeding in SG4 within ITU-T, which is addressing network management for IMT-2000 systems. SG7 is involved in security aspects of IMT-2000, and SG16 is involved in Internet protocol (IP) issues.

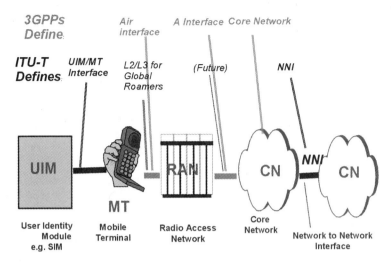

FIGURE 5.2   IMT-2000 interfaces and organizations defining the specifications for them.

Family of Systems Reference Model   The family of systems concept was developed in recognition of the fact that 3G systems would evolve from existing 2G systems, and that the existing 2G systems are based on a few different regional standards (Europe's GSM, U.S.'s ANSI-41 based, and Japan's PDC networks being the most predominant). A "green field" approach, where both GSM MAP and ANSI-41 standards are abandoned in 3G in favor of a new core network protocol is not practical given the huge investment in 2G networks around the world.

The family of systems concept was developed to allow existing operators to evolve their current networks to 3G, to reuse the existing features and services already available on their 2G networks while adding 3G capabilities, and to supplement early spotty 3G coverage with their 2G networks to provide ubiquitous service.

The ITU decided that it would *not* specify the protocol used within the core network, allowing instead the regional protocols (e.g., GSM MAP and ANSI-41) to be evolved independently. To be a "family member" of IMT-2000, a regional standard must:

- Support a set of service and network capabilities provided in the capability sets as defined in ITU-T (see ITU-T recommendation Q.1701).
- Support a set of IMT-2000 interfaces to be defined by the ITU (see IMT-2000 interfaces section below).
- Must have a core network that can interface with other "family member" networks to enable global roaming and interoperability.

IMT-2000 Interfaces   Figure 5.2 shows a simple network reference model that illustrates the interfaces that are being defined for IMT-2000 standards. The interfaces are as follows:

- User identity module (UIM) to mobile terminal (MT) interface will be specified by the ITU.

- MT to radio access network (RAN) interface is the air interface. The physical layer, or radio interface, is the subject of the radio transmission technologies defined by ITU-R TG8/1.

- RAN to core network (CN) interface is often referred to as the A interface. 3GPP and 3GPP2 are currently defining standards for this interface. The ITU may define standards for this interface in the future.

- The Network-to-Network Interface (NNI) is utilized in situations where end users roam between family member networks (e.g., between an evolved GSM system such as UMTS and an ANSI-41 system).

Challenges of Global Roaming Across Different Core Networks   There are two classifications for types of roaming, or mobility. There is *terminal mobility*, or the ability to roam with the same terminal, and there is *personal mobility*, which is the ability to roam and have the same end user services regardless of location but not necessarily the same terminal. Personal mobility requires some way of identifying the end-user to the network because the terminal device is not constant. For IMT-2000, the user identity module (UIM) performs that end-user identification function to the wireless network.

One of the challenges to achieving global roaming is the interoperability between the different family member core networks. The interface between core networks, as previously discussed, is via the NNI. 3G network interoperability within core networks (i.e., within a family member network) will be based on either ANSI-41 evolution or GSM MAP evolution. Therefore, the NNI is the bridge between these two family member networks. While these two protocols are functionally equivalent in what they provide, the implementation of the services or operations are different.

Evolving Packet Data Networking Standards   High-speed data service is a major driver for 3G standards. Providing high-speed data service over the air interface has received a tremendous amount of attention. But, the network side of data services is important as well. What are the services that need to be provided to the end user of packet data services? There are a few categories of services, each having different requirements, especially in regard to accounting and security:

- Access to the public Internet
- Access to private networks, or intranets

Network Management   The network management standardization activities of the ITU are focused on interfamily issues. The topics being addressed for network management in IMT-2000 include charging and accounting, management functions for roaming agreements, security management, service provisioning, configuration management, and fault and performance management.

5.3   3G STANDARDIZATION CONSORTIA

As mentioned previously, the demands of third-generation market situations have influenced new approaches for defining standards that will significantly accelerate the process and time to market. International consortia/special-interest groups have arisen to define specifications for 3G family members on a rapid timescale to allow manufacturers to get their products to market faster. The key bodies involved in these efforts are the third-generation partnership projects (3GPP and 3GPP2) and the Universal Wireless Communications Consortium (UWCC). Specifications developed by these consortia will in turn be adopted by regional standards development organizations (SDOs), including (but not limited to) TIA and T1P1 of the United States, ETSI of Europe, TTA of Korea, ARIB and TTC of Japan, and CWTS of China. With the creation of 3GPP and 3GPP2, the 3G standards development process becomes open to participation of standards organizations outside of Europe (GSM/ETSI's home) and the United States (ANSI-41's home), allowing participation to potentially all of the countries that will deploy the standards.

5.3.1 3GPP*

The widespread commercial success of GSM caused many operators and infrastructure manufacturers to question the commercial viability of planning 3G on the basis of a new core network architecture. During 1997 a decision was made that a UMTS (universal mobile telecommunications system) would be based upon an evolution of the GSM core network with the aim of having, as far as possible, a common core network for both GSM and UMTS.

In January 1998 ETSI agreed on wideband CDMA and time division CDMA as the radio technologies for the UMTS terrestrial radio access network (UTRAN). At the same time, Japan, Korea, and the United States were also developing wideband CDMA proposals, and China was developing a TD-CDMA proposal. There then followed a series of high-level discussions between the various regional standards organizations, and as a result 3GPP (Third-Generation Partnership Project) was formed at the end of 1998 with the aim of developing one UTRAN standard, based upon an evolved GSM core network.

3GPP has completed specification of the first release (Release 99) of the UMTS standard and is currently actively working on specifications for the Releases 4 and 5 of the standard. Separate publication of UMTS standard, variations on the generic standard produced by 3GPP, is planned by regional standards organizations to take account of regional requirements.

5.3.2 3GPP2

Another partnership program that has been established to develop regional standards for IMT-2000 is 3GPP2 (Third-Generation Partnership Project 2).[†] Whereas 3GPP is developing 3G standards for the evolution of GSM MAP and the UTRAN air interface, 3GPP2 is developing 3G standards for the evolution of ANSI-41-based networks, primarily involving the cdma2000 air interface.

*http://www.3gpp.org
†http://www.3gpp2.org.

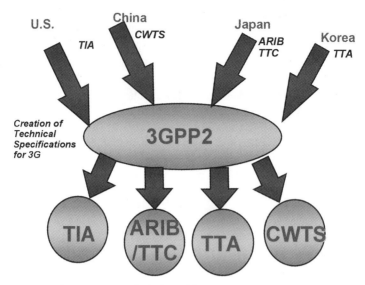

FIGURE 5.3 3GPP2 creates technical specifications to be standardized by partner SDOs.

Similar to 3GPP, 3GPP2 was formed to open up the standards development process to all the interested stakeholders, giving other countries* a voice in the creation and establishment of the standards. At the time of this writing 3GPP2 membership consists of TTA of Korea, ARIB and TTC of Japan, TIA of North America, and CWTS of China.

3GPP2 will create 3G specifications that will be standardized in the partner SDOs listed above (Fig. 5.3).

5.3.3 UWCC†

The Universal Wireless Communications (UWC) program was launched in 1995 as a collaborative effort among leading vendors and operators of wireless products and services to deliver an enhanced portfolio of global mobility services—across all spectral, market, and subscriber bands. Key goals of the UWCC are to promote deployment of IS-136 TDMA and IS-41 WIN products and services worldwide, stimulate subscriber and operator growth using TDMA/WIN on a global economic scale to optimize terminal costs; provide product availability and information to facilitate conversion of AMPS-based (Advanced Mobile Phone System) systems to TDMA/WIN, expedite deployment of TDMA/WIN into all global markets, and continue development of enhanced seamless TDMA/WIN services.

Since 1998, and with a recent direction of ANSI-136 to converge with GSM technology in the future, the UWCC has announced several initiatives related to roaming in cooperation with the U.S. GSM alliance. The UWCC is the driving industry body behind the UWC-136 proposal to the ITU for ANSI-136 evolution. Cooperation with the 3GPP has also been a priority of UWCC, since the 3GPP will be the key organization specifying the evolution of GSM and GPRS-based core networks.

*The standards development organizations (SDOs) of these countries are the actual partner organizations in both 3GPP and 3GPP2.
†http://www.uwcc.org

5.4 EVOLVING RADIO STANDARDS

5.4.1 CDMA-Based Standards

Cdma2000, or G3G MC   Cdma2000 is a third-generation RTT that is a direct evolution of the second-generation cdmaOne (IS-95) technology. It is designed to support all the capabilities prescribed by the ITU for IMT-2000 while providing a forward migration path for operators that have invested in cdmaOne technology. It is the only 3G RTT that can provide all IMT-2000 capabilities without requiring a significant overhaul of a carrier's 2G radio infrastructure. cdma2000 is defined by the TIA's IS-2000 specification, which was released in July 1999. 3GPP2 will define future releases of the standard. Cdma2000 forms the basis for the Multicarrier mode of the OHG's G3G proposal.

Cdma2000 is a wideband CDMA radio interface that offers significant advances over IS-95 CDMA (cdmaOne) to increase performance and capacity. Among these are a coherent pilot-based reverse radio interface (mobile to base station), a continuous reverse radio interface waveform, fast forward (base station to mobile) and reverse radio interface power control, transmit diversity, and an auxiliary pilot to support-beam forming applications and to increase capacity. The RTT supports data rates from 1.2 kbps to greater than 2 Mbps and a wide range of radio frequency (RF) channel bandwidths, ranging from 1.25 MHz (sometimes referred to as 1X) to 15 MHz (e.g., 12X). Early releases will concentrate on the 1.25- and 5-MHz deployment options (1X and 3X, respectively). The 1X option, although nominally supporting numerologies yielding data rates up to 307 kbps, will support practical data rates up to 144 kbps. The 3X option supports data rates up to 1 Mbps on a single supplemental channel, with 2 Mbps requiring allocation of two supplemental channels.

The cdma2000 carriers use chip rates that are multiples of the cdmaOne chip rate of 1.2288 Mcps. The structure of forward radio link RF carriers greater than 1X is achieved by multiplexing a direct sequence CDMA signal across multiple radio frequency carriers. This unique feature allows cdma2000 to coexist in the same spectrum as IS-95. It also permits wideband (3X+) carriers to spectrally overlay IS-95 and narrowband (1X) carriers as it enables the wideband signal to maintain orthogonality with the narrowband signals thereby reducing interference between the two signals to manageable levels.

In addition to the above-mentioned improvements to performance and capacity, cdma2000 introduces several new capabilities that are specifically designed to improve packet data performance. These include an advanced medium access control (MAC) for highly efficient high-speed packet data services, along with physical layer features such as a dedicated control channel (DCCH), a variable frame size packet data control channel operation (5, 10, and 20 ms), and enhanced paging and access channels for fast packet data service access control [common control channel (CCCH)]. Turbo codes enable higher transmission rates and increased capacity. The specifications will also incorporate advanced multimedia quality of service (QoS) control capabilities to enable scheduling and prioritization among competing services to implement negotiated QoS commitments.

Since standardization of the cdma2000 RTT, enhancements to the technology to substantially improve data traffic throughput and spectral efficiency have been introduced into discussions of 3GPP2 and have been designated as cdma2000 1XEV. Phase 1 is based on HDR, a technology developed by Qualcomm to enable peak data rates of up to 2.4 Mbps in a single, 1.25-MHz carrier. 1XEV leverages Qualcomm's HDR concepts and those of other companies into a standard approach to deliver these high data rates on cdma2000 networks. The radio design parameters of the initial version of 1XEV will be optimized for data traffic to maximize throughput. Future releases of the standard will investigate enhancements to support simultaneous voice and data traffic.

UTRA Wideband CDMA, or G3G DS The UTRA wideband CDMA interface* is the basis for the direct sequence mode of the OHG's global third-generation proposal, also known as G3G-DS. Circuit and packet-oriented services are supported and spectral efficiency is improved relative to second-generation radio systems. Other features of UTRA include seamless handover between cells and base stations that can interoperate asynchronously. From the operators' perspective an important aspect of UTRA is the configurability of the UTRA to facilitate the flexible and innovative introduction of services, according to market demand.

The UTRA W-CDMA physical layer[†] enables many users to be multiplexed onto the same spectrum, separated via orthogonal codes with the uplink (mobile to base station) separated from the downlink (base station to mobile) via different frequency bands. It requires adjacent channel spacing of 5 MHz and a channel center frequency raster of 200 kHz.

Information carried on the uplink and downlink in this mode is formatted into a 10-ms frame, which is subdivided into 16[‡] time slots of 625 µs. These time slot periods correspond to a power control period. There are seventy-two 10-ms frames in a super frame of 720 ms duration. Orthogonal codes are used to separate users that exist at the same time to minimize interference between them when recovered at the receiver.

In the uplink, the modulation of both dedicated physical channels (DPCH)—the dedicated physical control channel (DPCCH) and the dedicated physical data channel (DPDCH)—is BPSK. In the downlink QPSK modulation is used. Root-raised cosine (RRC) pulse-shaping filters are used to shape the modulated output.

The development of the air interface has had no constraints with regards to backward compatibility with second-generation systems, as this has not been required. This is due to the different frequency bands allocated for UMTS in Europe (1920–1980 MHz and 2110–2170 MHz paired, 1900–1920 MHz and 2010–2025 MHz unpaired).

UTRA's chip rate is 3.84 Mcps,[§] and the addition of a common, code division pilot channel was added in Release 99 of the specification. These changes will facilitate multimode terminal design for terminals incorporating all three modes of the G3G CDMA standard.

Formal support of the OHG's G3G CDMA recommendation within 3GPP was conferred during the July 1999 meeting of the 3GPP Project Coordination Group (PCG). The G3G TDD (time division duplex) mode uses the UTRA TD-CDMA mode as a basis for future harmonization with other CDMA-based TDD proposals to the ITU such as China's TD-SCDMA. This mode enables several users to be multiplexed onto the same spectrum via time slot allocation, and then different codes within a time slot enables simultaneous use of each timeslot by several users. Like the G3G-DS mode, it requires adjacent channel spacing of 5 MHz and has a channel center frequency raster of 200 KHz.

---

*3GPP TS 25.401 v1.0.0 UTRAN Overall Description.
†3GPP TS 25.213 v2.0.0 Spreading and modulation (FDD).
‡Harmonization of the chip rate to 3.84 Mcps will require changing this to 15 time slots.
§Prior to OHG proposal, the UTRA chip rate was specified as 4.096 Mcps.

TDD uses the same 625-μs time slot and 10-ms frame timing structure as G3G-DS. However, users are separated via a combination of 625-μs time slot and code allocation (within the time slot). Each 625-μs time slot may be uplink or downlink.

Harmonization of CDMA-based TDD technologies has not been as aggressively pursued as that for FDD (frequency division duplex) technologies at the time of this writing.

5.4.2 TDMA-Based standards

UWC-136 and EDGE The UWC-136 RTT proposal specifies an evolution to a global next-generation TDMA standard based on enhancements to the ANSI-136 TDMA standards that incorporate the ETSI-developed GSM EDGE and GPRS standards. The standardization effort has been spearheaded by the UWCC industry forum, with the RTT proposal submitted to the ITU by the TIA standards developing organization. The complete sets of specifications under development for UWC-136 will be contained in the ANSI-136 standards specifications.

To address the UWC-136 key characteristics, the ANSI-136, EDGE, and GPRS standards are evolving to support new or enhanced interfaces, capabilities, features, and services. Key to providing enhanced voice and data capacities and capabilities for next-generation TDMA systems is the ability to achieve greater spectral efficiency. Greater spectral efficiency is being achieved by a change to the modulation schemes, new slot formats, the addition of new interleaving and coding options, and other enhancements.

Developed by the TIA TR45 Technical Engineering Committee, the current version of ANSI-136 supports 30-kHz channel spacing, six time slots, and three calls per channel, circuit-switched data rates up to 28 kbps, and more. The UWC-136 compliant version of ANSI-136 will support 30-kHz, 200-kHz, and 1.6-MHz channel spacing, six time slots with either three or six calls per channel, high-speed packet data rates up to 384 kbps or 2 Mbps (depending on the environment), GSM interoperability, and more.

The GSM communications systems based on TDMA technology, currently support 200-kHz channel spacing, eight time slots and four or eight calls per channel, circuit-switched data rates up to 64 kbps, and more. Intended to augment GSM networks, enhanced data for GSM evolution (EDGE) is a digital air interface that supports circuit and packet data as well as voice. EDGE allows a threefold increase over current GSM data rates. The general packet radio service (GPRS) standards define a packet data network compatible with the GSM intersystem network that supports packet data up to 170 kbps. Both the EDGE and GPRS standards enhancements are under development in the European Telecommunications Standards Institute (ETSI) special mobile groups (SMG).

In addition to continual improvement of the ANSI-136 standard to enhance spectral efficiency, add new user services, and meet the ITU's IMT-2000 requirements, the evolution toward the convergence of ANSI-136 and GSM moves toward the goals of economies of scales of products and global roaming.

DECT* DECT (Digital Enhanced Cordless Telecommunications) is another IMT-2000 RTT submitted to the ITU as part of the IMT-2000 radio evaluation process. It is primarily

*http://www.dect.ch

intended for unlicensed spectrum bands, with low-power operation targeting cordless communications in residential, corporate, and public environments.

DECT provides for voice and multimedia traffic and contains many technical features that allow DECT-based cordless systems to play a central role in important new communications developments such as Internet access and interworking with other fixed and wireless services such as ISDN and GSM. The DECT standard makes use of several advanced digital radio techniques to achieve efficient use of the radio spectrum; it delivers high speech quality and security with low risk of radio interference and low power technology.

The DECT radio standard uses TDMA technology, and with the built-in interference management mechanisms, provides high system capacity to handle up to 100,000 users per km^2 floor space in an office environment. The standard employs ADPCM (adaptive differential pulse code modulation) speech encoding. DCS/DCA (dynamic channel selection/allocation) is a DECT capability that guarantees the best radio channels available to be used. This happens when a cordless phone is in standby mode and throughout a call. This capability ensures that DECT can coexist with other DECT applications and with other systems in the same frequency, with high-quality, robust, and secure communications for end users.

Other features of the DECT standard include encryption for maximum call security and optimized radio transmission for maximum battery life.

5.4.3 RAN Interconnection Methods

3GPP's RAN Interconnection—Iu*   The UTRAN consists of a set of radio network subsystems (RNSs) that are connected to the core network through Iu interfaces. Separate RNSs are interconnected to each other through Iur interfaces, and all cell-level mobility handling and resource control is managed within the UTRAN. Each RNS is comprised of radio node controllers (RNCs) responsible for handover decisions, which are connected through Iub interfaces to node Bs (radio base stations). It is intended that the Iu is a general-purpose interface, allowing connection not only to the UTRAN but also to other access networks. These could include the UMTS satellite radio access network (USRAN) and the broadband radio access network (BRAN) for example.

Additionally the Iu interface supports connection to both circuit-switched (CS) networks and packet-switched (PS) networks, as shown in Figure 5.4. To fully support this, the UTRAN can handle connection to either distributed CS and PS network nodes or a combined CS and PS network node.

The Iu protocol model contains three separate planes—the control plane, the transport network plane, and the user plane. The transport bearer across the Iu interface, and the other UTRAN interfaces, is based on an asynchronous transfer mode (ATM) bearer. The user plane utilizes the Radio Access Network Application Part (RANAP) protocol, which is capable of handling all procedures between the CN and the UTRAN. It is also capable of passing messages between the CN and the user equipment (UE), without interpretation by the UTRAN. The transport layer is administered by the Access Link Control Application Part (ALCAP) protocol, which configures and maintains the transport bearers. Specifically

---

*3GPP TS 25.410 v0.2.1 UTRAN Iu Interface: General Aspects and Principles.

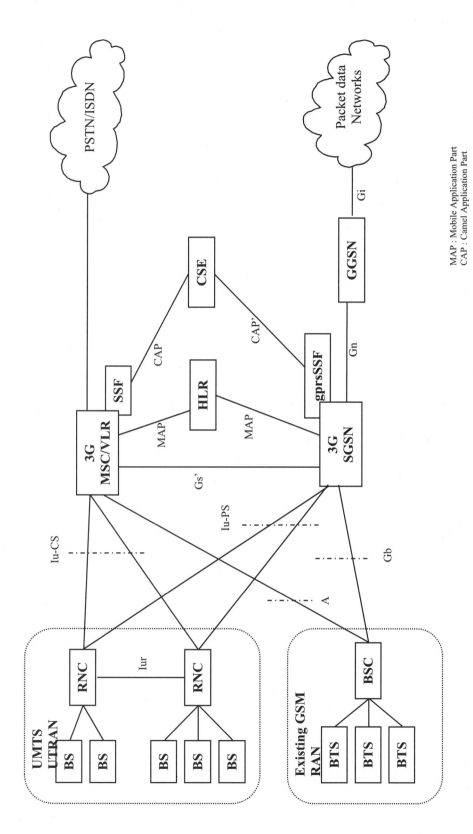

FIGURE 5.4 Release 99 network architecture.

MAP : Mobile Application Part
CAP : Camel Application Part

89

the ALCAP protocol dynamically configures AAL.2 services over the Iu within the CS domain, and AAL.5 is used for services over the Iu within the PS domain.

3GPP2's RAN Interconnection—IS-634/IOS   The specifications for interconnecting the cdma2000 radio access network are being developed on 3GPP2's TSG-A and are based on the IS-634 standard and are known as IOS (interoperability specification) specifications. IOS is the agreed-upon subset and evolution of IS-634 that manufacturers will build the products in conformance to specifications in order to assure intervendor interoperability among radio access networks. As of this writing, the latest version of this standard is in ballot, and will be published as TIA/EIA/IS-2001 (PN-4545).

The key elements in the IOS reference model include the mobile switching center (MSC); the base station (BS—includes radio equipment and associated control functions); the selection/distribution unit (SDU—responsible for identifying and routing traffic packets to appropriate network, e.g., voice or data; includes voice coders); the packet data service node (PDSN—responsible for interworking packet data with the Internet and/or other packet networks). The defined interfaces in IOS Version 5.0 are shown in Figure 5.5. Their functions are:

- A1—Signaling connection between the MSC and BS
- A2—64 kbps DS0 to carry full rate speech or circuit-oriented data
- A3—Carries user traffic (voice/data) using ATM AAL2 and signaling using ATM AAL5 between a BS and an SDU. Includes separate signaling and traffic subchannels.
- A5—User traffic connection that carries a circuit data call between the SDU and the IWF. This interface is only necessary when the SDU and IWF are in separate pieces of equipment.
- A7—Signaling connection using ATM AAL5 between two BSs to support efficient inter-BS packet mode soft handoff.
- A8—User traffic connection between source BSC and packet control function (PCF) for packet data services.
- A9—Signaling connection between the source BSC and PCF for packet data services.
- A10—User traffic path between a PCF and a packet data service node (PDSN).
- A11—Signaling connection between a PCF and a PDSN for packet data services.

5.5   EVOLVING NETWORK STANDARDS

5.5.1   GSM, UMTS, and CAMEL

Providing intelligent network (IN) services to GSM subscribers is a key element for operators to acquire competitive advantages by offering differentiation, reducing churn, and increasing revenue by providing fast time to market services. In the early years of GSM, operators were able to provide IN-based services using the existing ETSI IN capability sets. The disadvantage with this approach was that these services were only available in the home network.

The introduction of CAMEL (customized application for mobile enhanced logic) now allows GSM subscribers to have access to services provided by their home network when roaming to other networks. Associated with such subscribers are types of CAMEL

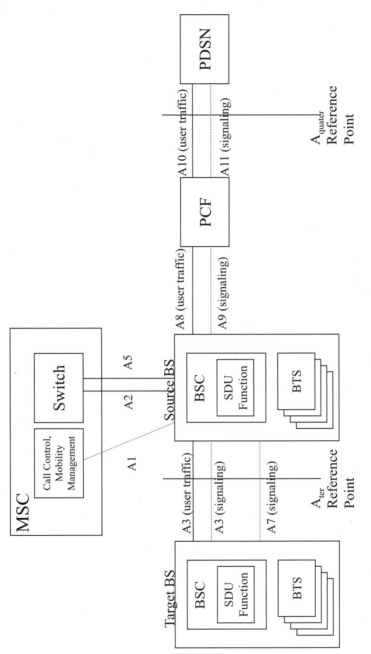

FIGURE 5.5 IS-634/IOS reference model.

subscription data, which is administered in the home location register (HLR). This information is sent from the HLR to the visited location register (VLR) associated with the region associated with the location of the subscriber. This subscription information identifies the CAMEL support required for the subscriber and identifies the home CAMEL service environment (CSE) used for execution of the necessary service logic. The features that form part of Release 99 provide:

- Functionality defined in ETSI IN capability set 2,* such as call party handling, service control point (SCP) initiated calls, and midcall event detection point.
- Mobility management triggers and mobility event notification to the CSE—this provides the potential to introduce location-based applications such as the delivery of location-specific information upon registration in a visited country.
- Support for data services such as GPRS and short message service (SMS)—limited in scope to flexible charging, usage monitoring screening, and authorization.
- Per call suppression of GSM supplementary services. This allows the settings of supplementary services such as call barring and call forwarding to be overridden on a per call basis by the IN service logic.
- Support of optimal routing by the CSE.
- Active location information retrieval allows the CSE to obtain accurate location information about the subscriber.

Packet-switched real-time applications are receiving considerable interest with respect to Releases 4 and 5. Although the focus is for new services and the debate of the value of IN for data services continues, there is a need in the near term to allows today's voice services to run and interoperate over a packet-based core network that is likely to be included as part of Releases 4 and 5.

5.5.2 ANSI-41 and WIN

ANSI-41 and WIN (wireless intelligent network) offer a comparable set of capabilities to GSM and CAMEL. The approach to providing IN services to wireless users was different in the WIN case to the initial GSM approach of providing ETSI IN services to GSM subscribers. The WIN approach was viewed as adding IN to a wireless network, instead of the inverse. Therefore, the virtual home environment was part of the initial WIN services. Although this may be academically interesting, both systems have evolved to support comparable sets of capabilities to the end users.

With the introduction of WIN, the wireless networks are expanded to include the IN network elements, namely the SCP, SN (service node), and IP (intelligent peripheral). IN service logic is executed in the SCP and the SN. The WIN call model is based in ITU IN capability set 2 and is executed in the services switching point, or the MSC. The WIN call model includes "triggers" at various points in call processing to execute service logic in the SCP or SN. The trigger points include points during call origination by the wireless subscriber, as well as call termination to the wireless subscriber, and points during system administrative activities such as mobility management or authentication.

*EN300-374-1, "Intelligent Network (IN); Intelligent Network Capability Set 2 (CS2)."

Phase 1 of WIN includes incoming call screening, calling name presentation, voice-activated dialing, voice-activated feature control, and voice-based user identification. Further enhancements to WIN include prepaid service, advice of charge, location-based services, reverse charging, Selective Call Rejection as well as other charging based features.

ANSI-41/WIN evolution is being driven by 3GPP2. Current priorities are focused on evolving ANSI-41 toward compliance with the ITU's goals for IMT-2000 services.* Key additions are targeted to address global roaming, virtual home environment enhancements, addition of quality of service capabilities for packet data, enhanced security procedures, and emergency and location-based services. Enhancements for support of the IS-2000 radio interface will also be included in the 1999 release of the standard.

5.5.3 GPRS

Evolution of GPRS   This development in GSM standards was started several years ago with the GPRS project. GPRS was a simple approach to allow packet access for GSM users, primarily aimed at Internet access. GPRS allows several users to share the same GSM time slot(s) via a link layer send/receive scheduling protocol. GPRS standards include both air interface and data networking interfaces. For 3G, the data networking interfaces are evolved for use in the UMTS core network as well as the 136-HS/EDGE core network.

In Release 99 further improvements are being added to support real-time QoS services.

The main goal of the GPRS-136 architecture is to integrate ANSI-136 and GSM GPRS as much as possible with minimum changes to either technology. To provide roaming between GPRS-136 and GSM GPRS networks, a separate functional GSM GPRS home location register (HLR) is incorporated into the architecture in addition to the ANSI-41 HLR. The general approach of the GPRS-136 data model is to overlay the circuit-switched network nodes with packet data network nodes for service provisioning, registration, mobility management, and accounting. Interworking is provided between the circuit-switched and packet data networks for mobiles capable of both services. With GPRS the user can access two forms of data network, X.25 and Internet protocol (IP) based. In addition, GPRS allows a user engaged in an active data transfer to suspend operation to make or receive a voice call. Depending on the data operation being performed, the data transaction may be resumed once the voice call is complete.

It is important to note that a new TDMA packet data channel (PDCH) has been defined to carry both user data and information. In addition, by using a concept called tunneling to pass ANSI-136 messages through the GPRS network elements, the existing features defined on the DCCH are maintained.

5.5.4 Mobile IP

Accessing corporate private networks is a key challenge for packet data networking standards, as corporate information technology professionals demand extremely secure access for their end users (and only their end users!). Although this capability is extremely challenging, the payoff is high when corporate end users can attach to their home network

---

*3G Capability Requirements Descriptions, Revision 4.1. June 2, 1999.

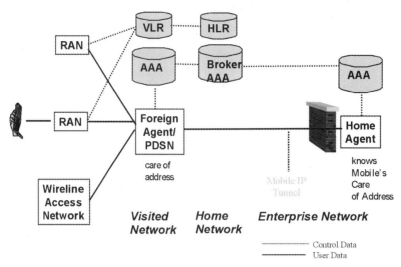

FIGURE 5.6 Providing intranet access to the end user.

while traveling and be able to receive e-mail and intranet services as if they were at their desk in the office (Fig. 5.6).

To provide seamless private network access to the wireless data user, the wireless network must interoperate with the data communications standards that are deployed on commercial router equipment. All equipment that provides Internet service conforms to the Internet Engineering Task Force (IETF) standards.*

Mobile IP is an IETF standard that allows users to change their point of attachment to the network, even during an active data session. Current enhancements being worked within IETF to mobile IP include the security support needed to support user access in public networks, such as wireless cellular networks. Originally, mobile IP was envisioned as a solution for providing mobility to end users within a private network, and it lacked many of the supporting structure to make it useful within public networks, such as accounting and security. Members of IETF Mobile IP Working Group are also working diligently, along with members of 3GPP2 and 3GPP, to solve the needs of wireless networks within the mobile IP standards.

Mobile IP entities included in Figure 5.6 are the foreign agent and the home agent. The foreign agent (FA) is a router that exists in the visited system and its function is to notify the home agent that the end user is attached to it, and the FA delivers data grams to the end users. The home Agent (HA) is a router that exists in the home system and its function is to maintain the users current location and to tunnel and deliver the user's data grams when the user is away from home.[†]

The other IETF entities that are included in Figure 5.6 are the authorization, authentication, and accounting (AAA) and the broker AAA servers. One AAA protocol that is being developed in IETF is called DIAMETER. Enhancements to mobile IP are currently being developed to interwork with DIAMETER to provide security and accounting functions.[‡]

---

*http://www.ietf.org.

†Charles Perkins. *Mobile IP Design Principles and Practices*. Addison-Wesley Longman, Reading, MA, 1998.

‡Pat Calhoun and Charles Perkins. "DIAMETER Mobile IP Extensions," IETF Draft, November 1998.

5.5.5 Network Interoperability Efforts

Several industry and standards initiatives related to aspects of network interoperability are underway. The International Telecommunications Union—Telecommunications Standardization Sector (ITU-T) supports the family of systems concept that allows multiple IMT-2000 systems to coexist and interoperate. The IMT-2000 family concept facilitates the realization of a global service offering among IMT-2000 systems. The evolved ANSI-41 and evolved GSM MAP are IMT-2000 "family member" core networks (CN). Q.FIN and Q.FSN are key ITU IMT-2000 requirement documents that address aspects of interoperability and roaming between IMT-2000 core networks. The framework for IMT-2000 networks (Q.FIN) provides an overall framework for the development of signaling requirements that supports the roaming between family members. Network-to-network interface (NNI) signaling requirements are defined in the signaling network (Q.FSN) document. NNI is the interface between different core networks of IMT-2000 systems. Support for IMT-2000 capabilities and interfaces will facilitate roaming between family members.

To facilitate roaming between ANSI-136/ANSI-41 networks and GSM networks prior to availability of the ITU's NNI, the UWCC and the GSM alliance industry bodies established the GSM/ANSI-136 Interoperability Team (GAIT), in February 1999. The working groups (WGs) of the GAIT include network interoperability, handset and SIM (subscriber identity module), and interoperability testing. The scope of the GAIT activities are to develop mobile station [including subscriber identity module (SIM)] and network functional requirements and specifications for the interoperability and interworking of GSM and ANSI-136-based cellular and PCS systems. In addition, the scope of the activities is to work with other industry bodies (such as the 3GPPs) and the appropriate standards bodies to standardize the GSM and ANSI-136 interoperability service.

The GAIT has defined two phases of interoperability. Phase I addresses basic features and functionality for intertechnology roaming while minimizing impacts to existing standards. Phase II addresses advance features such as data services, features requiring more time to develop such as GSM frequency bands, and features requiring additional standardization. In phase I the network interoperability WG is developing network interoperability and internetworking function (IIF) specifications and the handset and SIM WG is developing mobile stations and SIM cards specifications. The IIF specifications initiative identifies recommended updates and modifications to the existing TIA TR46 IS-129 standard. Published July 1996, the IS-129 standard defines "Interworking/Interoperability Between DCS 1900 and IS 41 Based MAPs for 1800 MHz Personal Communications Systems Phase I."

5.5.6 Migrating to IP-based Core Networks

Interest in IP technology and IP telephony with respect to cellular systems has increased considerably during the last couple of years. This can be attributed to:

- Rapid growth of Internet traffic and in particular the rapid growth of services and applications developed for Internet
- Rapid decline in the cost/performance ratio of Internet technology compared to other telecommunications technologies (such as that traditionally deployed by PTTs in providing PSTN, ISDN, or B-ISDN)

- Interest in integrating telephony service with other applications such as Web-based directory services
- The opportunity for converged core network for all services, voice, multimedia, and data, providing service platforms easier and quicker service creation

As a result there is an effort to evolve the existing core network to a packet-based core network. Various standards organizations and industry forums are now actively working on IP in relation to wireless.

On June 10, 1999, AT&T Wireless Services, Inc., British Telecommunications Plc, Rogers Cantel Inc., Ericsson, Lucent Technologies, Nokia Corporation, Nortel Networks, Telenor AS, and Telecom Italia Mobile announced the formation of a new focus group, named 3G.IP, to promote an IP-based wireless system for 3G mobile communications technology. The 3G.IP Focus Group plans to support the development of next-generation wireless services such as voice, high-speed data and Internet access, imaging, and video conferencing on an all-IP-based network architecture using a common core network based on evolved general packet radio system (GPRS). 3G.IP functions as a special-interest group to get work on the all-IP core standardized in 3GPP and ETSI. In addition to 3G.IP, the Mobile Wireless Internet Forum (MWIF) has been formed to investigate the utility of IETF protocols into the wireless IP core network. Active efforts are currently underway in both 3GPP and 3GPP2 to incorporate the ideas initiated in 3G.IP and MWIF into their reference models for next-generation services. A high-level representation of a potential unified reference model is shown in Figure 5.7.

Voice-over IP technology is increasingly viewed as the way to achieve a converged core network for all services and to provide a service platform for easier and quicker service creation. However, the application of voice-over IP in a wireless environment provides added challenges such as mobility, radio resource efficiency considerations, and quality of service support. Existing control protocols synonymous with IP telephony, H.323 (ITU-T), and session initiation protocol (SIP) (IETF) are also being investigated as potential candidates for the provision of multimedia application in UMTS. The adoption of media gateway function (MGF) and media gateway control functions (MGCF) as show in Figure 5.7 allow the internetwork operability with legacy circuit-switched networks. In the near term, existing IN-based value-added services such as prepaid and advanced routing can continue to be used for IP telephony. Obvious benefits can be reaped if the existing portfolio of IN services used in a circuit-switched environment could be made available to IP telephony.

Standards consortia focusing on wireless IP core network evolution include the IETF, 3GPP, 3GPP2, MWIF, and ETSI's TIPHON project. This common interest in leveraging IP as a core network for wireless may indicate that there is potential for a single converged wireless core network in the future built around IP. Perhaps this is the path to the single, global network envisioned by the ITU in its IMT-2000 goals.

5.6 RELATED STANDARDIZATION EFFORTS

In addition to the standards referred to in the previous sections, there is a great deal of activity and interest in standards to improve terminal and user interface operation.

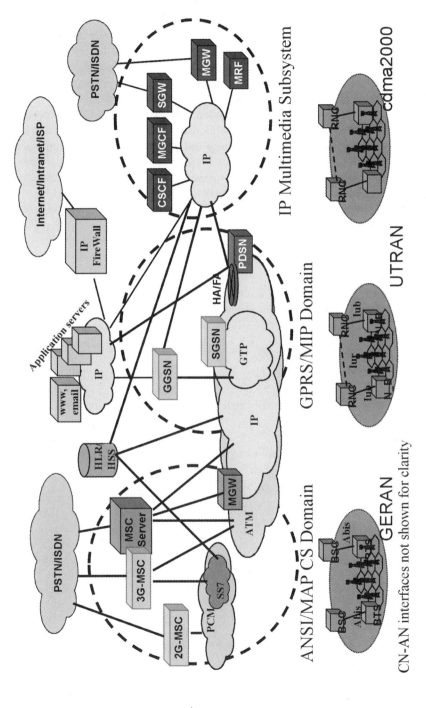

FIGURE 5.7 All-IP core network architecture.

Bluetooth* is a low-power radio interface with upper layer application protocols that is primarily designed to interoperate portable computing devices such as laptops and personal digital assistants (PDAs) with mobile terminals. While this is the most obvious application, there are many other possibilities that are envisioned for its use. The interface standard eliminates a significant source of confusion for users in finding compatible mobiles and cellular modems, as well as making sure that they have the right cable to interconnect them. This will be an important enabler for wireless data, significantly improving ease of use. It has gained a great deal of industry support among mobile phone as well as portable electronics manufacturers.

WAP[†] (wireless application protocol) is another industry effort to facilitate ease of use; in particular to help solve issues associated with interconnecting small-screen wireless data devices to information services using relatively low data bandwidth when compared with traditional TCP/IP.

In addition, there are three primary PDA operating systems emerging as contenders for the next wave of mobile computing—Windows CE, Symbian, and Palm OS.

5.7 SUMMARY

As indicated in this chapter, wireless standards organizations have been extremely active in defining the technologies that will form the basis of wireless communications as we move into the 21st century. The challenges of managing an extremely competitive environment with the desire to converge wireless, wireline, voice, and data technologies to provide users with the kind of services that they will need are overwhelming. Operators and manufacturers are placing a lot of emphasis on protecting their current investment while migrating their equipment to serve these users. While there is a great deal of activity, it is still unclear which of the above technologies, if any, will emerge as dominant in the next generation. Regardless of what actually materializes, 3G wireless standardization is a very exciting ride.

Acknowledgments The authors would like to acknowledge the contributions of the following people in preparation of this material: Yunsong Yang, Roger Guenther, Ian Cordon, Gordon Young, Nigel Berry, Michel Grech, and Jin Yang.

REFERENCES

Given the rapid changes to standards as wireless networks evolve to support third- and even fourth-generation services, the best approach to providing references is to cite the Web sites of the relevant standards fora. The following are some of the key standards Web sites:

1. Third-Generation Partnership Project: http://www.3gpp.org
2. Third-Generation Partnership Project 2: http://www.3gpp2.org
3. International Telecommunications Union's IMT-2000 site: http://www.itu.int/imt/
4. United States' TIA standards: http://www.tiaonline.org/standards

*http://www.bluetooth.com
†http://www.wapforum.org

5. United States' T1 standards: http://www.t1.org
6. Europe's ETSI standards: http://www.etsi.org
7. Japan's ARIB standards: http://www.arib.or.jp
8. Japan's TTC standards: http://www.ttc.or.jp
9. China's CWTS standards: http://www.cwts.org/index_eng.html
10. Korea's TTA standards: http://www.tta.or.kr
11. Internet Engineering Task Force: http://www.ietf.org
12. Universal Wireless Communications Consortium (UWCC): http://www.uwcc.org
13. CDMA Development Group (CDG): http://www.cdg.org
14. UMTS Forum: http://www.umts-forum.org
15. cdma2000 standards: http://www.tiaonline.org/standards/sfg/imt2k/cdma2000/cdma2000table.cfm

CHAPTER 6

Standardization on Broadband Wireless Access: Wireless ATM and IP

MASAHIRO UMEHIRA

6.1 INTRODUCTION

Significant advantages of wireless access are "tetherlessness" and "mobility" because it provides users with ubiquitous access to telecommunications services. This has been proven by the wide acceptance of cellular/cordless telephone services. Aiming at the N-ISDN-equivalent services for the mobile users, a great deal of efforts have been made for standardizing the third-generation mobile communications system, for example, the international mobile telecommunications 2000 (IMT-2000). IMT-2000 will offer full mobility with a service bit rate of 144 kbits/s for mass-market applications and support 2 Mbits/s with restricted mobility and for favorable (indoor) propagation conditions only. The first IMT-2000 was introduced in Japan in October 2001. It should be noted that it took more than 10 years for IMT-2000 global standardization. Global standardization is truly important in the context of global roaming with terminal interoperability, and inexpensive terminal cost taking advantage of the scale merits.

With significant growth of the Internet and intranets, the demand for bandwidth is increasing. In the fixed networks, asynchronous transfer mode (ATM), ISDN, and x digital subscriber line (xDSL) (e.g., ADSL for asymmetric) are expected to offer broadband services. With the rapid growth of the wireless mobile services, broadband mobile wireless access to enable multimedia services is drawing a great deal of attentions. However, it is a challenge for wireless access systems to enable broadband access in mobile environments since multipath fading makes broadband radio transmission much more difficult than in line-of-sight environments.

Aiming at broadband services for wireless mobile users, many research projects have been launched. One of the promising candidates for broadband wireless networks is wireless ATM, which is intended to provide the mobile end users with a variety of broadband ATM services, for example, constant bit rate (CBR), variable bit rate (VBR), and unspecified bit rate (UBR) for both isochronous and asynchronous applications [1, 2]. However, we should note that current end-user terminal for multimedia is mostly the PC

Wireless Communications in the 21st Century, Edited by Shafi, Ogose, and Hattori.
ISBN 0-471-155041-X © 2002 by the IEEE.

(personal computer), which talks IP (Internet protocol) but not ATM. In addition, IEEE802-based Ethernet is widely deployed in current office/
enterprise local area networks (LANs), and ATM-to-the-end-terminal concept has not been widely accepted at this moment. Therefore, end-to-end ATM scenario of wireless ATM does not seem attractive in the near-future market.

On the other hand, IP has been rapidly expanding over both private and public network domains since IP is the only way to make it possible to interconnect PCs over heterogeneous networks, for example, Ethernet LAN, ATM, ISDN, and public switched telephone network (PSTN), including conventional cellular networks. Considering that IP is used to provide multimedia services for the wireless mobile users, broadband wireless access networks will be connected to either mobile ATM networks or mobile IP networks, for IP services.

With the above-mentioned backgrounds, this chapter overviews the standardization on broadband wireless access, including both wireless ATM and IP. This chapter first overviews the standardization efforts on broadband wireless access in four key forums, that is, wireless ATM working group (WATM-WG) of the ATM Forum and IEEE802.11 in United States, the European Telecommunications Standards Institute (ETSI) project BRAN (Broadband Radio Access Networks) in Europe, and MMAC (Multimedia Mobile Access Communication Systems Promotion Council) in Japan. This chapter also describes the update of technical specifications developed so far by IEEE802.11, ETSI-BRAN, and MMAC on the radio access layer of the 5-GHz band broadband wireless access systems. Most remarkably, they have successfully achieved the harmonized radio physical layer specifications. It also overviews the network architectures for broadband wireless access systems and describes the mobile ATM protocols developed by the WATM-WG of the ATM Forum.

6.2 STANDARDIZATION EFFORTS RELATED TO BROADBAND WIRELESS ACCESS

6.2.1 ATM Forum WATM-WG [1–4]

Outline of the WATM-WG   The WATM-WG of the ATM Forum was established in the middle of 1996, after having extensive discussions on the charter and the work plan in Birds of a Feather (BoF) meetings of the ATM Forum. The deployment scenarios of WATM are shown in Figure 6.1. A broad range of wireless network access scenarios for both private and public services are included, that is, fixed wireless access, microcellular-based mobile systems that support end-user mobility, mobile infrastructure where ATM node mobility is supported, satellite links and ad hoc networks. In addition, WATM capability for end-user mobility support can be used for a backbone network for personal communication systems (PCS) networks. This PCS-ATM interworking scenario has been also considered by the WATM-WG. The first release of the WATM specifications, WATM1.0, will be dedicated to the microcellular-based mobile system scenario.

As the charter of the WATM-WG indicates, there are two major technical areas—"mobile ATM" protocol for mobility support within an ATM network and "radio ATM" for ATM-based wireless access. The WATM protocol architecture compared with the standard ATM protocol stuck is shown in Figure 6.2. It should be noted that there is a user plane for user applications and a control plane for control signaling. In the case of WATM,

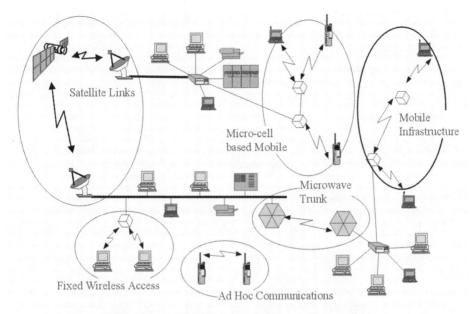

FIGURE 6.1   Deployment scenarios of wirelss ATM.

the "radio access layer" needs to be enhanced to transport ATM cells reliably over air and to support quality of service (QoS) control required for ATM services. For this purpose, the medium access control (MAC) layer and the data link control (DLC) layer are added to the radio physical (PHY) layer. Furthermore, wireless control protocol is necessary for radio resource management, for example. In addition to radio access layer enhancement, the control plane needs to be enhanced to support the mobility of the wireless mobile terminal including handoff, location management, routing for mobile connections, traffic/QoS management, and wireless network control. "Mobile ATM" protocol extension needs to be added to the current user-network interface (UNI) and network-to-network interface (NNI) specifications for mobility support.

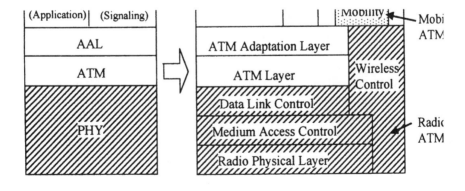

FIGURE 6.2   WATM protocol architecture.

As mentioned, WATM includes a broad range of technical work items. Therefore, WATM-WG has been cooperating with radio groups such as ETSI-BRAN and MMAC for WATM specification development. They share the work items of WATM. Currently, the WATM-WG is responsible for the mobile ATM issues, while ETSI-BRAN and MMAC have been dedicated to radio access layer specification development. These three groups have a liaison relationship to facilitate the mutual cooperation on the WATM specification development.

Technical Issues of WATM As mentioned, there are two technical issues to be considered, that is, radio ATM and mobile ATM. Each is briefly described below.

Radio ATM Radio ATM includes the radio PHY layer, the MAC layer, the DLC layer, and wireless control protocol. The first target of radio ATM specifications is WATM using the 5-GHz band. Currently, ETSI-BRAN and MMAC are developing the radio ATM specifications.

The WATM radio physical layer specification will be varied based on the performance requirements of the future broadband wireless access scenario. Typical WATM PHY requirements for microcell-based mobile systems are: range of 100 to 500 m based on microcellular arrangement, bit rates of 25 Mbps or higher, short burst preamble overhead, efficient frequency utilization, low error floor, and low required C/N in multipath fading environments. There are many issues to be considered, including applicable frequency bands, antennas, radio cell design, basic modulation method and bit rate, countermeasures for multipath fading such as diversity, equalization and FEC (forward error correction), coding, security, and encryption. Regulatory matters concerning the frequency spectrum must be taken into account.

The WATM MAC layer is to support shared use of the radio channel by multiple WATM terminals. It needs to be able to support standard ATM services such as CBR, VBR, and UBR traffic classes with associated QoS controls at reasonable QoS levels while maintaining reasonably high radio channel efficiency. In addition, since end-to-end ATM performance is sensitive to cell loss frequently encountered in radio channels, WATM DLC layer including powerful error control capability is indispensable for WATM radio access layer. Error detection/retransmission protocols and FEC are employed for DLC protocol.

Wireless control is to support control plane functions related to the radio access layer for radio resource control and management functions for PHY, MAC, and DLC layers. Moreover, wireless control functions are necessary to support registration/authentication and handoff.

Mobile ATM The advantage of wireless systems is definitely mobility resulting from its tetherless feature. To enable nomadic access to ATM networks, additional mobility enhancement is required in ATM networks, that is, mobile ATM. They include handoff control, location management, mobile user authentication and registration, routing of mobile connections, and so on. The WATM-WG of the ATM Forum is dedicating itself to the mobile ATM specification development.

Handoff is a basic mobile network capability to maintain the end-to-end connection while the mobile terminal changes its point of attachment within the ATM network. UNI and NNI signaling needs to be extended because dynamic rerouting of a set of VCs from one to another is indispensable. The handoff process involves handoff initiation, rerouting mechanism including path extension and/or reestablishing subpaths, traffic control, and renegotiation during handoff, operation, administration, and management (OAM) cell option for seamless handover, and NNI extension for route optimization.

Location management means the capability of mapping from "mobile terminal ID" to "routing ID" used to locate the current ATM endpoint. Location management issues include ATM addressing principles, mobility management protocols for address updates, queries, mobile user authentication, and registration.

Routing function means the capability of mapping "mobile terminal routing IDs" to "paths" in the network and route identification and optimization for handoff. ATM routing algorithms need to be upgraded to support these additional mobility-related functions. Impact of terminal mobility on traffic control and QoS management also needs to be considered, for example, the call admission control (CAC) in mobile ATM and dynamic QoS renegotiation after handoff. Wireless network management issues include performance management, fault identification and isolation, dynamic network reconfiguration, and network and user administration.

6.2.2 ETSI-BRAN [5, 6]

ETSI work on HIPERLANs (high performance radio local area networks) for private and business radio networks was initiated in the former Radio Equipment and Systems (RES10) Technical Committee. The RES10 developed the HIPERLAN Type 1 Functional Specification. HIPERLAN Type 1 is an ISO8802-compatible wireless LAN just like the one developed by the IEEE802.11 group. The differences between them are the operating frequency and the transmission speed. HIPERLAN Type 1 operates in the 5-GHz band at the rate of 23.5 Mbps, while IEEE802.11 wireless LAN operates in 2.4 GHz at the rate of 1 to 2 Mbps. Both can be used as an ad hoc network without wired infrastructure and as an extension of a wired LAN. Note that no commercial HIPERLAN Type 1 devices are available at this moment.

ETSI-BRAN is currently developing the standards for three types of broadband radio access networks: HIPERLAN Type 2, HIPERACCESS, and HIPERLINK. The deployment scenarios of BRAN systems are illustrated in Figure 6.3. ETSI-BRAN is now focusing on the work of HIPERLAN Type 2 and HIPERACCESS. The basic sets of the HIPERLAN Type2 specifications have been approved. The HIPERACCESS specifications are expected to be approved by the end of 2001.

HIPERLAN Type 2 is a short-range wireless access system intended for complementary access mechanism for universal mobile telecommunication system (UMTS) as well as for private use as a wireless LAN-type system. HIPERLAN Type 2 gives consumers in corporate, public, and home environments wireless access to the Internet and future multimedia, as well as real-time video services at speeds of up to 54 Mbps. The HIPERLAN Type 2 also provides interworking with several core networks including the Ethernet, IEEE 1394, ATM, and UMTS. Spectrum has been allocated for HIPERLANs in the 5-GHz range. It should be noted that functional enhancement for home link applications are also the scope of HIPERLAN Type 2.

HIPERACCESS is a long-range, point-to-multipoint high-speed fixed wireless access system operating at the typical rate of 25 Mbps, intended for residential and small business users to provide a flexible and competitive alternative to wired access networks. HIPERACCESS is targeting high frequency bands; for example, the 40.5–43.5-GHz band. For HIPERACCESS, TDMA will be used as multiple access scheme and a single carrier modulation scheme will be used.

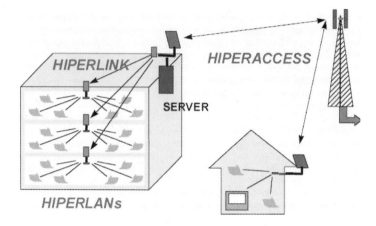

FIGURE 6.3   Deployment scenarios of ETSI-BRAN's HIPERLANs HIPERACCESS and HIPER-LINK.

HIPERLINK is a short-range very high speed interconnection of HIPERLANs and HIPERACCESS, up to 155 Mbps over distances up to 150 meters. Spectrum for HIPERLINK is available in the 17-GHz range.

6.2.3   MMAC [7, 8]

The Japanese Ministry of Post and Telecommunications (MPT) formed a study committee on multimedia mobile access to conduct the study of future mobile communications from July, 1995, to May, 1996. The report issued by the committee suggests the following:

1. High speed and high quality will be necessary in future mobile communications for multimedia services.
2. It is important to enable seamless wireless access to multimedia backbone networks, for example, ATM-network.
3. Early deployment of MMAC is desirable to supplement IMT-2000, around 2002.

In addition, the report encouraged creating a forum to promote the MMAC system development. Following the recommendation in the MMAC report, MMAC Promotion Council was established in December 1996 to promote the development of MMAC systems in cooperation with the Association of Radio Industries and Businesses (ARIB) of Japan.

MMAC-PC consists of a technical committee and a promotion committee. The MMAC technical committee comprises the High Speed Wireless Access Subcommittee, Ultra High Speed Wireless LAN Subcommittee, and two ad hoc committees, that is, the 5 GHz Band Mobile Access Ad hoc Committee and the Wireless Home-Link Ad-hoc Committee, which were established in December 1998. the 5 GHz Band Mobile Access ad-hoc Committee has two working groups, one is ATM-WG working with High Speed Wireless Access Subcommittee and Ethernet-WG. The Wireless Home-Link ad-hoc Committee is developing the specifications for IEEE1394-based wireless LAN. The targets of the above-mentioned MMAC systems are summarized in Table 6.1. Those MMAC family systems

TABLE 6.1 Target Specifications of MMAC Systems

	High-Speed Wireless Access	Ultra-High-Speed Wireless LAN	5-GHz Band Mobile Access	Wireless Home Link
Service area	Public space (outdoor, indoor) Private space (indoor, premises)	Private space (indoor)	Public space (outdoor, indoor) Private space (indoor, premises)	Private space (indoor)
Connected networks, interface	Public network (ATM, IP) Private network; (ATM, IP, Ethernet)	Private network (ATM)	Public network (ATM, IP) Private network (ATM, IP, Ethernet)	IEEE1394, etc.
Information rate	30 Mbps	156 Mbps	20 to 25 Mbps	30 to 100 Mbps
Terminal equipment	Notebook-type PCs, etc.	Desktop PCs and WSs, etc.	Notebook-type PCs, handy terminals, etc.	PCs, audio visual equipment, etc.
Mobility	Stationary or pedestrian	Stationary	Stationary or pedestrian	Stationary or pedestrian
Frequency bands	25/40/60 GHz	60-GHz band	5 GHz band	5/25/40/60 GHz
Bandwidth	500 to 1000 MHz	1 to 2 GHz	>100 MHz	>100 MHz
Bit error rate	$<10^{-6}$	Equivalent to wired networks (10^{-8} to 10^{-10})	$<10^{-6}$	Equivalent to wired networks (10^{-8} to 10^{-10})

will enable wireless access in various environments, that is, home (indoor), office (indoor), premises, and public spaces (outdoors and indoors). MMAC is aiming at mobile communication systems, which enable high-speed wireless access to multimedia information "anytime and anywhere" with seamless connections to optical fiber networks. The service scenarios of MMAC systems are shown in Figure 6.4. The stable draft specifications of the MMAC systems are expected to appear by the end of 2000. In addition, field trial of the MMAC systems is scheduled to confirm the developed specifications from late 2000 to early 2001. Their commercial launch target date is 2002.

High-speed wireless access is a semimobile communication system operating at the rate of 30 Mbps using the 25/40/60 GHz. It has been decided to adopt the same specifications for both ATM-type wireless assess using 5 GHz and high-speed wireless access using 25 GHz. Thus, they shall be essentially the same wireless access systems. They will be deployed both outdoors and indoors, for public services as well as private services. The High Speed Wireless Access Subcommittee has a close liaison relationship with WATM-WG of the ATM Forum concerning the mobile ATM specifications, as well as with ETSI-BRAN's HIPERLAN Type 2 group concerning the radio access layer specifications in the 5-GHz band. The specifications of High Speed Wireless Access Network type a (HiSWANa) was completed and approved in December 2000.

Ultra-high-speed wireless LAN is a wireless LAN that can transmit up to 156 Mbps using the millimeter wave radio band (30 to 300 GHz) in indoor office environments. High-end business users are envisaged for the ultra-high-speed wireless LAN.

Ethernet-WG of the 5-GHz Band Mobile Access Ad-Hoc Committee is developing the specifications of Ethernet-type wireless LAN using the 5-GHz band. This group has a close liaison relationship with the IEEE802.11 group. The group adopted the same PHY specifications for their Ethernet-type wireless LAN using the 5-GHz band as those adopted by IEEE802.11 Task Group a (TGa). The specifications of Ethernet-type wireless access system were approved in 2000.

The wireless Home-Link Ad-hoc Committee is responsible for the specification development of wireless home link, which can transmit up to 100 Mbps using 5/25/40/60 GHz. The specifications of the wireless 1394 system, which is dedicated for IEEE 1394 interface, were completed and approved in March 2001.

6.2.4 IEEE802.11 [9]

The IEEE802.11 group has been developing the Ethernet-compatible wireless LAN standards. Two types of network configurations are supported; one is independent configuration where the stations communicate directly to each other with no infrastructure, and the other is infrastructure configuration where the stations communicate to access points, a part of a distribution system. The MAC protocol is carrier sense multiple access with collision avoidance (CSMA/CA). The IEEE802.11 standard provides 2 physical layer specifications for radio, operating in the 2- to 4-GHz band at the rate of 1 to 2 Mbps.

In IEEE802.11, TGa was established to develop high-speed PHY specifications for wireless LAN operating in the 5-GHz band, where 300 MHz is available in the United States. The 5-GHz PHY is aiming at the rate of 20 Mbps or higher. TGa completed the specifications of 5-GHz PHY. In parallel with TGa, TGb was established to develop high-speed PHY specifications for wireless LAN operating in the 2.4-GHz band, targeting the data rates of 11 Mbps for DS-PHY. TGb also completed the specifications of 2.4-GHz

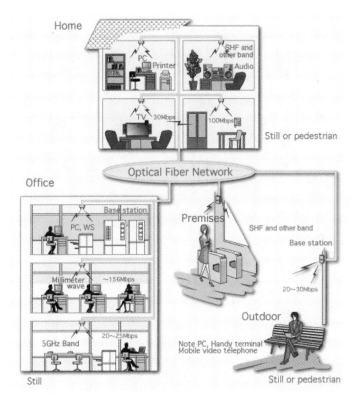

FIGURE 6.4 Service scenarios of MMAC.

high-speed PHY. Both new PHY specifications have been already approved as a formal standard.

6.2.5 Mutual Cooperation and Achievement

As mentioned so far, there are two broadband wireless access standardization bodies in United States, the WATM-WG of the ATM Forum and IEEE802.11. There is ETSI-BRAN in Europe and MMAC in Japan, aiming at the broadband wireless access.

Though they initiated their activities on the broadband wireless access systems independently, they quickly established mutual liaison relationships among them. Their mutual relationship is illustrated in Figure 6.5. Concerning the radio PHY specification of the radio access layer, three groups of IEEE802.11, ETSI-BRAN, and MMAC sent liaisons to one after another to let other standardization bodies know the update of the work. Especially, the results of the selection of modulation scheme and channel spacing are remarkable. When the work started, each group received various radio PHY proposals. There were two main streams, the single-carrier modulation with equalizer and the coded orthogonal frequency division multiplexing (OFDM). There were extensive discussions on the advantages and disadvantages of the proposed modulation schemes, and finally all of the three groups reached the same conclusions, that is, coded OFDM. However, another discrepancy among the groups included the detailed system parameters, for example, the number of subcarriers, subcarrier modulation scheme, detection scheme, and the like. This

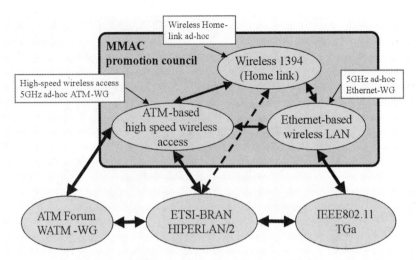

FIGURE 6.5 Liasion relationships among 5-GHz band broadband wireless related standardization bodies.

discrepancy was mainly caused by differences on the assumed frequency spacing and the propagation environments related to the deployment scenarios. In fact, IEEE802.11 assumed 15-MHz spacing and both MMAC and ETSI-BRAN assumed 25-MHz spacing. Once they reached the consensus on the frequency spacing, that is, 20-MHz spacing, they achieved the well-harmonized coded OFDM parameters. The details will be given in Section 6.3.

Concerning the mobile ATM specifications, requirements for the mobile ATM specifications were sent to the WATM-WG from ETSI-BRAN and MMAC. In addition, MMAC and ETSI-BRAN successfully coordinated their radio access network architecture as well as the basic payload data unit (PDU) format. The success of harmonized specifications will result in the possible global roaming of the wireless terminal and reasonable wireless terminal cost.

6.3 STANDARDIZATION ON RADIO ACCESS LAYER [10]

6.3.1 Radio Physical Layer

In the standardization process in IEEE802.11, ETSI-BRAN, and MMAC, two candidates were proposed in each group. One is single-carrier modulation with adaptive equalizer and the other is coded OFDM to reduce the intersymbol interference due to the delayed signals by using guard interval. In the IEEE802.11 TGa, the comparison criteria was the required C/N to achieve the packet error rate of 10^{-1} when the packet is 1536 bytes long. In the case of ETSI-BRAN and MMAC, PDU is segmented to a short packet at the transmitting side and reassembled to the original PDU at the receiving side. Thus, the comparison criteria was the required C/N to achieve the packet error rate of 10^{-2} when the packet is 64 bytes long. In addition to the required C/N, the PER (packet error rate) performance in various propagation environments was compared. Implementation complexity was also a controversial issue since it depends on the robustness against severe multipath fading or delay spread. Major points for the performance comparison between two candidates were the required back-off of the amplifier and the propagation environments. In the case of ETSI-BRAN and MMAC, the root-mean-square (rms) delay spread for the performance comparison was ranged from 50 to 250 ns, taking the outdoor or wide-open space operation into account. The disadvantage of the OFDM is definitely large back-off required in the power amplifier. However, this disadvantage is not significant when powerful FEC such as convolutional-coding–Viterbi-decoding is used to OFDM since the coded OFDM can achieve large coding gain in combination with interleaving even in multipath environments. First, IEEE802.11 TGa decided to adopt coded OFDM for its modulation scheme, based on the joint proposal of Nippon Telegraph and Telephone (NTT) and Lucent. Afterwards, ETSI-BRAN and MMAC made a decision to adopt the coded OFDM scheme.

The comparison between single carrier modulation and OFDM is shown in Figure 6.6. The major reasons why coded OFDM was selected are as follows:

- Reliable transmission in severe multipath fading environments
  - Complicated adaptive equalizer is not required for OFDM since guard interval (GI) reduces the intersymbol interference.
  - Interleaving in the frequency domain for OFDM makes it possible to avoid the burst error. This randomized error feature is advantageous enough to achieve high coding gain, especially if the powerful FEC such as convolutional-coding–Viterbi-decoding is used.
  - It is possible to achieve large diversity gain if the subcarrier-based diversity is employed for OFDM.
- Scalability and flexibility according to the propagation environments
  - It is possible to provide a wide range of bit rates by changing the combination of sub-carrier modulation scheme and coding rate. This feature enables link adaptation, that is, high bit rate transmission in less severe propagation environments, and high-quality transmission can be achieved in severe propagation environments at the sacrifice of reduced transmission speed.

Once IEEE802.11, ETSI-BRAN, and MMAC reached the same solution for broadband wireless access, they tried to harmonize the system parameters of coded OFDM. Finally,

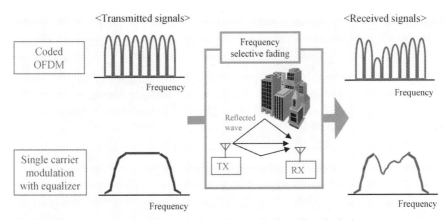

FIGURE 6.6 Comparison of modulation schemes for broadband wireless access systems.

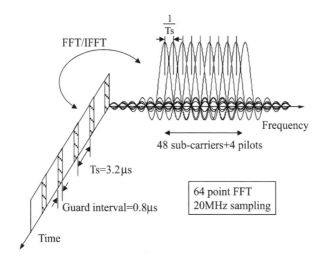

FIGURE 6.7 Major parameters of coded OFDM.

they achieved the well-harmonized OFDM specifications, that is, 64-point Fast Fourier Transform/Inverse FFT (FFT/IFFT) for OFDM, 20-MHz channel spacing, 800-ns guard interval, and 4 μs OFDM symbol period. Coherent detection using four pilots for automatic frequency control (AFC) was selected as a detection scheme. The 48 subcarriers are used for information. Major OFDM system parameters are shown in Figure 6.7. Table 6.2 shows the available bit rate of the coded OFDM, by changing the combination of subcarrier modulation scheme and FEC coding rate. Variable coding rate of FEC is

TABLE 6.2 Major Specifications on Physical Layer of Broadband Wireless Access Systems in 5-GHz Band

	ETSI-BRAN/MMAC(ATM-WG)	IEEE802.11a/MMAC(Ethernet-WG)
MAC	TDMA-TDD/DSA	CSMA-CA
Modulation and demodulation	Coded OFDM/coherent detection BPSK, QPSK, 16QAM, 64QAM	Coded OFDM/coherent detection BPSK, QPSK, 16QAM, 64QAM
Transmission rate	6 Mbps (BPSK, $R = \frac{1}{2}$)	6 Mbps (BPSK, $R = \frac{1}{2}$)
	9 Mbps (BPSK, $R = \frac{3}{4}$)	9 Mbps (BPSK, $R = \frac{3}{4}$)
	12 Mbps (QPSK, $R = \frac{1}{2}$)	12 Mbps (QPSK, $R = \frac{1}{2}$)
	18 Mbps (QPSK, $R = \frac{3}{4}$)	18 Mbps (QPSK, $R = \frac{3}{4}$)
	27 Mbps (16QAM, $R = \frac{9}{16}$)	24 Mbps (16QAM, $R = \frac{1}{2}$)
	36 Mbps (16QAM, $R = \frac{3}{4}$)	36 Mbps (16QAM, $R = \frac{3}{4}$)
	54 Mbps (64QAM, $R = \frac{3}{4}$)	48 Mbps (64QAM, $R = \frac{2}{3}$)
		54 Mbps (64QAM, $R = \frac{3}{4}$)
Forward error correction	Convolutional-coding–Viterbi-decoding with intrasymbol interleaving	Convolutional-coding–Viterbi-decoding with intra-symbol interleaving
Channel spacing	20 MHz	20 MHz

implemented employing the punctured code of the $R = \frac{1}{2}$, $K = 7$ convolutional code, where R is FEC coding rate, and K is constraint length. The combination of BPSK and $R = \frac{1}{2}$ provides the transmission rate of 6 Mbits/s with robustness against severe multipath fading and interference. The combination of 64 QAM and $R = \frac{3}{4}$ can provide very high speed transmission rate of 54 Mbps, even though the propagation condition needs to be less severe and the wireless terminal is near to the access point. Link adaptation will provide both flexibility and robustness according to the propagation conditions. Performance enhancement using turbo coding will be one of the issues for further study.

6.3.2 MAC and DLC Layer

MAC protocol of Ethernet-based wireless LAN developed by IEEE802.11 is CSMA/CA (carrier sense multiple access with collision avoidance). Basically, best effort service is available for data services when CSMA/CA is used. Note that it is difficult to provide QoS control when distributed control such as CSMA/CA is used for MAC protocol. On the other hand, HIPERLAN Type 2 of ETSI-BRAN and HiSWANa of MMAC are aiming at QoS-controlled services. Therefore, they employ centralized control at the AP (access point) for MAC protocols. It is a mandatory requirement for multimedia services to support asymmetric traffic efficiently and to carry both burst data traffic and circuit-switched-type traffic efficiently. Therefore, DSA (dynamic slot assignment) based on TDMA-TDD structure was selected for ESTI-BRAN's HIPERLAN Type 2 and MMAC's HiSWANa. The feature of DSA is that the AP transmits FCCH (frame control channel) every frame. FCCH includes the information of the frame structure in a frame period, for example, receiving timing and transmitting timing on a per mobile terminal basis. The frame structure for DSA is shown in Figure 6.8. Frame period is 2 ms. Each MAC frame includes BCCH (broadcast control channel) to broadcast only the necessary information to all mobile terminals (e.g., access point ID). It also includes RACH (random access channel), which is based on contention-based slotted ALOHA with some priority mechanism and is used for time slot request, for example.

In ESTI-BRAN's HIPERLAN Type 2 and MMAC's HiSWANa, original PDU is segmented to 48-bytes-long packets at the transmitting side. This segmentation to short packets is to enhance the robustness against the multipath fading and impulsive noise coming from the radar. When a DLC-SDU is erred, it is retransmitted by using ARQ (automatic repeat request) protocol. Each DLC-SDU needs CRC for error detection, sequence number for retransmission control, and SAR (segmentation and reassembly) control information. Resultant length of the DLC-SDU is 54 bytes long. Since the PDU

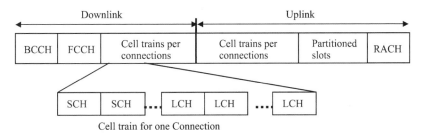

FIGURE 6.8 MAC frame structure of HIPERLAN/2 and MMAC's HiSWANa.

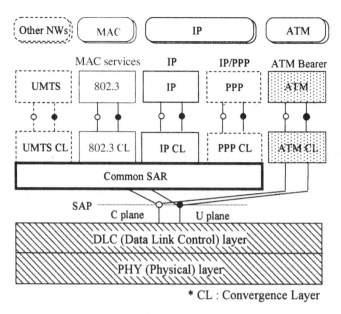

FIGURE 6.9 Architecture of ETSI-BRAN HIPERLAN/2 and MMAC HiSWANa.

length for user information of an ATM cell is the same as that of DLC-SDU, the DLC architecture can support both IP and ATM. Furthermore, common SAR mechanisms can be used to connect it to various types of networks, for example, 802-based Ethernet, IP over PPP, and the like. This architecture is flexible enough to support multiprotocols of the core networks as shown in Figure 6.9, where a convergence layer is necessary for interworking between the wireless access network and the core network. With this architecture, it also will be possible to connect HIPERLAN Type 2 and MMAC's HiSWANa to the core network for IMT-2000 or UMTS.

6.3.3 Frequency Spectrum for Broadband Wireless Access [11, 12]

Since broadband wireless access systems such as HIPERLAN Type 2 and MMAC's HiSWANa can provide much higher bit rates than that of IMT-2000, that is, 2 Mbps at the highest, they need a wide frequency spectrum. At this moment, the 5-GHz band is a promising candidate for broadband wireless access by sharing with other radio locations and navigation services. The current status of the 5-GHz band allocation in the United States, Europe and Japan is shown in Figure 6.10.

In the United States, in 1995, the Wireless Information Networks Forum (WINForum) and Apple Computer, Inc. filed a petition for rule making requesting that the Federal Communications Commission (FCC) should allocate 250 to 300 MHz of spectrum in the 5-GHz band to establish a new unlicensed wireless radio service to promote the full deployment of the National Information Infrastructure (NII). Accordingly, the FCC proposed to make 350 MHz of spectrum at 5.15 to 5.35 GHz and 5.725 to 5.875 GHz available for unlicensed national information infrastructure (U-NII) devices, which will provide short-range, high-speed wireless digital communications. Currently, 300 MHz of spectrum at 5.15 to 5.35 GHz and 5.725 to 5.825 GHz are available for use by U-NII devices in the United States. However, since 100 MHz at 5.15 to 5.25 GHz is shared with

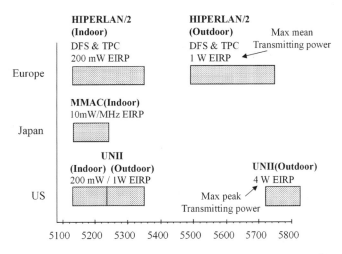

FIGURE 6.10 5-GHz band availability in United States, Europe, and Japan.

the feeder link of mobile satellite services, the U-NII devices using 5.15 to 5.25 GHz have to operate indoors.

In Europe, the 5.15- to 5.25-GHz band was allocated for HIPERLANs. Recently, ERC decided to allocate more 5-GHz band spectrum for HIPERLANs. Consequently, 200 MHz at 5.15 to 5.35 GHz will be made available for indoor use of HIPERLAN devices. In addition, 255 MHz at 5.47 to 5.725 GHz will be available for HIPERLAN Type 2. However, DFS (dynamic frequency selection) and TPC (transmitting power control) are mandatory requirements to facilitate the frequency sharing with other services such as radiolocation and radio navigation, without frequency coordination. Currently, 455 MHz in the 5-GHz band are available in Europe to HIPERLAN$_S$. Note that 200 MHz at 5.15 to 5.35 GHz is commonly available in the United States and Europe.

In Japan, 100 MHz at 5.15 to 5.25 GHz have been recently made available for MMAC devices including wireless home link applications. However, since these 100 MHz are shared with the feeder link (uplink) of mobile satellite services using nongeostationary satellites, MMAC devices are limited to operate indoors only as far as they use 100 MHz at 5.15 to 5.25 GHz. Since no spectrum is allocated for outdoor use in Japan, more work on frequency sharing is necessary to identify the frequency band candidate for outdoor use.

6.4 STANDARDIZATION ON MOBILE ATM [13–15]

6.4.1 Outline of Mobile ATM Specifications

Figure 6.11 shows the architecture of wireless ATM adopted by WATM-WG of the ATM Forum; one is integrated architecture where the AP is embedded in EMAS-E (end-user mobility supporting ATM switch at the edge) and the other is modular architecture where AP is physically separated from EMAS-E. In the case of integrated architecture, switching functions and control functions are located in EMAS-E and RT (radio transmitter-receiver) is attached to the EMAS-E. In the case of modular architecture, access point control protocol (APCP) is used to exchange the signaling message for handover and registration,

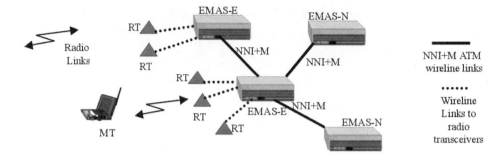

(a) Integrated architecture.

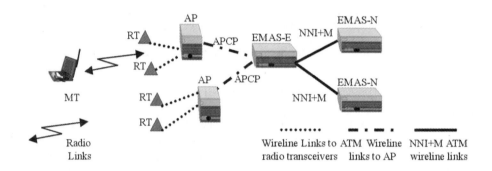

(b) Modular architecture.

FIGURE 6.11   WATM network architectures.

for example, between AP and EMAS-E. The advantage of integrated architecture is implementation simplicity. However, since both radio network control and mobile ATM network control functions are implemented in a single switch entity, everything needs to be modified if radio specifications change. On the other hand, the advantage of modular architecture is that the radio access network, comprising AP and MT, and the mobile ATM network can evolve independently since AP is responsible for radio network control and the EMAS-E is responsible for wired network control. APCP is interworking between two entities. Thus, as far as APCP is maintained, no modification is necessary in AP if EMAS-E is modified.

Figure 6.12 shows the location server and authentication server in mobile ATM. There are two approaches for implementation. In the integrated approach, either the location server or the authentication server is embedded in the EMAS-E. In the modular approach, the location server and the authentication server are located in a physically different location from the EMAS-E. Both architectures need to be supported.

Concerning the addressing in mobile ATM specifications, it is decided not to separate the address field for the mobile terminal identifier. This means that MT always needs two ATM addresses, that is, the "permanent (home) ATM address" when the MT is attached to home EMAS-E and the "temporary (visited) ATM address" when the MT is attached to the foreign EMAS-E. This approach is similar to mobile IP protocol developed by Internet

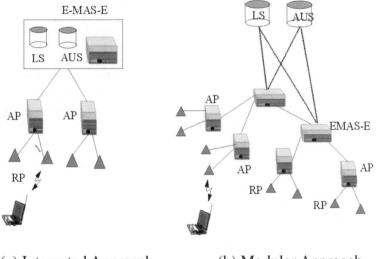

(a) Integrated Approach   (b) Modular Approach

FIGURE 6.12 Location server and authentication server in WATM networks.

Engineering Task Force (IETF). Assuming this principle, location management is carried out in the following way. Figure 6.13 illustrates the location management according to the mobile ATM specifications of WATM-WG. The address of the calling party is A.1.1.0 and the called party is mobile whose home address is C.2.1.1. Currently, the mobile terminal is attached to the foreign EMAS-E of C.1.1. This location information (C.2.1.1 to C.1.1) needs to be stored in the home location server. Since the caller does not know C.2.1.1 is mobile, the setup message destined to C.2.1.1 is sent to the home EMAS-E. As the location server connected to the home EMAS-E knows C.2.1.1 is attached to C.1.1, it

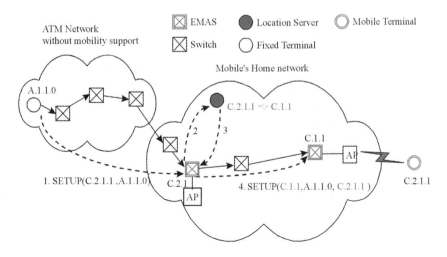

FIGURE 6.13 Location management in WATM-WG of the ATM Forum.

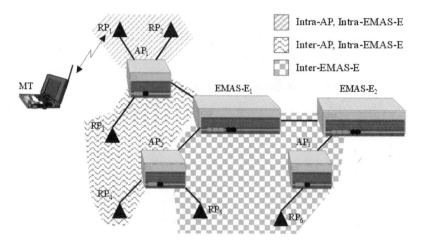

FIGURE 6.14 Handover supported by WATM specifications.

translates the destination address from C.2.1.1 to C.1.1. As the foreign EMAS-E knows the foreign mobile terminal is attached to C.1.1, it sends the setup message (A.1.1.0 to C.2.1.1 via C.1.1) to the called mobile terminal. Finally, the setup message reaches the called mobile terminal and a connection is successfully established. The disadvantage of this approach is that routing will not be optimized. To solve this problem, routing optimization using crankback capability of private network-node interface (PNNI) is proposed.

Concerning the handover, WATM1.0 will support both backward handover and forward handover. Backward handover is preferable to minimize the break time during the handover process, however, forward handover is indispensable if the backward handover process fails. The WATM1.0 supports hard handover, but not soft handover. Further consideration will be necessary for lossless handover utilizing a buffer in the ATM switch and soft handover. There are three handover scenarios, which are (1) intra-AP, intra-EMAS handover, (2) inter-AP, intra-EMAS handover, and (3) inter-AP, inter-EMAS handover. They are illustrated in Figure 6.14.

6.4.2 IP Mobility Support Options

There are three architecture options for IP mobility support. The first option is a gateway architecture, where a wireless network is transparent for IP and the gateway node (GN) is located between two networks. It is assumed that the wireless network is a mobility capable network like cellular networks. A logical connection is established between the GN and the MT. In the case of dial-up services, the GN works as a remote access server for Internet access. To set up a connection for incoming IP packets, the GN needs to translate the IP address of the destined IP address to the node address of the MT. The gateway architecture is shown in Figure 6.15.

The second option is an overlay network architecture. The feature of the overlay network architecture is that a connectionless IP network is overlaid on a connection-oriented mobile network. The GN is used to connect a connection-oriented mobile network to external IP networks. The MT needs to have a node address, for example, ATM address, in addition to the IP address assigned to the mobile host (MH). The IP address of the MH

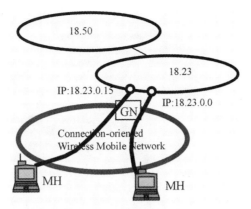

FIGURE 6.15   Gateway architecture.

is translated to the node address of the MT for routing control of a connection. The overlay network architecture is illustrated in Figure 6.16. In the case of mobile ATM as a connection-oriented mobile network protocol, the MT needs a permanent home ATM address and a temporary ATM address to indicate the current network point of attachment. A permanent home IP address is also assigned to the MH connected to the MT. It is assumed that the binding of the home IP address and the home ATM address is never changed. When an IP packet destined to the MH comes in, ATM-ARP (address resolution protocol) is used to map the IP address of the MH to the home ATM address of the MT. The home ATM address is further translated to the temporary ATM address. In this way, an ATM connection according to the destination IP address is set up and maintained. Note that the IP to ATM address translation function is necessary at the GN.

The third option is integrated network architecture, where the backbone network is an IP subnet and mobility is supported on IP layer instead of a connection level. Mobile IP has been developed for this purpose by IETF. In the current mobile IPv4, an HA (home

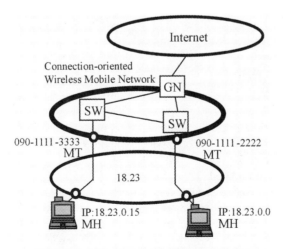

FIGURE 6.16   Overlay network architecture.

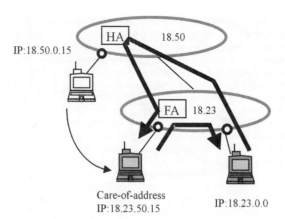

FIGURE 6.17 Routing mechanism in mobile IP.

agent) in a home network and an FA (foreign agent) in a visited network are assumed. The MH lets the HA know its current network point of attachment, that is, care-of-address given by the FA. IP packets destined to the MH are first sent to the HA. They are forwarded from the HA to the FA, and further forwarded to the destination MH. On the other hand, the MH sends IP packet directly to the destination host. Therefore, a forward path is different from a backward path. This is called triangle routing of the mobile IP, which is illustrated in Figure 6.17. A logical connection is set up between the MT and the HA or FA. In the case of Ethernet-based wireless access network, ARP is used to map IP address of the MH within a subnet to the MAC address of the MT. In this way, an IP packet encapsulated in an Ethernet packet can reach the destination MH.

The overlay network architecture based on mobile ATM is effective as far as the mobile ATM networks are interconnected one after another over ATM backbone networks. In this case, the MT can roam over interconnected mobile ATM networks and multiple protocols such as IP and X.25 can be supported. However, to make roaming over heterogeneous networks possible, it needs IP mobility support in addition to ATM mobility support. On the other hand, the integrated architecture needs IP mobility support only. In addition, it should be noted that the current mobile IP does not support route optimization, thus throughput performance can be degraded by the transfer delay due to the triangle routing, that is, IP router processing delay and path delay.

6.5 CONCLUSIONS

This chapter has overviewed the standardization on broadband wireless access, including radio access layer issues and network issues on both wireless ATM and IP. First, the standardization efforts in the WATM-WG of the ATM Forum, IEEE802.11, ETSI-BRAN, and MMAC have been outlined. Though each group has a slightly different scope, the basic system concept of broadband wireless access in the 5-GHz band is quite similar among them, that is, aiming at 20 Mbps or higher, short-range microcellular based system, and the like. This common system concept resulted in the successful harmonization on the radio physical layer specifications for the 5-GHz band broadband wireless access systems among IEEE802.11, ETSI-BRAN, and MMAC. The harmonized radio physical layer specifications will help to reduce the wireless terminal cost and would enable global roaming of the wireless terminal. This chapter has also outlined the mobile ATM specifications developed by the WATM-WG of the ATM Forum. Location management and handover support will be available in WATM1.0. It also briefly described the issues on IP mobility support. As mentioned before, the advantage of IP is to interconnect the PCs over heterogeneous networks. Thus, IP mobility is becoming more important.

IEEE802.11, ETSI-BRAN, and MMAC have completed the specifications of their broadband wireless access systems using 5-GHz band. With the previously mentioned specifications, some prototype developments were reported and demonstrated. A commercial product will appear very soon.

REFERENCES

1. "Charter, Scope and Work Plan for Proposed Wireless ATM Working Group," ATM Forum, Anchorage, April 1996, ATM Forum/96-0530/PLEN.

2. DEMO'98 Workshop proceedings, http://comet.columbia.edu/demo98.

3. D. Raychaudhuri, "Broadband Wireless Networks: Current Status and Future Directions," Proc. 1st Workshop on Wireless Broadband Testbed (DEMO'98), Berlin, Germany, Oct. 15, 1998.

4. J. Mikkonen, C. Corrado, C. Evci, and M. Prögler, "Emerging Wireless Broadband Networks," *IEEE Commun. Mag.*, pp.112–117, Feb. 1988.

5. http://www.etsi.org/bran/.

6. Jamshid Khun-Jush, "A Short Range Broadband Radio Access System for IP and Multimedia Applications," panel session on broadband wireless, PIMRC'99, September 14, 1999, Osaka, Japan.

7. "Multimedia Mobile Access Communication," MMAC systems promotion council activity brochure, Association of Radio Industries and Businesses (ARIB), Tokyo, Japan, 1997.

8. M. Umehira, "Japanese View: Multimedia Mobile Access Communication Systems—MMAC," Proc. 1st Workshop on Wireless Broadband Testbed (DEMO'98), Berlin, Germany, Oct. 15, 1998.

9. http://grouper.ieee.org/groups/802/11/main.html.

10. http://www.etsi.org/technicalactiv/h2tech.htm.

11. FCC, ET Docket No. 96-102 Amendment of the Commission's Rules to Provide for Operation of Unlicensed NII Devices in the 5 GHz Frequency Range, January 9, 1997.

12. ITU-R Document 8A-9B/168-E, "Report on the Compatibility Studies Related to the Possible Extension Band for HIPERLAN at 5 GHz," June 1999.

13. M. Umehira et al., "Wireless and IP Integrated System Architectures for Broadband Mobile Multimedia Services," Proc. WCNC'99, New Orleans, Sept. 1999.

14. ATM Forum BTD-WATM-01.12, "Draft Wireless ATM Capability Set 1 Specification."

15. IETF, "IP Mobility Support," RFC2002, Oct. 1996.

PART 3

PROPAGATION ISSUES

CHAPTER 7

Multipath Effects Observed for the Radio Channel

HENRY L. BERTONI

7.1 INTRODUCTION

Starting from the earliest measurements of the spatial variation of radio communication signals to terrestrial terminals, large variations were found over distances whose minimum scale length is about one half the wavelength λ. At the smallest scale, the variations are sensitive to the frequency and show variation with time as objects and people move in the vicinity of the radio link. When measured by a moving receiver, these rapid spatial variations are perceived as fast time variations, leading to the term *fast fading*. These variations have been understood in terms of the existence of multiple paths by which the signals can propagate between the antennas forming the link. Variations over larger scales show only a gradual variation with frequency and are thought to result from changes in the shadowing by buildings and from the overall decrease of received signal with distance. Because of the larger spatial scale, these variations have sometimes been called *slow fading*.

This chapter reviews the multipath characteristics of the radio channel that have been observed under different conditions. Because the radio links in modern wireless systems are to low base station antennas over relatively short paths from subscribers in cities, or even inside buildings, the buildings have a major influence on the received signal. Modern systems operate at frequencies f above 300 MHz, so that the wavelength λ is less than a meter. Because the buildings are large compared to wavelength, it is appropriate to think of the radio waves as traveling along rays between the base station and the subscriber. Rays are used in this chapter to provide a conceptual picture for discussing and classifying the observed characteristics. Ray methods have also been used by various authors to make quantitative predictions of the propagation, as discussed by Bertoni [1].

Rays connecting the base station and the subscriber can be multiply reflected, or otherwise scattered by the buildings, and diffracted around them, thus allowing signals to propagate between the base station and subscriber even when they are not visible to each

Wireless Communications in the 21st Century, Edited by Shafi, Ogose, and Hattori.
ISBN 0-471-155041-X © 2002 by the IEEE.

other. Because the propagation is reciprocal, the same ray paths are involved no matter if the base station is transmitting and the subscriber is receiving or if the subscriber is transmitting. Some possible ray paths from the base station to a street-level subscriber are suggested in the cartoon of Figure 7.1 for the case when there is no direct path between them. Thinking of the base station as being the transmitter, a small fan of rays illuminates the buildings in the vicinity of the subscriber. These rays then reach the subscriber as a result of reflection and diffraction.

The ray segments arriving at the subscriber in Figure 7.1 appear to come from a 360° range of angles in the horizontal plane and a wedge of angles in the vertical direction. Because the differences in path length are on the order of the street width, which is around 30 m, the time difference for this cluster of rays is on the order of 100 ns. In addition, signals from the base station may be reflected from large structures at a greater distance before arriving at the building in the vicinity of the subscriber. Such paths may have delays on the order of 1 μs, followed by additional delays due to scattering in the vicinity of the subscriber, which results in another cluster of rays arriving at the subscriber. While the cartoon in Figure 7.1 has been drawn to illustrate the grouping of ray paths described above, there may be a continuum of scatterers at all distances from the base station and subscriber. A similar ray description can also be used for indoor propagation, although the differences in path length and time delay are smaller by an order of magnitude.

Early measurements of channel characteristics were made for high base station antennas using very narrow band or continuous wave (CW) sources and dipole antennas having uniform (omnidirectional) radiation patterns in the horizontal plane [2–7], while later measurements employed low base station antennas [8–15]. Other narrowband measurements were made using directional antennas [16]. To assist with the design of digital

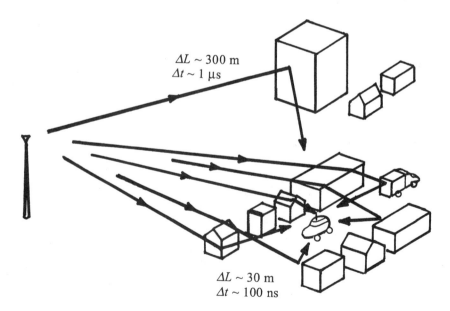

FIGURE 7.1 Scattering from objects in the vicinity of the mobile result in signals arriving at the subscriber from many directions. Scattering by more distant objects leads to additional clusters of multipath arrivals.

systems, broadband pulsed excitation measurements were carried out [17–22]. Most recently, high-resolution space–time signal processing has been employed on pulsed measurements to directly separate out the rays at the base station and subscriber by direction of arrival and time delay [23, 24]. Depending on the type of measurements, the presence of multiple ray paths is observed through different effects.

In addition to the multipath effects, the received signal amplitude for outdoor links shows variations due to shadowing by buildings, which occurs over a distance scale of 5 to 10 m. Finally, the signal amplitude shows a systematic dependence on the distance R between base station and subscriber. Other chapters in this book discuss the nature of the signal variations observed over these large scales.

7.2 MEASUREMENT OF MULTIPATH ARRIVALS

Measurements capable of space and time resolution of the signal received at street level were made using pulsed transmission and a synthetic aperture antenna mounted on a van [23]. The measurements were made in downtown Paris at 890 MHz, and had a time resolution of 0.1 μs. The synthetic aperture, covering 1 by 2 m in the horizontal plane, allowed determination of the direction of arrival in both the vertical and horizontal planes. Figure 7.2(a) shows the time-dependent power for a single received pulse at one location of the omnidirectional receiving antenna. From a synthesis of received pulses as the antenna was moved over the aperture, the direction of arrival of the rays was found using the Unitary ESPRIT algorithm.

The time delay and azimuth direction of arrival of the rays at a midblock location is shown in Figure 7.2(b). Each vertical spike represents an individual ray, whose azimuth angle is shown in a polar format, and whose time delay is indicated by radial position, with the outer circle representing the latest arrivals. At this location the rays arrive from a broad range of angles centered about the directions along the street, which runs north and south, and are spread over a time interval of about 5 μs. At other locations of the van, the rays were more uniformly distributed over the entire 360° in the horizontal plane [23]. The distribution in elevation angles of the rays extended to about 50° [23]. The number of individual arrivals identified in Figure 7.2 is limited by the temporal and spatial resolution of the measuring systems and would be higher for a system with greater resolution.

Superresolution processing was applied to 1800-MHz pulsed signals arriving at an elevated base station in Aalborg, Denmark, from a street-level mobile. The results of the processing are shown in Figure 7.3 [24]. In these measurements a one-dimensional array was employed, so that only the direction of arrival in the horizontal plane could be distinguished. At the base station the rays are spread over a limited range of angles, which in the case of Figure 7.3 is about 60°. The relative delay time is $T_s = 0.923$ μs, so that the arriving rays can be divided into five clusters separated by intervals on the order of 1 μs. Based on the cartoon in Figure 7.1, the first cluster may represent rays scattered in the vicinity of the mobile that subsequently propagate directly to the base station, while later clusters represent rays that undergo both local scattering as well as scattering from distant objects. Placing the base station antenna below the rooftops increases the angular spread [24]. Viewing the scattering as occurring in a region of fixed size about the mobile [25], the angular spread will decrease with distance R. The opposite dependence on distance has also been observed [26]. The dependence of angular spread and time-delay spread on the propagation environment and path geometry is discussed further in later sections.

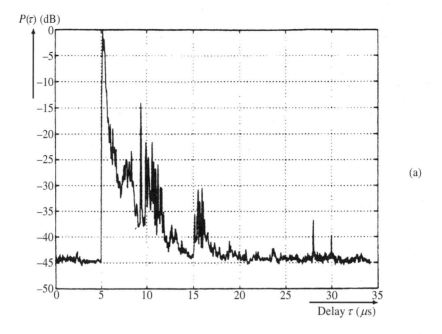

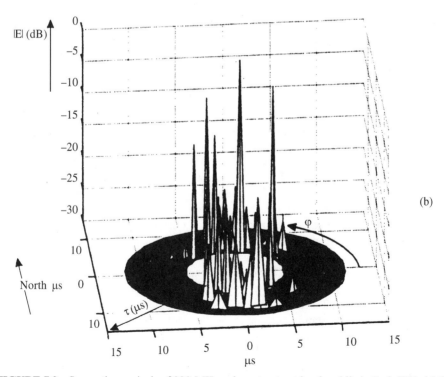

FIGURE 7.2 Space–time arrivals of 890-MHz pulses at a street-level mobile in Paris [23]: (a) time delay profile received by an omnidirectional antenna at one position in a synthetic array and (b) azimuth–time distribution of rays obtained using 2-DUnitary EXPRIT.

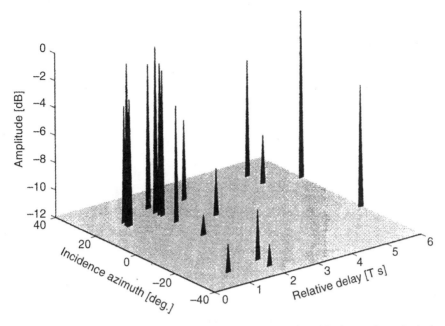

FIGURE 7.3 Estimated angle of arrival and time delay of individual rays for pulsed signals received at an elevated base station antenna in Aalborg, Denmark [24].

Measurements of the rays arriving at a mobile have been made with a CW system, which has the effect of integrating the space–time rays over time [16]. These measurements were made at 900 MHz in Tokyo using a 22° spot beam antenna that was mechanically rotated. For midblock locations, arriving signals were distributed over all directions in the horizontal plane and up to at least 45° in elevation, as found in Paris [23].

Pulsed measurement made with omnidirectional antennas integrate the received signals over the direction of arrival, leaving the profile of the received echoes over time, as in Figure 7.2(a). The received power as a function of time is referred to as the power-delay profile. In Figure 7.2(a) it is possible to identify four, and possibly five, individual clusters of arriving rays. While many pulsed measurements have been made using omnidirectional antennas [17–22], one set made in Redbank, New Jersey [27], shows an unusually clear example of the arriving ray clusters. The power-delay profile for one subscriber location in Redbank is shown in Figure 7.4. To explain similar time-delay profiles for indoor paths, Saleh and Valenzuela [28] originally proposed the cluster model. In their model the arrivals within each cluster decay rapidly with time, while the amplitudes of the individual clusters decay more slowly in time. In mountainous regions, scattering from the mountains can result in clusters arriving with delays on the order of 10 µs [29].

7.3 MULTIPATH PHENOMENA FOR NARROWBAND EXCITATION

The extensive measurements made to support the design and installation of first-generation cellular mobile radio systems employed narrowband or CW excitation and omnidirectional antennas. In such a measurement system, the multipath arrivals discussed above are integrated both over time and direction of arrival. The total signal is thus a superposition of

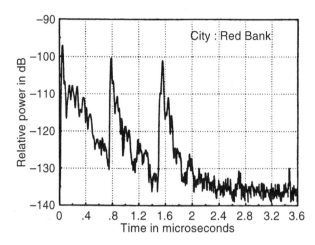

FIGURE 7.4   Power delay profile measured in Red Bank, New Jersey, using a 850-MHz channel sounder with the equivalent of a 50-ns pulse width [27].

many CW contributions that travel along different paths or rays. For a subscriber located at position x, the contributions combine to give the total received voltage, whose complex amplitude $V_C(x)$ is given by

$$V_C(x) = \sum_i V_i(x) e^{j\phi_i(x)} e^{-jkL_i(x)} \tag{7.1}$$

Here $k = 2\pi/\lambda = 2\pi f/c$ is the wavenumber and $L_i(x)$ is the path length of the individual ray. The ray amplitudes $V_i(x)$ will vary relatively slowly over distances on the order of a wavelength, but the phases will show significant variation. Because the rays arrive from different directions, the path lengths $L_i(x)$ of the rays will undergo different variations with x, some increasing and some decreasing, and all at different rates. These variations will result in relative phase changes of the different rays leading to rapid amplitude variations of $V(x)$ in space. In addition, $V_C(x)$ will exhibit variations as a function of frequency, time variations on a scale of seconds, and polarization effects, discussed below.

7.3.1   Spatial Fading over Small Areas

In a typical scenario, measurements of the signal are made as a vehicle carrying the receiving equipment travels along a street. On occasional streets, the base station is visible from the vehicle, and the propagation path is line-of-sight (LOS). However, on most streets the vehicle is obscured from the base station by the surrounding buildings, and the path is non-line-of-sight (NLOS). Figure 7.5 shows the magnitude $V(x) = |V_c(x)|$ of the received signal as a function of the distance traveled by a vehicle that is obscured from the base station for 910-MHz transmission [30]. The amplitude is seen to undergo variations of up to 20 dB over distances on the order of one half the wavelength λ, which at this frequency is about $\frac{1}{3}$ m. In a moving vehicle, this spatial variation is perceived as a rapid time variation, which has led to the term fast fading. Various studies have shown that the signal variation can be explained in terms of the spatial interference pattern set up by the rays arriving from all directions [31–33]. As a consequence, the interference pattern observed

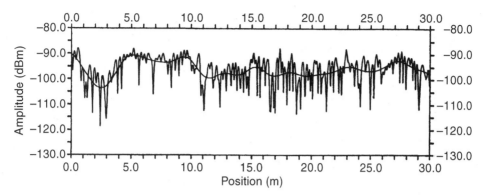

FIGURE 7.5 Narrowband (CW) signal variation in dBm measured along a street and the sliding average (shadow fading) as a function of position for non-LOS conditions [30].

by a moving subscriber will have the same characteristics for any direction of travel in the horizontal plane, that is, for a subscriber crossing the street or moving along a street.

In addition to the rapid fluctuations in Figure 7.5, the signal has a slower variation of about 10 dB. The slow variation, or fading, can be separated from the fast fading using a sliding average over a window of length 2W that is several meters long [25, 30]. The sliding average $\overline{V(x)}$ is shown in Figure 7.5 and represents the slow-fading variation. Slow fading is thought to arise from shadowing by buildings and other objects, and is often referred to as shadow fading or shadow loss.

Whereas $V(x)$ represents the fast fading in combination with the slow fading, the normalized voltage $V(x)/\overline{V(x)}$ gives the fast fading alone. Because the fast fading has an irregular appearance, its statistical properties have been used to anticipate system performance. Let $r_i = V(x_i)/\overline{V(x_i)}$ be the sampled value of the normalized voltage at a point x_i of the fast fading. For locations that are heavily shadowed by surrounding buildings, it is typically found that the probability distribution function (PDF) $p(r)$ for r_i approximates a Rayleigh distribution [25], [30–33]:

$$p(r) = \frac{r}{\rho^2} \exp\left(-\frac{r^2}{2\rho^2}\right) \tag{7.2}$$

The mean value of r can readily be shown to be $\langle r \rangle = \rho\sqrt{\pi/2} \approx 1.25\rho$. Because the mean value of $V(x)$ is $\overline{V(x_i)}$, the mean value $\langle r \rangle = \langle V(x)/\overline{V(x)} \rangle = 1$ and therefore $\rho = \sqrt{2/\pi} \approx 0.80$. Thus for heavily shadowed environments, knowing the small area average voltage $\overline{V(x_i)}$ is sufficient to completely characterize the statistical properties of the signal.

For LOS or other paths where one ray makes a dominant contribution to the received voltage, the distribution function is typically found to be that of a Rician distribution [34], whose probability density function $p(r)$ is defined for $r \geq 0$, and is given by

$$p(r) = \frac{r}{\rho^2} \exp\left[-\left(\frac{r^2 + r_0^2}{2\rho^2}\right)\right] I_0\left(\frac{rr_0}{\rho^2}\right) \tag{7.3}$$

Here $I_0(\cdot)$ is a modified Bessel function of the first kind and zero order [35], $r_0^2/2$ is proportional to the power of the dominant wave signal, and ρ^2 is proportional to the net power of all the other waves [34]. The relative amplitude of the dominate signal is often measured by means of the parameter $K = r_0^2/(2\rho^2)$. When r_0 vanishes, $I_0(0) = 1$ and the distribution reduces to the Rayleigh distribution (7.3). When r_0 is large, the distribution approaches a Gaussian distribution centered at r_0. It is possible to solve for ρ in terms of K by requiring that $\langle r \rangle = \langle V(x)/\overline{V(x)} \rangle = 1$. The Rician has the property that the distribution of r is more concentrated about the mean value. Thus a voltage with Rician distribution will exhibit smaller fluctuations about its mean value than will a voltage with Rayleigh distribution.

7.3.2 High-Order Fast-Fading Statistics

In addition to the distribution of the narrowband signal about its mean value, system designers are concerned with other statistical parameters of the multipath fading, such as the correlation distance of the fading pattern, the average fade duration, and the level crossing rate [25, 31, 34]. The correlation distance is found from the normalized autocorrelation function $C(s)$ found from the complex voltage envelope $V_C(x)$ using

$$C(s) = \int_{-W}^{W} V_C(x)V_C^*(x - s)\, dx \bigg/ \int_{-W}^{W} |V_C(x)|^2 \, dx \qquad (7.4)$$

Here it is assumed that the subscriber antenna is moved over a window $2W$ that is 50 m or more. An example of a normalized autocorrelation function obtained from measurements at 821 MHz [36] is shown in Figure 7.6. In Figure 7.6, the full scale of 300 sample points corresponds to a spatial offset s of approximately 4 m. The autocorrelation function is seen to go to zero at about 10 sample points, or a distance $s = 0.13$ m. For comparison, the wavelength at 821 MHz is $\lambda = 0.365$ m. The decorrelation distance is the offset s at which the normalized autocorrelation function drops below some value, frequently taken as 0.5. From Figure 7.6 it is seen that the received voltage is decorrelated after a distance that is less than $\lambda/3$. One implication of this result is that diversity reception at the subscriber set can be achieved using antennas separated by a fraction of a wavelength. Thus for frequencies of 1800 MHz and above, even in a handset the two antennas can be placed far enough apart to achieve diversity reception.

   Multipath fading is also observed at elevated base stations. As seen in Figure 7.3, and by reversing the arrows in Figures 7.1, the rays arrive over a limited range of angles $\Delta\theta$. In narrowband systems, these rays give rise to spatial fading at the base station, although the scale length over which it occurs is much greater than λ. To overcome this fading, two or more antennas separated by 10λ or more are used for diversity reception [36]. Measurements of spatial correlation at base station sites in cities indicate that the multipath arrivals come from a region about the subscriber whose radial extent is about 100 m [37]. At a distance $R = 1$ km from the base station, such a region appears to have angular width $\Delta\theta = 6°$, which is consistent with the spread of the earliest arrivals in Figure 7.3. Theoretical arguments for this case give the decorrelation distance as $\lambda/\Delta\theta$[1], with $\Delta\theta$ in radians, which for an angular width of 6°, is 9.55λ.

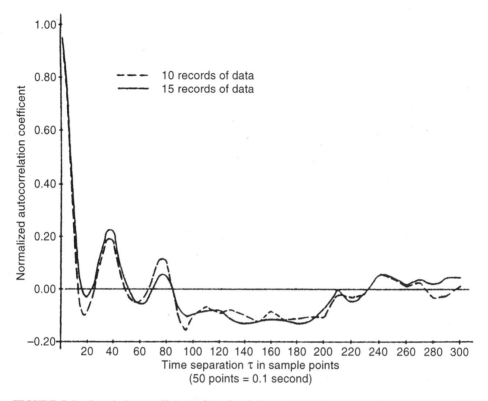

FIGURE 7.6 Correlation coefficient of the fast fading at 821 MHz vs. sample point number. The full scale of 300 sample points corresponds to a spatial offset s of approximately 4 m, (10 sample points $= 0.13$ m) [36].

7.3.3 Doppler Spread

When the vehicle is moving at road or train speeds, the signals arriving from various directions exhibit a range of Doppler frequency shifts that also effect receiver performance. Signals propagation directly toward or away from the vehicle will experience the maximum Doppler shift of $v_d = \pm v_s/\lambda$, where v_s is the vehicle speed. The power spectral density can be found by replacing the offset s in the autocorrelation function by $v_s t$, and taking its Fourier transform [38]. An example of the Doppler power spectral density obtained from measurements at 1800 MHz is shown in Figure 7.7 [24], where the frequency shift is normalized to v_d. The measured spectrum is seen to vary about the classical spectrum, which is obtained by assuming a continuous and uniform distribution of rays arriving in the horizontal plane [31].

7.3.4 Fading for Indoor and Mobile-to-Mobile Communications

When both the base station and portable are located inside a building, as suggested in Figure 7.8, fast fading is experienced when either end of the link is moved [39, 40]. For mobile-to-mobile communications, the link is also symmetric in the sense that both ends are located in a strongly scattering environment. As a result, fast fading occurs when either

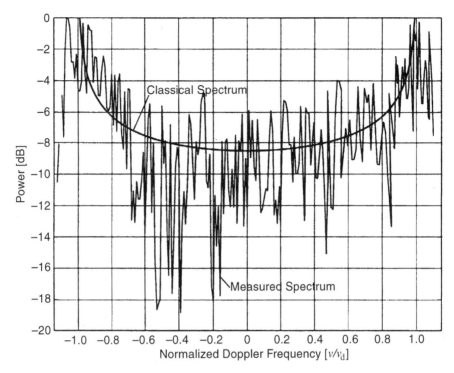

FIGURE 7.7   Comparison of the received power spectral density due to Doppler spread measured at 1800 MHz with a simple theoretical curve [24]. The maximum Doppler shift is v/λ for a subscriber velocity v.

end of the link is moved. This behavior is in contrast to the case of high outdoor base stations antennas, which receive multipath arising from scattering in a region about the subscriber. Because both ends of the link experience fading for in-building and mobile-to-mobile propagation, the average obtained by spatially averaging over one end of the link will change if the other end is moved a distance on the order of a wavelength [41, 42].

An example of the multiple rays generated by reflections at both ends of an indoor link is indicated in Figure 7.8. Two rays that are nearly parallel when they arrive at the receiver may leave the transmitter at widely different directions. When the receiver is moved to obtain the small area average, the sum of the contributions from these two rays will be nearly constant. However, the sum of the two ray contributions will depend strongly on the exact position of the transmitter. Thus the average obtained by moving the receiver will depend on the position of the transmitter, as indicated for the positions x_{T1} and x_{T2} in the cartoon of Figure 7.8. Conversely, rays that are nearly parallel when they leave the transmitter may arrive at very different angles at the receiver. To obtain a true small area to small area path loss, it is therefore necessary to average over both ends of the link.

Actual measurements of the foregoing effect are shown in Figure 7.9 for transmitter and receiver on different floors of a two-story building [42]. Here the average obtained by moving one end of the link around a circular path several wavelengths in diameter is compared to the average obtained when both ends are moved around circular paths. In Figure 7.9 the horizontal axis gives the location along a corridor of the center of the

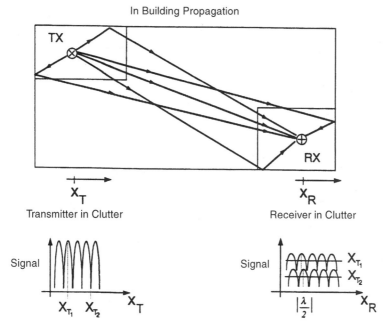

FIGURE 7.8 Multipath at both ends of an indoor link will result in fast fading of the received signal when either end of the link is moved [41].

receiver circle. The curve obtained by averaging over both ends of the link gives smoother distance dependence than obtained when averaging only at the transmitter end. The difference between the two curves has a random appearance, with peak differences of ±5 dB.

7.3.5 Frequency Dependence and Slow Time Fading

Because the phases of the individual ray paths in (7.1) depend on frequency through the term kL_i, the relative phases of paths with different lengths will change with frequency. As a result, the amplitude of the received voltage will also vary with frequency. An example of the frequency dependence of the radio channel about the center frequency of 910 MHz is shown in Figure 7.10 for a 1.2-km link in downtown Toronto [21]. For two rays of length L_i and L_j, the difference in phase is $(L_j - L_i)2\pi f/c$. If $L_j - L_i = 1$ km, then phase difference will change by 2π when f changes by 0.3 MHz. In other words, the two-ray contributions can go from constructive to destructive and back to constructive addition, or vice versa, in 0.3 MHz, which is consistent with the spacing between minima in Figure 7.10. For indoor links the differences in path length will be much smaller, requiring a greater change in frequency to produce the same cycle in the interference [43].

For an individual ray path, the motion of scattering objects, such as vehicles, pedestrians, and trees, causes a change in L_i even when the subscriber is stationary. The resulting change in the phases of the individual rays will therefore change with time over a

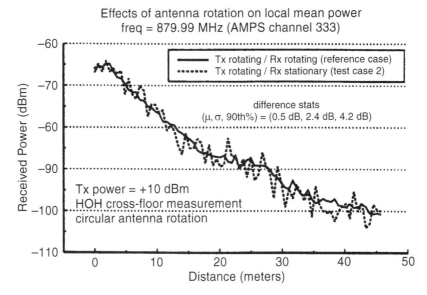

FIGURE 7.9   Comparison of the small area averages obtained on indoor links when small area averaging is taken around a circle at both ends of the link or only on a circle at one end of the link [42]. Transmitter and receiver were on different floors of a two-story building. The horizontal axis represents the horizontal separation between the centers of the averaging circles.

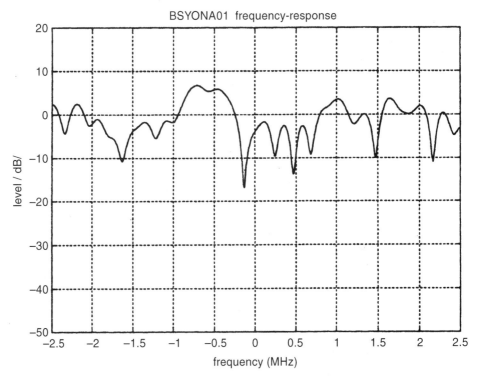

FIGURE 7.10   Frequency variation about the center frequency of 910 MHz for a 1.2-km radio link in downtown Toronto [21].

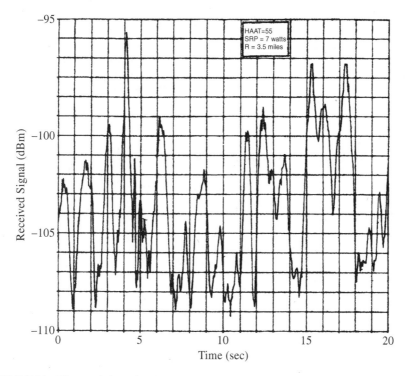

FIGURE 7.11 Time variation of a 900-MHz CW signal received at a vehicle parked on a suburban street [44].

scale of seconds, causing the amplitude of the received signal to change. Figure 7.11 shows time variation observed in the amplitude of the received voltage for a 900-MHz signal measured at a vehicle parked in a suburban environment [44]. It is seen that variations in the voltage amplitude occur over a time scale of 0.5 s or less. Pedestrian walking speed is around 4 km/h (2.4 miles/h) or 1.1 m/s, so that in 0.5 s a person walks about 0.5 m, which is slightly larger than λ at 900 MHz.

7.3.6 Polarization Coupling

The diffraction and scattering processes that cause multipath arrivals can also result in depolarization of the waves. Measurements have been made of the co-polarized (vertical) and cross-polarized (horizontal) electric fields received at the base station [45–49]. Table 7.1, which is taken from Lotse et al. [48], lists ratios between the small area average signals received in the vertical and horizontal polarizations. For transmission from a vertically polarized antenna mounted on top of a vehicle, the ratio ranges from about 7 dB in cities to 13 dB in the far suburbs. Surprisingly, there is significantly more power in the vertical polarization for hand-held portable terminals in suburban Veddesta. Other measurements using hand-held terminals operating inside buildings found equal power in the two polarizations received at an outdoor base station [50]. While both polarizations exhibit fast fading as the subscriber antenna is moved, the fading patterns are uncorrelated [49], suggesting the use of polarization diversity in place of spatial diversity at the base station.

TABLE 7.1   Small Area Averages of the Cross Polarization Ratio Measured at 1800 MHz in Sweden [48][a]

Environment	Mobile Configuration	Horizontal-to-Vertical Power Ratio (dB)
Kungshholmen, urban	Roof-mounted	-7 ± 2
	protable, outdoors	-4 ± 2
	portable, inside van	-3 ± 2
	portable, indoors	-1 ± 4
Kista, suburban	Roof-mounted	-8 ± 2
	portable, outdoors	-2 ± 1
	portable, inside van	-1 ± 1
	portable, indoors	-3 ± 1
Veddesta, suburban	Roof-mounted	-13 ± 1
	portable, outdoors	-6 ± 1
	portable, inside van	-7 ± 1
	portable, indoors	-7 ± 1

[a]The error limits indicate one standard deviation about the mean value.

7.4   MULTIPATH PHENOMENA FOR BROADBAND EXCITATION

With the advent of digital systems, the response to pulsed excitation has been the subject of many measurements [17–23, 27, 43, 51, 52]. Since the bandwidth of these systems is typically 5 MHz or less, it is the finite bandwidth channel response that is of importance. For center frequency near 1 GHz, such radiated pulses have a small fractional bandwidth, and hence each pulse contains many radio frequency (RF) cycles. Thus the received pulses will exhibit some of the characteristics found for CW excitation. Viewed another way, the duration of pulses having a 5-MHz bandwidth is about 200 ns, which is less than the differences in the delay between many ray paths. What may appear to be an individual received pulse is therefore a composite of multiple ray arrivals.

To set these multipath effects in a framework, suppose that $w(t)e^{j\omega t}$ is the voltage, normalized to the free space path-loss, that would be received when the transmitter and receiver are in free space with a small separation to avoid time delays. In the presence of multipath, the received voltage at position, $\underline{r}$ and time, $\underline{t}$ will be $V_C(\underline{r}, t)e^{j\omega t}$, where $V_C(\underline{r}, t)$ is the complex amplitude. Neglecting distortion of the individual pulses as a result of reflection, diffraction, or scattering, the complex received voltage amplitude is found by summing the contributions from the individual ray paths as

$$V_C(\underline{r}, t) = \sum_i w[t - \tau_i(\underline{r})] V_i(\underline{r}) e^{j\phi_i(\underline{r})} e^{-jkL_i} \qquad (7.5)$$

Here $\tau_i(\underline{r}) = L_i/c$ is the ray time delay. At any point $\underline{r}$, the received voltage envelope is $|V_C(\underline{r}, t)|$, while the envelope power $P(\underline{r}, t)$ is proportional to $|V_C(\underline{r}, t)|^2$. When plotted as a function of time, the envelope power is referred to as the power-delay profile, as seen in Figures 7.2(a) and 7.4.

7.4.1 Fading of Individual Pulses

Because many of the individual received pulses will partially overlap in time, received pulses will appear broadened in time, and will have amplitudes that depend on the relative phases of the individual ray contributions. Thus the amplitudes of the received pulses will exhibit fading when the subscriber antenna is moved over distances that are a fraction of a wavelength, as in the case of fast fading observed for CW signals. This effect is demonstrated by the measured pulse response shown in Figure 7.12 observed on indoor LOS and obstructed links using a 5-ns-wide pulse at 2.4 GHz [51, 52]. For each type of path, there are 16 power-delay profiles $|V_C(x_k, t)|^2$ taken at locations x_k ($k = 1, 2, \ldots, 16$) that are separated by $\lambda/4$. For the LOS case of Figure 7.12(a), the amplitude of the first

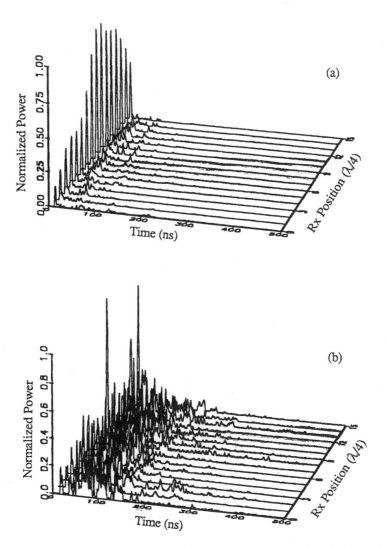

FIGURE 7.12 Power delay profiles measured at 16 locations separated by $\lambda/4$ on indoor links using a 5-ns wide pulse at 2.4 GHz: (a) LOS path and (b) obstructed path [51].

arrival shows dependence on the subscriber location. This is thought to result from interference between the direct ray and the ray reflected from the floor. For the obstructed paths, the individual peaks in the power-delay profiles of Figure 7.12(b) exhibit much more severe fading. Statistical analysis of the measurements has shown that the fading of individual peaks is nearly Rayleigh for obstructed paths and Rician for LOS paths [51].

7.4.2   rms Time-Delay Spread

The root-mean-square (rms) delay spread τ_{rms} is commonly used as a measure of pulse spreading. For a given receiver and transmitter location, the rms delay spread is defined as

$$\tau_{\mathrm{rms}}^2(\underline{r}) = \frac{\displaystyle\int_0^\infty (t - T_0)^2 |V_c(\underline{r}, t)|^2 \, dt}{\displaystyle\int_0^\infty |V_c(\underline{r}, t)|^2 \, dt} \tag{7.6}$$

where T_0 is the mean excess delay. In the case of LOS paths, where the fading of the main peak greatly influences the integrals in (7.6), $T_0(\underline{r})$ and $\tau_{\mathrm{rms}}(\underline{r})$ can have considerable variations with position. For heavily obstructed paths, where no one peak dominates, the variations of $T_0(\underline{r})$ and $\tau_{\mathrm{rms}}(\underline{r})$ are smaller than for LOS paths, but their mean values are greater [51].

Figure 7.13 shows a cumulative distribution function of the rms delay spread measured at 1800 MHz on outdoor paths in urban and suburban Toronto [21]. The median value of τ_{rms} for these measurements was 0.71 μs for urban cells and 0.31 μs for suburban cells. These values are consistent with measurements reported in several other cities [17, 19, 22, 53–55], although some authors have observed smaller median values [27, 56]. As previously noted, very hilly areas give even larger delay spreads [29]. For links inside office buildings, the delay spreads τ_{rms} is typically less than about 50 ns [57–59], which is an order of magnitude smaller than found for outdoor links.

Greenstein et al. [60] have reviewed the published measurements of delay spread for outdoor systems and have proposed a statistical model for τ_{rms} in the form

$$\tau_{\mathrm{rms}} = T_1 R^\varepsilon y \tag{7.7}$$

where R is the distance from the base station in kilometers. The exponent ε has value $\varepsilon = 0.5$ in urban, suburban, and rural areas, while in mountainous areas $\varepsilon = 1$. The average delay spread T_1 at $R = 1$ km ranges from 0.1 μs in rural areas up to 1 μs in urban macrocells, while in mountainous areas $T_1 = 0.5$ μs. The term y in (7.7) is a random variable such that $10 \log(y)$ is a normal distribution with standard deviation of 2 to 6. Recently, simulations of delay spread in cities have been made using a ray-tracing computer code [61]. The simulations are in agreement with the Greenstein model (7.7). Cheon et al. [61] further show that the delay spread is not sensitive to the distribution of building heights but does increase somewhat when the base station is lowered below the rooftops.

To define a measure of the pulse spread that is independent of the exact position within a small area, Devastrvatham [58, 59] has averaged the power-delay profiles over a small

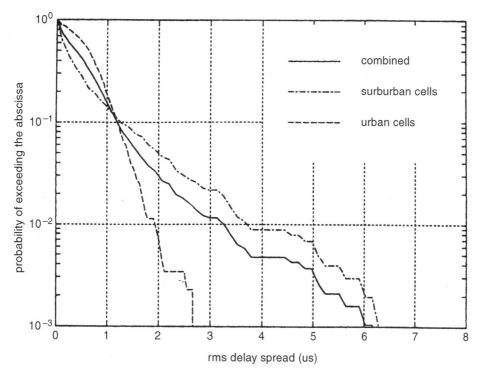

FIGURE 7.13 Cumulative distribution function of rms delay spreads measured for urban and suburban environments in Toronto [21].

area, and from this average profile defined T_0 and τ_{rms}. The significance of averaging can be understood by using (75) to express the power-delay profile in the form

$$|V_c(\underline{r}, t)|^2 = \sum_i \sum_j V_i(\underline{r}) V_j(\underline{r}) w(t - \tau_i) w^*(t - \tau_j) e^{j[\phi_i(\underline{r}) - \phi_j(\underline{r})]} e^{-jk(L_i - L_j)} \qquad (7.8)$$

For small displacements, $V_i(\underline{r})$ will be nearly constant, and the change in the pulse delay t_i will be small compared to the pulse width. However, $k(L_i - L_j)$ for $i \neq j$ will vary by 2π or more, so that the spatial average of the corresponding exponential terms in (7.8) will be small. As a result, spatial averaging reduces the double summation in (7.8) to the single sum

$$\langle |V_c(\underline{r}, t)|^2 \rangle = \sum_i |V_i(\underline{r})|^2 |w(t - \tau_i)|^2 \qquad (7.9)$$

The spatial average (7.9) of the time-delay profiles is representative of the local environment of the subscriber. For indoor links, it is necessary to average over both ends of the link, as discussed in Section 7.3.4. The values of T_0 and τ_{rms} obtained from using (7.9) in (7.6) are found to be close to the values obtained by averaging $T_0(\underline{r})$ and

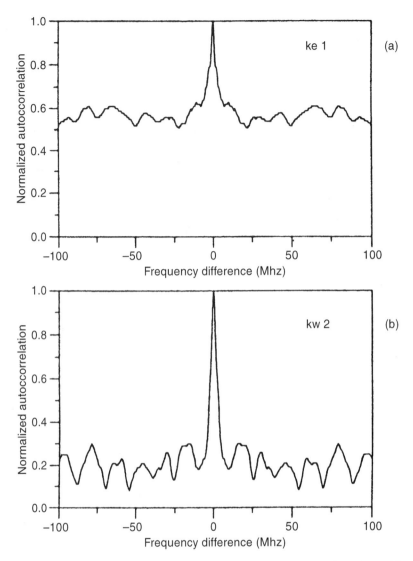

FIGURE 7.14 Normalized coherence function obtained on indoor links using a 5-ns wide pulse at 2.4 GHz: (a) LOS path and (b) obstructed path [51].

$\tau_{rms}(\underline{r})$ for the individual profiles [51]. If the spatial average time-delay profile is integrated over time to find the total energy W, it is seen from (7.9) that

$$W = \int_0^\infty \langle |V_c(\underline{r}, t)|^2 \rangle \, dt = \left[\int_0^\infty |w(t)|^2 \, dt \right] \sum_i |V_i(\underline{r})|^2 \qquad (7.10)$$

The summation in (7.10) represents the spatial average of the received power for narrowband excitation, which is seen to be a direct measure of the received energy for pulsed sources.

7.4.3 Coherence Bandwidth

Since the pulse response of a radio channel contains information over a band of frequencies, pulse measurements can be used to characterize the frequency dependence via Fourier transforms [21]. The transform requires knowledge of the phase of the received voltage, which is more difficult to measure than the amplitude. The coherence function $R_c(\omega)$ is an alternative way of studying the frequency dependence of a channel base on power measurements only. The coherence function is obtained by taking the Fourier transform of the average power-delay profile [31, Ch. 1; 62], or

$$R_c(\Delta\omega) = \int_{-\infty}^{\infty} \langle |V_c(\underline{r}, t)|^2\rangle e^{-j\Delta\omega t}\, dt \tag{7.11}$$

Figure 7.14 shows two examples of normalized coherence functions obtained from measurements on indoor links using a 5-ns-wide pulse at 2.4 GHz [51]. For the LOS path the coherence function is seen to be greater than 0.5 over a bandwidth of about 50 MHz, while for the obscured path it remains above 0.5 only over the bandwidth of about 5 MHz.

To understand its meaning of the coherence function, recognize that for free space propagation, where there is no multipath interference, the received signal will show no spreading in time. In this case the channel is independent of frequency and the coherence bandwidth is infinite. For LOS paths, where the direct ray gives the dominant contribution to the signal, the signal will show only small spreading, there will be only weak frequency dependence, and the coherence bandwidth is large. On obstructed paths many multipath arrivals having roughly equal amplitudes will spread the received signal in time. In this case the frequency dependence for CW excitation will show rapid Rayleigh-like variations, as seen in Figure 7.11, and coherence bandwidth will be small. The coherence bandwidth is therefore inversely related to the rms delay spread.

7.5 ANGULAR SPREAD FOR SPACE–TIME SIGNAL PROCESSING

Development of multiple antenna technologies, which seeks to take advantage of the multipath to increase channel capacity, have given additional reasons for studying the angle of arrival spread at elevated base stations. Wideband measurements are being made using array antennas [24, 26, 56, 63] to find the individual ray arrivals in space and time, such as those shown in Figure 7.3. To characterized the angle spread it is common to employ the second moment of the angular deviation of the arriving rays. For a given subscriber location, let θ_i be the angle of arrival of the ith ray, as measured from some reference direction. The rms angle spread θ_{rms} is commonly taken to be

$$\theta_{\text{rms}}^2 = \frac{\sum_i (\theta_i - \bar{\theta})^2 V_i^2}{\sum_i V_i^2} \tag{12}$$

where $\bar{\theta}$ is the mean value of arrival angles θ_i.

When measurements are made at the base station for many subscriber locations, a cumulative distribution function of θ_{rms} can be constructed. The distribution for measurements made in Aarhus, Denmark, using base station antenna heights of 20, 26, and 32 m

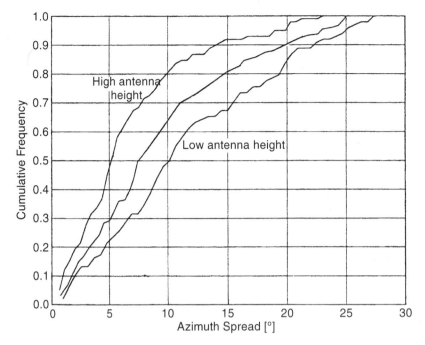

FIGURE 7.15   Cumulative distribution function of azimuth angle spread for subscriber locations in Aahus, Denmark, as measured at 1800 MHz for base station antenna heights of 20, 26, and 32 m [26].

[24, 26] is shown in Figure 7.15. Because the buildings are 4 to 7 stories, the 20 m antenna height is below the highest rooftops, the 26-m height is near to or above the highest buildings, while the 32-m-high antenna is well above the tallest buildings. The median value of the distribution is seen to be smallest for the highest antenna, where it has a value of 5°.

Little information is available about the influence of distance or building height distribution on angle spread. Measurements of angle spread θ_{rms} in urban and suburban environments in Sweden found that the median value was several times greater in urban areas than in suburban areas when the base station antenna was located well above the buildings [63]. A related set of measurements [56] found that the cross-polarized signals had the same distribution of delay spread and angle spread as the co-polarized signals, although they were lower in amplitude. Simulations carried out using a ray tracing computer code [61] have found no systematic variation of the median value of θ_{rms} on distance from the base station. However, it did show significant dependence of median value on the building height distribution and on the height of the base station antenna relative to the building.

7.6   SUMMARY

Various measurements demonstrate that propagation between a base station and a street-level mobile takes place along multiple ray paths. Due to scattering by obstacles, such as buildings and vehicles, in the vicinity of the mobile, the rays at the mobile come from all directions in the horizontal plane and over a range of angles in the vertical plane. As seen at an elevated base station, the rays resulting from scattering near the mobile occupy a small range of angles. However, scattering by large buildings results in additional clusters of rays that can have a wider angular spread.

The multiple ray paths lead to a wealth of observed phenomena for different types of transmitted signals. For narrowband or CW signals, the phase of the individual ray fields depends on the position of the receiver and the frequency, so that changing either parameter results in fading as the ray fields go in and out of phase. For radio links that do not have line of sight between the antennas, the fading can be 20 dB or more and has statistical distribution that is close to Rayleigh. Because some of the scatterers move in time, the phase of some of the rays change in time, leading to slow time fading at a stationary subscriber. Scattering that produces the multiple rays also produces cross-polarization coupling and Doppler spread when the subscriber is moving.

Because of the large number of ray paths, whose differences in length cover many scales, the reception of pulsed signals having finite duration will also involve multiple ray arrivals that overlap in time. As a result, what appears at the receiver as a distinct pulse is in fact the sum of many arrivals with small delay spread, so that the apparent pulses will exhibit fading similar to that of narrowband signals. Propagation inside buildings exhibits the same effects observed on outdoor links, except that the differences in path length are smaller.

REFERENCES

1. H. L. Bertoni, *Radio Propagation for Modern Wireless Applications*, Prentice Hall PTR, Upper Saddle River, NJ, 2000.

2. Y. Okumura, E. Ohmori, T. Kawano, and K. Fukuda, "Field Strength and Its Variability in VHF and UHF Land-Mobile Radio Service," *Re. Elec. Comm. Lab.*, Vol. 16, pp. 825–873, 1968.

3. D. O. Reudink, "Properties of Mobile Radio Propagation Above 400 MHz," *IEEE Trans. Veh. Tech.*, Vol. VT-23, pp. 143–159, 1974.

4. K. Allsebrook and J. D. Parson, "Mobile Radio Propagation in British Cities at Frequencies in the VHF and UHF Bands," *IEEE Trans. Veh. Tech.*, Vol. VT-26, pp. 313–322, 1977.

5. G. D. Ott and A. Plitkins, "Urban Path-Loss Characteristics at 820 MHz," *IEEE Trans. Veh. Tech.*, Vol. VT-27, pp. 189–197, 1978.

6. K. K. Kelly, "Flat Suburban Area Propagation at 820 MHz," *IEEE Trans. Veh. Tech.*, Vol. VT-27, pp. 198–204, 1978.

7. N. H. Shepherd et al., "Coverage Prediction for Mobile Radio Systems Operating in the 800/900 MHz Frequency Range," Special Issue *IEEE Trans. Veh. Tech.*, Vol. VT-37, pp. 3–72, 1988.

8. S. T. S. Chia, R. Steele, E. Green, and A. Baran, "Propagation and Bit Error Ratio Measurements for a Microcellular System," *IRE*, Vol. 57, pp. S255–S266, 1987.

9. J. H. Whitteker, "Measurements of Path Loss at 910 MHz for Proposed Microcell Urban Mobile Systems," *IEEE Trans. Veh. Tech.*, Vol. VT-37, pp. 125–129, 1988.

10. P. Harley, "Short Distance Attenuation Measurements at 900 MHz and 1.8 GHz Using Low Antenna Heights for Microcells," *IEEE J. Selected Areas in Comm.*, Vol. 7, pp. 5–11, 1989.

11. A. J. Rustako, Jr., N. Amitay, G. J. Owens, and R. S. Roman, "Radio Propagation at Microwave Frequencies for Line-of-Sight Microcellular Mobile and Personal Communications," *IEEE Trans. Veh. Tech.*, Vol. 40, pp. 203–210, 1991.

12. P. E. Mogensen, P. Eggers, C. Jensen, and J. B. Andersen, "Urban Area Radio Propagation Measurements at 955 and 1845 MHz for Small and Micro Cells," Proc of GLOBECOM'91, Phoenix, AZ, pp. 1297–1302, 1991.

13. A. J. Goldsmith and L. J. Greenstein, "A Measurement-Based Model for Predicting Coverage Areas of Urban Microcells," *IEEE J. Selected Areas in Comm.*, Vol. 11, pp. 1013–1023, 1993.

14. H. H. Xia, H. L. Bertoni, L. R. Maciel, A. Lindsay-Stewart and R. Rowe, "Radio Propagation Characteristics for Line-of-Sight Microcellular and Personal Communications," IEEE Trans. Ant. and Prop., Vol. 41, no. 10, pp. 1439–1447, 1993.

15. H. H. Xia, H. L. Bertoni, L. R. Maciel, A. Lindsay-Stewart, and R. Rowe, "Microcellular Propagation Characteristics for Personal Communications in Urban and Suburban Environments," *IEEE Trans. Veh. Tech.*, Vol. 43, no. 3, pp. 743–752, 1994.

16. T. Taga, "Analysis for Mean Effective Gain of Mobile Antennas in Land Mobile Radio Environments," *IEEE Trans. Veh. Tech.*, Vol. VT-39, pp. 117–131, 1990.

17. D. C. Cox and R. P. Leck, "Correlation Bandwidth and Delay Spread in Multipath Propagation Statistics for 910 MHz Urban Radio Channels," *IEEE Trans. Comm.*, Vol. COM-23, pp. 1271–1280, 1975.

18. D. M. J. Devasirvatham, "Time Delay Spread and Signal Level Measurements of 850 MHz Radio Waves in Building Environments," *IEEE Trans. Ant. Prop.*, Vol. AP-34, pp. 1300–1305, 1986.

19. S. Y. Seidel, T. S. Rappaport, S. Jain, M. L. Lord, and R. Singh, "Path Loss, Scattering and Multipath Delay Statistics in Four European Cities for Digital Cellular and Microcellular Radiotelephone," *IEEE Trans. Veh. Tech.*, Vol. VT-40, pp. 721–730, 1991.

20. S. Kozono and A. Taguchi, "Mobile Propagation Loss and Delay Spread Characteristics with a Low Base Station antenna on an Urban Road," *IEEE Trans. Veh. Tech.*, Vol. VT-42, pp. 103–108, 1993.

21. E. S. Sousa, V. M. Jovanovic and C. Daigneault, "Delay Spread Measurements for the Digital Cellular Channel in Toronto," *IEEE Trans. Veh. Tech.*, Vol. VT-43, pp. 837–847, 1994.

22. J. A. Wepman, J. R. Hoffman and L. H. Loew, "Analysis of Impulse Response Measurements for PCS Channel Modeling Applications," *IEEE Trans. Veh. Tech.*, Vol. VT-44, pp. 613–618, 1995.

23. J. Fuhl, J-P. Rossi, and E. Bonek, "High-Resolution 3-D Direction-of-Arrival Determination for Urban Mobile Radio," *IEEE Trans. Ant. Prop.*, Vol. 45, pp. 672–682, 1997.

24. K. I. Pedersen, P. E. Mogensen, B. H. Fleury, F. Frederiksen, K. Olesen, and S. L. Larsen, "Analysis of Time, Azimuth and Doppler Dispersion in Outdoor Radio Channels," Proc. ACTS, 1997.

25. W. C. Y. Lee, *Mobile Communications Engineering*, McGraw-Hill, New York, 1982.

26. K. I. Pedersen, P. E. Mogensen, and B. Fleury, "Spatial Channel Characteristics in Outdoor Environments and Their Impact on BS Antenna System Performance, *Proc. IEEE Veh. Tech. Conf.*, 1998.

27. D. M. J. Devasirvatham, "Radio Propagation Studies in a Small City for Universal Portable Communications," *Proc. IEEE Veh. Tech. Conf.*, pp. 100–104, 1988.

28. A. A. M. Saleh and R. A. Valenzuela, "A Statistical Model for Indoor Multipath Propagation," *IEEE Selected Areas in Commun.*, Vol. 5, pp. 128–137, 1987.

29. A. Zogg, "Multipath Delay Spread in a Hilly Region at 210 MHz," *IEEE Trans. Veh. Tech.*, Vol. VT-36, pp. 184–187, 1987.

30. M. Lecours, I. Y. Chouinard, G. Y. Delisle and J. Roy, "Statistical Modeling of the Received Signal Envelope in a Mobile Radio Channel," *IEEE Trans. Veh. Tech.*, Vol. VT-37, pp. 204–212, 1988.

31. W. C. Jakes, Ed., *Microwave Mobile Communications*, IEEE Press: Piscataway, NJ, Chapter 1, 1974.

32. H. W. Nylund, "Characteristics of Small-Area Signal Fading on Mobile Circuits in the 150 MHz Band," *IEEE Trans. Veh. Tech.*, Vol. VT-17, pp. 24–30,1968.

33. G. L. Turin, F. D. Clapp, T. L. Johnston, S. B. Fine, and D. Lavry, "A Statistical Model of Urban Multipath Propagation," *IEEE Trans. Veh. Tech.*, Vol. VT-21, pp. 1–9, 1972.

34. D. Parsons, *The Mobile Radio Propagation Channel*, Wiley, Chichester, England, 1996, pp. 134–135.

35. M. Abramowitz and I. A. Stegun, *Handbook of Mathematical Functions*, Dover: New York, 1965, pp. 376, 487.

36. S-B. Rhee and G. I. Zysman, "Results of Suburban Base Station Spatial Diversity Measurements in the UHF Band," *IEEE Trans. Comm.*, Vol. COM-22, pp. 1630–1636, 1974.

37. F. Adachi, M. T. Feeney, A. G. Williamson, and J. D. Parsons, "Crosscorrelation between the Envelopes of 900 MHz Signals Received at a Mobile Radio Base Station Site," *IEE Proc.*, Vol. 133, Pt. F, pp. 506–512, 1980.

38. A. Papoulis, *Probability, Random Variables, and Stochastic Processes*, 3rd ed., McGraw-Hill, New York, 1991, pp. 319–332.

39. R. J. C. Bultitude, S. Mahmoud, and W. Sullivan, "A Comparison of Indoor Radio Propagation Characteristics at 910 MHz and 1.75 GHz," *IEEE J. Selected Areas in Commun.*, Vol. 7, pp. 20–30, 1989.

40. T. S. Rappaport and C. D. McGillen, "UHF Fading in Factories," *IEEE J. on Selected Areas in Commun.*, Vol. 7, pp. 40–48, 1989.

41. W. Honcharenko, H. L. Bertoni, and J. Dailing, "Bi-Lateral Averaging over Receiving and Transmitting Areas for Accurate Measurements of Sector Average Signal Strength inside Buildings," *IEEE Trans. Ant. Prop.*, AP-43, pp. 508–512, 1995.

42. R. A. Valenzuela, O. Landron, and D. L. Jacobs, "Estimating Local Mean Signal Strength of Indoor Multipath Propagation," *IEEE Trans. Veh. Tech.*, Vol. VT-46, pp. 203–212, 1997.

43. H. Hashemi and D. Tholl, "Statistical Modeling and Simulation of the RMS Delay Spread of Indoor Radio Propagation Channels," *IEEE Trans. Veh. Tech.*, Vol. VT-43, pp. 110–120, 1994.

44. N. H. Shepherd et al., "Special Issue on Radio Propagation," *IEEE Trans. Veh. Tech.*, Vol. VT-37, p. 45, 1988.

45. W. C. Y. Lee and Y. S. Yeh, "Polarization Diversity System for Mobile Radio," *IEEE Trans. Comm.*, Vol. COM-20, pp. 912–913, 1972,

46. S. A. Bergmann and H. W. Arnold, "Polarization Diversity in Portable Communications Environments," *Elect. Lett.*, Vol. 22, pp. 609–610, 1986.

47. R. G. Vaughn, "Polarization Diversity in Mobile Communications," *IEEE Trans. Veh. Tech.*, Vol. VT-39, pp. 177–186, 1990.

48. F. Lotse, J.-E., Berg, U. Forssen, and P. Idahl, "Base Station Polarization Diversity Reception in Macrocellular Systems at 1800 MHz," *Proc. IEEE Veh. Tech. Conf.*, pp. 1643–1646, 1996.

49. J. J. A. Lempiainen, J. K. Laiho-Steffens, and A. F. Wacker, "Experimental Results of Cross Polarization Discrimination and Signal Correlation Values for a Polarization Diversity Scheme," *Proc. VTC 97*, pp. 1498–1502.

50. P. C. F. Eggers, I. Z. Kovacs, and K. Olesen, "Penetration Effects on XPD with GSM 1800 Handset Antennas, Relevant for BS Polarization Diversilty for Indoor Coverage," *Proc. IEEE Veh. Tech. Conf.*, pp. 1959–1964, 1998.

51. S. Kim, H. L. Bertoni, and M. Stern, "Pulse Propagation Characteristics at 2.4 GHz Inside Buildings," *IEEE Trans. Veh. Tech.*, Vol. 45, pp. 579–592, 1996.

52. H. L. Bertoni, S. Kim, and W. Honcharenko, "Review of In-Building Propagation Phenomena at UHF Frequencies," *Proc. IEEE ASILOMAR-29*, pp. 761–765, 1996.

53. R. J. C. Bultitude and G. K. Bedal, "Propagation Characteristics on Microcellular Urban Mobile Radio Channels at 910 MHz," *IEEE J. Selected Areas in Commun.*, Vol. 7, pp. 31–39, 1989.

54. A. M. D. Turkmani, D. A. Demery, and J. D. Parsons, "Measurement and Modeling of Wideband Mobile Radio Channels at 900 MHz," *IEE Proc. 1*, Vol. 138, pp. 447–457, 1991.

55. H. Asplund, A. A. Glazunov, and J. E. Berg, "An Investigation of Measured and Simulated Wideband Channels with Applications to 1. 25 and 5 MHz CDMA Systems," *Proc. IEEE Veh. Tech. Conf.*, pp. 562–566, 1998.

56. M. Nilsson, B. Lindmark, M. Ahlberg, M. Larsson, and C. Beckmanm, "Measurements of the Spatio-Temporal Polarization Characteristics of a Radio Channel at 1800 MHz," *Proc. IEEE Veh. Tech. Conf.*, 1999.

57. D. M. J. Devasirvatham, "Time Delay Spread Measurements of Wideband Radio Signals Within a Building," *Elect. Lett.*, Vol. 20, pp. 950–951, 1984.

58. D. M. J. Devasirvatham, "Multipath Time Delay Spread in the Digital Portable Radio Environment," *IEEE Commun. Mag.*, Vol. 25, pp. 13–21, 1987.

59. D. M. J. Devasirvatham, "Multipath Time Delay Jitter Measured at 850 MHz in the Portable Radio Environment," *IEEE J. Selected Area in Commun.*, Vol. 5, pp. 855–861, 1987.

60. L. J. Greenstein, V. Erceg, Y. S. Yeh and M. V. Clark, "A New Path-Gain/Delay-Spread Propagation Model for Digital Cellular Channels," *IEEE Trans. on Veh. Tech.*, Vol. 46, pp. 477–485, 1997.

61. C. Cheon, G. Liang, and H. L. Bertoni, "Monte Carlo Simulation of Delay and Angle Spread in Different Building Environments," *Proc. IEEE Veh. Tech. Conf.*, Sept., 2000.

62. J. G. Proakis, *Digital Communications*, McGraw-Hill, New York, 1989, pp. 702–719.

63. M. Larsson, "Spatio-Temporal Channel Measurements at 1800 MHz for Adaptive Antennas," *Proc. IEEE Veh. Tech. Conf.*, 1999.

CHAPTER 8

Indoor Propagation Modeling

HOMAYOUN HASHEMI

8.1 INTRODUCTION

The international research and development organizations and regulatory bodies are facing great challenges in designing and implementing the third and fourth generations of wireless communications systems. The ultimate goal of personal communication services (PCS) is to provide instant communications between individuals located anywhere in the world and at any time. Realization of futuristic pocket-size multimedia transceivers and subsequent Dick Tracy wrist watch phones are major communication frontiers. An important consideration in successful implementation of the currently evolving PCS is indoor communication, that is, transmission of voice, data, and motion video to people on the move inside buildings. Indoor communication covers a wide variety of situations ranging from communication with individuals walking in residential or office buildings, supermarkets or shopping malls, and the like, to fixed stations sending messages to robots in motion in assembly lines and factory environments of the future, to transmission of high bit rate data and motion video to carriers of laptop and palmtop transceivers.

Multipath fading seriously degrades the performance of communication systems. Unfortunately, one can do little to eliminate multipath disturbances. However, if the channel is well characterized, the effect of these disturbances can be reduced by proper design of the transmitter and receiver. Detailed characterization of radio propagation is therefore a major requirement for successful design of indoor communication systems. Such characterization is particularly useful if it leads to a reliable simulation model.

The goal of this presentation is to provide a tutorial coverage of the indoor radio and infrared propagation channels. Since the multipath medium can be fully described by its time and space varying impulse response, an emphasis is placed on characterization of the channel's impulse response. After proper mathematical (the impulse response) formulation of the channel, other related topics such as radio channel's spatial and temporal variations, large-scale path losses, mean excess delay and root-mean-square (rms) delay spread, are

Wireless Communications in the 21st Century, Edited by Shafi, Ogose, and Hattori.
ISBN 0-471-155041-X © 2002 by the IEEE.

addressed. Finally, the indoor infrared channel and its properties are described. The interested reader is also referred to the previous tutorial presentations by the author [1–4].

8.2 TYPES OF VARIATIONS IN THE CHANNEL

The signal transmitted from the fixed base station in an indoor environment reaches the portable radio receivers via one or more *main waves*. These main waves consist of a line-of-sight (LOS) ray and several rays reflected or scattered by main structures such as outer walls, floors, and ceilings. The LOS wave may be attenuated by the intervening structure to an extent that makes it undetectable. The main waves are random upon arrival in the local area of the portable receivers. They break up in the environment of the portable receivers due to scattering by local structure and furniture. The resulting paths for each main wave arrive with very close delays, experience about the same attenuation, but have different phase values due to different path lengths. The individual multipath components are added according to their relative arrival times, amplitudes, and phases, and their random envelope sum is observed by the portable receivers. The number of distinguishable multipath components recorded in a given measurement and at a given point in space depends on the shape and structure of the building and on the resolution of the measurement setup [1].

Let X_{ijk} ($i = 1, 2, \ldots, N$; $j = 1, 2, \ldots, L$; $k = 1, 2, \ldots, M$) be a random variable representing a parameter of the channel at a fixed point in the three-dimensional space. For example, X_{ijk} may represent amplitude of a multipath component at a fixed delay, amplitude of a narrowband fading signal, the number of detectable multipath components, mean excess delay, or delay spread. The index k in X_{ijk} numbers spatially adjacent points in a given portable site of radius 1 to 2 m. These points are very close (in the order of several centimeters or less). The index j numbers different sites with the same base-portable antenna separations, and the index i numbers groups of sites with different antenna separations. This scenario is illustrated in Figure 8.1.

With the above notations there are three types of variations in the channel [1]. The degree of these variations depends on the type of environment, distance between samples, and on the specific parameter under consideration. For some parameters one or more of these variations may be negligible.

8.2.1 Small-Scale Variations

A number of impulse response profiles collected in the same "local area" or site are grossly similar since the channel's structure does not change appreciably over short distances (Fig. 8.1). Therefore, impulse responses in the same site exhibit only variations in fine details. With fixed i and j, X_{ijk} ($k = 1, 2, \ldots, M$) are correlated random variables for close values of k. This is equivalent to the correlated fading experienced in the mobile channel for close sampling distances.

8.2.2 Midscale Variations

This is a variation in the statistics for local areas with the same antenna separation (Fig. 8.1). As an example, two sets of data collected inside a room and in a hallway, both having the same antenna separation, may exhibit great differences. If μ_{ij} denotes the mathematical expectation of X_{ijk} [i.e., $\mu_{ij} = E_k(X_{ijk})$, where E_k denotes expectation with respect to k],

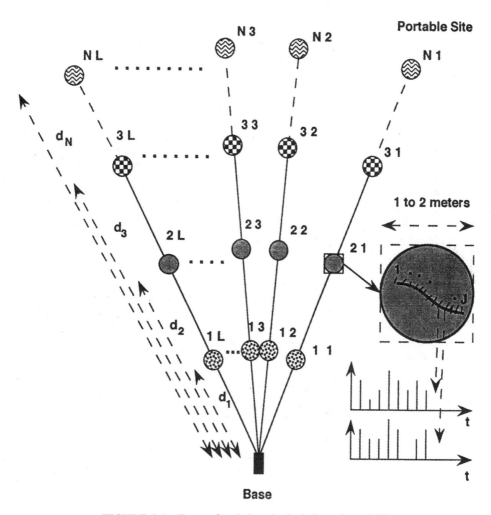

FIGURE 8.1   Types of variations in the indoor channel [1].

then for fixed i, μ_{ij} is a random variable. For amplitude fading this type of variation is equivalent to the shadowing effects experienced in the mobile environment. Different indoor sites correspond to intersections of streets, as compared to midblocks.

8.2.3   Large-Scale Variations

The channel's structure may change drastically, when the base-portable distance increases, among other reasons due to an increase in the number of intervening obstacles (Fig. 8.1). As an example, for amplitude fading, increasing the antenna separation normally results in an increase in path loss. Using the previous terminology $\xi(d_i) = E_{jk}(X_{ijk}) = E_j\,(\mu_{ij})$ is different for different d_i's. If X_{ijk} denotes the amplitude, this type of variation is equivalent to the distance-dependent path loss experienced in the mobile environment. For the mobile channel $\xi(d)$ is proportional to d^{-n}, where d is the base-mobile distance and n is a constant). Different path-loss models for the indoor channel will be discussed in a subsequent section.

More details about these three types of variations are included in [1]. Extensive measurements of the channel's frequency response based on this picture of the channel is reported in [5].

8.3 WIDEBAND CHANNEL

8.3.1 Mathematical Model

The complicated random and time-varying indoor radio propagation channel can be modeled in the following manner: For each point in the three-dimensional space, the channel is a linear time-varying filter with the impulse response given by:

$$h(t, \tau) = \sum_{k=0}^{N(\tau)-1} a_k(t)\delta[\tau - \tau_k(t)]e^{j\theta_k(t)} \tag{8.1}$$

where t and τ are the observation time instant and application time of the impulse, respectively, $N(\tau)$ is the number of multipath components, $\{a_k(t)\}$, $\{\tau_k(t)\}$, $\{\theta_k(t)\}$ are the random time-varying amplitude, arrival-time, and phase sequences, respectively, and δ is the delta function. The channel is completely characterized by these path variables [1]. This is a wideband model that has the advantage that, because of its generality, it can be used to obtain the response of the channel to the transmission of *any* transmitted signal $s(t)$ by convolving $s(t)$ with $h(t)$ and adding noise.

The time-invariant version of this model, first suggested by Turin of the University of California at Berkeley to describe multipath fading channels, has been successfully used in mobile radio applications [6–13]. For the stationary (time-invariant) channel, Eq. (8.1) reduces to:

$$h(t) = \sum_{k=0}^{N-1} a_k\delta(t - t_k)e^{j\theta_k} \tag{8.2}$$

The output $y(t)$ of the channel to a transmitted signal $s(t)$ is therefore given by

$$y(t) = \int_{-\infty}^{\infty} s(\tau)h(t - \tau)\,d\tau + n(t) \tag{8.3}$$

where $n(t)$ is the low-pass, complex-valued additive Gaussian noise.

With the above mathematical model, if the signal $x(t) = \text{Re}\{s(t)\exp[j\omega_0 t]\}$ is transmitted through this channel environment [where $s(t)$ is any low-pass signal and ω_0 is the carrier frequency], the signal $y(t) = \text{Re}\{\rho(t)\exp[j\omega_0 t]\}$ is received where

$$\rho(t) = \sum_{k=0}^{N-1} a_k s(t - t_k)e^{j\theta_k} + n(t) \tag{8.4}$$

In a real-life situation a portable receiver moving through the channel experiences a space-varying fading phenomenon. One can therefore associate an impulse response "profile" with each point in space. It should be noted that profiles corresponding to points close in space are expected to be grossly similar because principle reflectors and scatterers that give rise to the multipath structures remain approximately the same over short distances. A number of closely spaced impulse response profiles of the channel are illustrated in Figure 8.2.

8.3.2 Statistical Characterization

Extensive sets of measurements were carried out for statistical modeling of the channel's impulse response [5, 14–20]. The goal of measurements was to investigate the small-scale, midscale, and large-scale variations in the wideband statistics of the channel.

In a typical wireless indoor radio communication system, two-way transmission takes place between a fixed station and a moving unit. Correspondingly, the following measurement plan was devised: Four transmitter-receiver antenna separations of 5, 10, 20, and 30 m were considered. For each antenna separation several places for the fixed antenna ("base" unit) were selected. The selection was made on the basis of what was considered to be typical positions for base antenna in future systems. For each antenna separation a total of 20 small 1.5- to 2-m areas ("locations") were selected for the moving ("portable" unit) antenna position. The selections were made on the basis of good variation of typical conditions within the buildings. Each location was carefully chosen, with both line-of-sight (LOS) and non-line-of-sight (NLOS) (obstructed) topographies included. The number of locations with obstructed paths between the transmitter and receiver was higher for larger antenna separations, consistent with conditions encountered in real-life wireless indoor communication systems. For each location 75 frequency response estimates of the channel were recorded by displacing the portable antenna in steps of 2 cm using a step motor. These measurements were based on the scenario illustrated in Fig. 8.1 with $N = 4$, $L = 20$, and $M = 75$.

The above measurement plan was executed at two dissimilar office buildings (A and B). The result is a large database of 12,000 frequency response functions (75 samples per location at 2-cm spacing$\times$20 locations per antenna separation$\times$4 antenna separations per building$\times$2 buildings). The frequency response data were then converted to the time domain using classical Fourier analysis, resulting in an equal number of impulse response estimates [14, 15].

Due to the motion of people and equipment indoor propagation channels are, in general, time varying. Since the purpose of these measurements was to investigate the channel's variations in space (not in time), it was essential to keep the channel stationary during the measurements. All measurements were therefore performed at night or on weekends when there were few, if any, other personnel in the vicinity of the measurement setup.

The 12,000 measured frequency response data were converted to time domain by the inverse fast Fourier transform algorithm. A resolution of 5 ns for the impulse response data was therefore obtained. Details of the measurement plan and procedure are reported in [14, 15]. Reference [14] also reports detailed wideband statistical modeling of the channel's impulse response based on this database.

Statistical analysis of the 12,000 estimates of the channel's impulse response collected in the above measurement campaign are reported in [5, 14]. The results are summarized below:

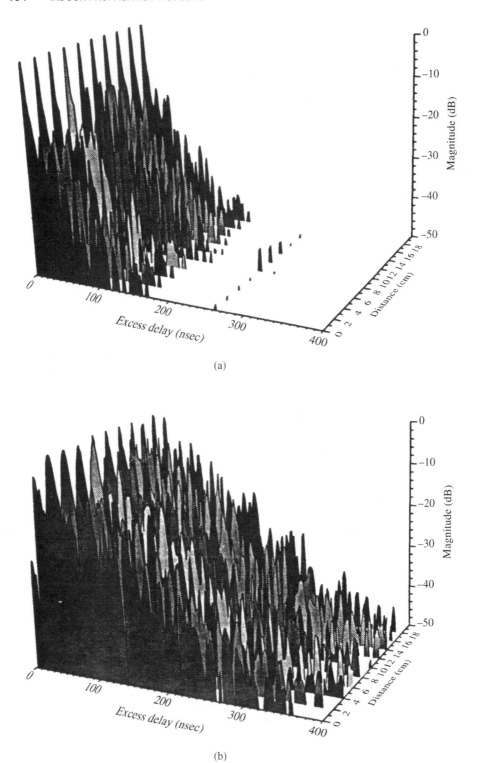

(a)

(b)

FIGURE 8.2   Sequences of spatially-adjacent impulse response profiles for a medium-size office building. Center frequency is 1.1 GHz. Base-portable antenna separation is 5 m. (a) A line-of-sight location and (b) a non-line-of-sight location [15].

The number of multipath components in each profile N was found to be a Gaussian-distributed random variable with a mean value that increases with increasing the antenna separation. When all components of a profile with amplitudes within 30 dB of the peak amplitude were included, mean of N increased from 13 to 24 for building A and from 13 to 19 for building B, as the antenna separation increased from 5 to 30 m. Standard deviations were between 3 and 7, also increasing with distance.

A Poisson fit to the arrival times of multipath components proved to be unsatisfactory. A modified Poisson process (the Δ-K model) provided good fits to the data for both buildings and all antenna separations. Most K values were in the 0.5 to 0.8 range, indicating that paths are often less clustered than what a standard Poisson process dictates.

The distribution of the amplitude of individual multipath components was obtained for both global data (1500 profiles representing one antenna separation) and local data (75 profiles representing one location). The lognormal distribution passed the Kolmogorov–Smirnov test of goodness of fit with a 90% confidence level for a majority of cases tested (87% for global data and 75% for local data). For global data standard deviations were in the 4- to 10-dB range, decreasing with increasing the excess delay. Local data showed smaller standard deviations of 3 to 5 dB. Mean amplitudes of multipath components (in decibels, relative to the peak amplitude of each profile) decreases almost linearly with increasing the excess delay for both local and global data, although only the results for global data were presented. A more elaborate analysis of the amplitude distributions is reported in [16].

Correlation on log amplitudes for impulse response estimates of spatially adjacent points were investigated. The average correlation coefficient ρ_s is between 0.7 and 0.9 for the portable antenna displacement of 2 cm, but it drops fairly rapidly for larger displacements; ρ_s is about the same for all antenna separations with the exception of the 5-m data, which showed higher correlations. The value of ρ_s is greater for initial paths than for later paths.

Amplitude of adjacent multipath components of the same profile showed small correlation for both buildings. Typical correlation coefficients are between 0.2 and 0.3 [5].

Performance of digital indoor communication systems is very sensitive to statistical properties of the phase sequence $\{\theta_k\}_0^\infty$. The signal phase is critically sensitive to path length and changes by the order of 2π as the path length changes by a wavelength (30 cm at 1 GHz). Considering the geometry of the paths, moderate changes (in the order of meters) in the position of the portable receiver results in a great change in phase. When one considers an ensemble of points, therefore, it is reasonable to expect a Uniform[0,2π) distribution; that is, on a global basis, θ_k has a $U[0, 2\pi)$ distribution. This phenomenologically reasonable assumption can be taken as a fact with no need for empirical verification. For small sampling distances, however, great deviations from uniformity may occur. Furthermore, phase values are strongly correlated if the channel's response is sampled at the symbol rates (tens to hundreds of kilobits per second). Phase values at a fixed delay for a given site are, therefore, correlated. Adjacent detectable multipath components of the same profile, on the other hand, have independent phases since their excess range (excess delay multiplied by the speed of light) is larger than a wavelength, even for very high resolution (a few nanoseconds) measurements.

Taking these into consideration, it is accurate to say that the absolute phase value of a multipath component at a fixed point in space is not important; emphasis of the modeling should be placed on *changes* in phase as the portable moves through the channel.

An elaborate study of the phase of individual multipath components was carried out in a separate work [17, 18] based on guidelines provided in [1]. Two phase models for the wideband indoor radio propagation channels were proposed and studied in detail. In model I, the phase of each path is updated deterministically using several random independently located scatterers for each multipath component. In model II the phase of each path is updated with random independent Gaussian increments whose standard deviations change with distance. Performance of these models were evaluated by means of extensive computer simulations, and by utilization of a large database of 12,000 impulse response profiles of the channel reported above. First- and second-order statistics of the narrowband continuous wave (CW) fading waveforms were obtained using simulated phases, and compared with those of empirical data. It was shown that model I with five independent scatterers for each path and model II with appropriate choice of standard deviation of Gaussian increments both provide results consistent with empirical results. Multiple reflectors are considered explicitly in model I, while in model II it is implicitly included. Furthermore, the two phase models were compared with each other. Statistical properties of phase increments in model I, particularly when the number of scatterers increases were studied, were simulated and compared with increments of model II. It was shown that with five scatterers, phase increments of model I agree well with the normal distribution of phase increments of model II. Standard deviation of phase increments in model I shows good agreement with the standard deviation of phase increments of model II. Comparison of these models was also carried out from computational efficiency. It was shown that with five scatterers in model I, model II is 1.7 times faster (more efficient) than model I.

The results reported above can be used in the performance analysis and design of indoor radio communication systems, either directly or through an elaborate simulation model based on the reported statistical characterization.

8.4   LARGE-SCALE PATH LOSSES

While the impulse response approach is useful in characterization of the channel at a microscopic level, path-loss models describe the channel at a macroscopic level. Path-loss information in indoor environments are essential in determination of size of the coverage area for radio communication systems, and in selecting optimum locations for base antennas. Obtaining three-dimensional propagation contour plots using a building's blue-print and the knowledge of its construction material is a challenging job that requires detailed and reliable path-loss models.

The indoor channel exhibits much larger path losses as compared to the outdoor mobile channel. Furthermore, large variations in the path loss are possible over very short distances. The propagation environment is very complicated and a universally accepted path-loss model is not yet available. A review of indoor propagation measurements, however, indicate that there are four distinct path-loss models. These models are briefly reviewed in this section. Although in principle one can obtain path losses using the wideband pulse transmission techniques, a number of available models were derived from narrowband CW measurements.

According to model I the received signal power follows an inverse exponent law with the distance between antennas; that is, $P(d) = P_0 d^{-n}$, where $P(d)$ is the power received at a distance d from the transmitter, and P_0 is the power at $d = 1$ m; P_0 depends on the transmitted power, frequency, and antenna heights and gains. Path loss is therefore proportional to d^{-n} where n depends on the environment. On a logarithmic scale this corresponds to a straight-line path loss with a slope of $10n \log(d)$. A number of investigators have used this model because of the simplicity of the model and its previous successful application to the mobile channel. This model provides the mean value of random path loss. A standard deviation σ is also associated with each measurement.

In model II the received power follows a d^{-n} law as in model I. The exponent n, however, changes with distance. A distance-dependent exponent that increases from 2 to 12 with increasing d has been reported for indoor measurements carried out in a multistory office building. In the measurements the fixed transmitter was located in the middle of a corridor and the portable receiver was placed inside rooms or along other corridors, on the same floor and on other floors. The value of n estimated from the data collected by Ericsson of Sweden was 2 (for $1 < d < 10$ m), 3 (for $10 < d < 20$ m), 6 (for $20 < d < 40$ m), and 12 (for $d > 40$ m). The large values of n are probably due to an increase in the number of walls and partitions between the transmitter and receiver when d increases.

In model III logarithmic attenuations are associated with various types of structure between the transmitter and receiver antennas. Adding these individual attenuations, the total path loss in decibels can be calculated.

In model IV a decibel per unit of distance (antenna separations) path loss is assumed within a building.

Details of these models and an associated survey of the literature is reported in [1].

8.5 RMS DELAY SPREAD

The random and complicated indoor radio propagation channel can be characterized using the impulse response approach described in the previous section. Since the response of the channel to the transmission of any signal can be obtained from the impulse response through the process of convolution, full characterization of the channel can be based on $h(t)$.

Partial characterization of the channel is possible with a one-number representation of an impulse response profile, the rms delay spread τ_{rms}:

$$\tau_{\text{rms}} = \sqrt{\frac{\int_{-\infty}^{\infty} (t - \tau_m)^2 |h(t)|^2 \, dt}{\int_{-\infty}^{\infty} |h(t)|^2 \, dt}} \tag{8.5}$$

where the time axis is scaled such that the first path in a profile arrives at $t = 0$; τ_m is the mean excess delay defined as:

$$\tau_m = \frac{\int_{-\infty}^{\infty} t |h(t)|^2 \, dt}{\int_{-\infty}^{\infty} |h(t)|^2 \, dt} \tag{8.6}$$

The above expressions show that τ_m is the first moment of the power-delay profile ($|h(t)|^2$) with respect to the first arriving path, and the delay spread τ_{rms} is the square root of the second central moment of a power-delay profile. It has been shown that τ_{rms} is the ratio of the power in the second to the power in the first term of the Taylor series expansion of the fading channel's transfer function. It is a good measure of multipath spread; it gives an indication of the potential for intersymbol interference. Strong echoes (relative to the LOS path) with long delays contribute significantly to τ_{rms}. Performance of communication systems operating in multipath environments are very sensitive to the value of τ_{rms}. More specifically, analysis and simulation studies have shown that with no diversity or equalization, maximum rate of data transmission in an indoor environment is a few percent of $(\tau_{rms})^{-1}$. Detailed analysis of τ_{rms} can therefore provide valuable information to designers of indoor radio communication systems.

Analysis of the rms delay spread for two dissimilar office buildings, based on extensive measurements of [5], are reported in [19, 20]. The database included 12,000 estimates of the channel's impulse response (75 samples per small location×20 locations per transmitter-receiver antenna separation×4 antenna separations of 5, 10, 20, and 30 m per building×2 buildings). The rms delay spread τ_{rms} was calculated for each impulse response, and the resulting 12,000 values were used to determine statistical properties of τ_{rms}.

The τ_{rms} values were typically between 10 and 50 ns, with the mean value is in the 20- to 30-ns range. The mean τ_{rms} increases with increasing dynamic range; τ_{rms} also showed a clear dependence on transmitter-receiver antenna separation. For the dynamic range of 30 dB mean τ_{rms} increases from 16.9 to 35.1 ns for building A and from 17.5 to 26.6 ns for building B, as antenna separation increases from 5 to 30 m. Standard deviations for τ_{rms} values corresponding to each antenna separation (1500 values) also increase with increasing antenna separation (ranging from 4 to 11 ns for building A and from 4 to 7 ns for building B) [19].

Distribution of τ_{rms} was obtained for both global data (1500 values corresponding to one antenna separation and 6000 values representing data for one building), and local data (75 values representing one location). Normal distributions passed the Kolmogorov–Smirnov test of goodness of fit with a 90% confidence level for the global data, both buildings, and for all antenna separations. Visual inspection of the local distributions also showed good normal fit for the majority of cases. Quantitatively, for 93% of locations the mean square error between theoretical and empirical distributions was smaller for normal distribution, as compared to several other known distributions [19].

The rms delay spread for spatially adjacent impulse responses were found to be correlated. Correlation coefficient function had a value of approximately 0.9 at sampling distance of 2 cm, dropping to less than 0.3 at 10 cm [19]. Mean τ_{rms} for each location showed great linear dependence with average path loss for that location.

On the basis of the results of the statistical analysis a simulation model capable of generating a consistent set of τ_{rms} values for spatially adjacent points has been developed. In this model joint distribution of τ_{rms} for two spatially adjacent points is assumed to be bivariate normal with means, standard deviations, and correlation coefficients estimated from the data. Application of this model, with examples, has been presented and discussed [19].

8.6   SPATIAL VARIATIONS OF THE CHANNEL

When a single unmodulated carrier (constant envelope) is transmitted in a multipath environment, due to vector addition of the individual multipath components, a rapidly fluctuating CW envelope is experienced by a receiver in motion. To deduce this narrowband result from the wideband impulse response model, a constant input ($=1$) is convolved with the channel's impulse response [1]. Excluding noise, the resultant CW envelope R and phase ϕ *for a single point in space* are given by:

$$\mathrm{Re}^{j\phi} = \sum_{k=0}^{\infty} a_k e^{j\theta_k} \tag{8.7}$$

where a_k and θ_k are the envelope and phase of the kth multipath component of an impulse response, respectively. Sampling the channel's impulse response frequently enough, one should be able to generate the narrowband CW fading results for the receiver in motion, using the wideband impulse response model.

Using the above approach, each set of 75 impulse response profiles collected in one location of the large database described in the previous section was converted to one segment of 75 spatial fading data covering a length of 1.5 m in space. As a result, 160 segments of 1.5 m narrowband CW fading data were obtained [21, 22]. Figure 8.3 shows samples of spatial variations data.

Statistical analysis of the data indicates that: (i) CW fading characteristics depend on the local environment of the portable (low) antenna, and are very similar at both buildings and for all antenna separations. (ii) There is higher degree of fluctuations for NLOS topographies. (iii) Average dynamic range of fluctuations is around 25 dB for both buildings and all antenna separations; deep fades of 30 to 40 dB below mean values have been observed. (iv) Signal envelope becomes essentially uncorrelated after a separation of 10 cm. (v) Level crossing rates are on the average 4.2, 2.2, 0.8, and 0.4 per 150 cm for threshold levels of 5, 10, 15, and 20 dB below the mean value, respectively. Percentages of data below threshold (i.e., duration of fades) are on the average 2.6 and 1.8% for threshold levels of 5 and 10 dB below the mean value, respectively. A major conclusion is that statistical properties of the narrowband CW fading waveforms (spatial variations) are about the same for both buildings, and is almost independent of base-portable antenna separation [21].

8.7 TEMPORAL VARIATIONS OF THE CHANNEL

When both antennas (fixed, or base, and mobile, or portable) are stationary, motion of people and equipment around the antennas result in multipath disturbances and fading effects in an indoor environment [1]. This temporal fading phenomenon degrades performance of high bit rate transmission in indoor environments such as computer communication typical in wireless local area networks (LANs).

Extensive CW temporal fading measurements at 1100 MHz in a modern office environment under diversified sets of conditions were carried out [22, 23]. Four antenna separations of 5, 10, 20, and 30 m were considered. For each antenna separation four different base-portable antenna positions (configurations) were selected. In positioning base and portable antennas typical locations that included both LOS and NLOS topographies were chosen.

For each one of the 16 resulted cases (4 antenna separations×4 configurations per antenna separation), three types of motion were considered: motion around base antenna only, motion around portable antenna only, and motion around both antennas. The 12 recordings labeled (1B 0P), (2B 0P), (3B 0P), (4B 0P), (0B 1P), (0B 2P), (0B 3P), (0B 4P), (1B 1P), (2B 2P), (1B 2P), (1B 3P), where (iB jP) represents motion of i persons in the

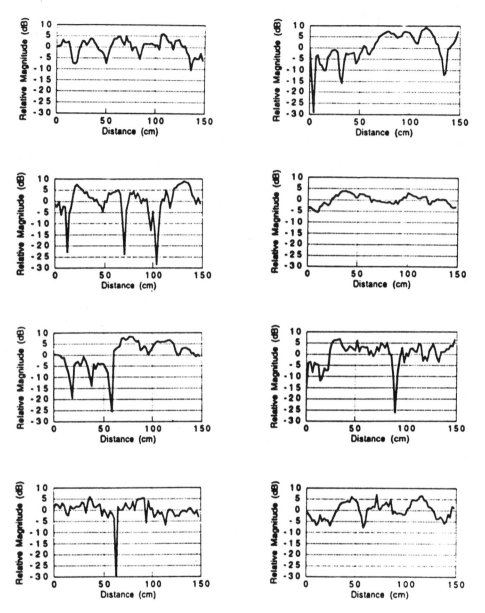

FIGURE 8.3 Spatial variations envelop fading data in an office environment. Center frequency is 1.1 GHz. Plots in each row represent from top to bottom base-portable antenna separations of 5, 10, 20, and 30 m, respectively [21].

vicinity of the high (base) antenna and j persons in the vicinity of the low (portable) antenna.

Each recording described above corresponds to 60 s of channel's temporal variations. During each 60 s care was taken to have continuous motion of the same nature. This means that if three people moved around the portable unit, they continued to do so throughout the whole 60 s. Therefore, each 60-s recording represents a sample from a stationary stochastic process. During each 1-min recording care was also taken to eliminate or minimize all other unplanned motion in the environment. To satisfy this latter requirement, most measurements were carried out late at night. Details of the measurement plan, measurement technique, equipment used, building structure, processing of data, and the like are reported in [23]. Typical temporal variations data are shown in Figure 8.4.

The resulting large database of 11,520 s of narrowband CW envelope fading data were analyzed. The analysis included both qualitative evaluation of the variations and quantitative analysis consisting of the amplitude fading distributions, correlation properties, level crossing rates, fade durations, and frequency domain properties [23]. A summary of the results are as follows: (i) Standard deviation and dynamic range of fluctuations are greater for motion around the portable antenna, as compared to motion around the base antenna. (ii) The Weibull and Nakagami distributions provide the best fit to envelope fading for most data records. (iii) Envelope values become essentially uncorrelated for temporal separations greater than 1 s. (iv) Level crossing rates and duration of fades are both higher for motion around the portable, as compared to motion around the base. (v) The mean spectrum widths of all data records for -3, -5, -10, and -20 dB threshold crossings are 0.35, 0.48, 1.06, and 4.54 Hz, respectively [23].

These results are particularly useful for high data rate wireless computer communications applications where both terminals are stationary but motion of people results in multipath distortions in the received signal.

8.8 COMPARISON BETWEEN INDOOR AND OUTDOOR RADIO CHANNELS

The indoor and outdoor channels are similar in their basic features: They both experience multipath dispersions caused by a large number of reflectors and scatterers. They can both be described using the same mathematical model. However, there are also major differences, briefly described in this section.

The *conventional* outdoor mobile channel (with an elevated base antenna and low-level mobile antennas) is stationary in time and nonstationary in space. Temporal stationarity is due to the fact that signal dispersion is mainly caused by large fixed objects (buildings). In comparison, the effect of people and vehicles in motion are negligible. The indoor channel, on the other hand, is stationary neither in space nor in time. Temporal variations in the indoor channel's statistics are due to the motion of people and equipment around the low-level portable antennas.

The indoor channel is characterized by higher path losses and sharper changes in the mean signal level, as compared to the mobile channel. Furthermore, applicability of a simple negative-exponent distance-dependent path-loss model, well established for the outdoor mobile channel, is not universally accepted for the indoor channel.

Rapid motions and high velocities typical of the outdoor mobile users are absent in the indoor environment. The indoor channel's Doppler shift is therefore negligible.

Maximum excess delay for the outdoor mobile channel is typically several microseconds if only the local environment of the mobile is considered [7] and more than 100 µs if reflection from distant objects such as hills, mountains, and city skylines are taken into account [10, 11]. The outdoor rms delay spreads are of the order of several microseconds without distant reflectors, and 10 to 20 µs with distant reflectors. The indoor channel, on the other hand, is characterized by excess delays of less than 1 µs and rms delay spreads in

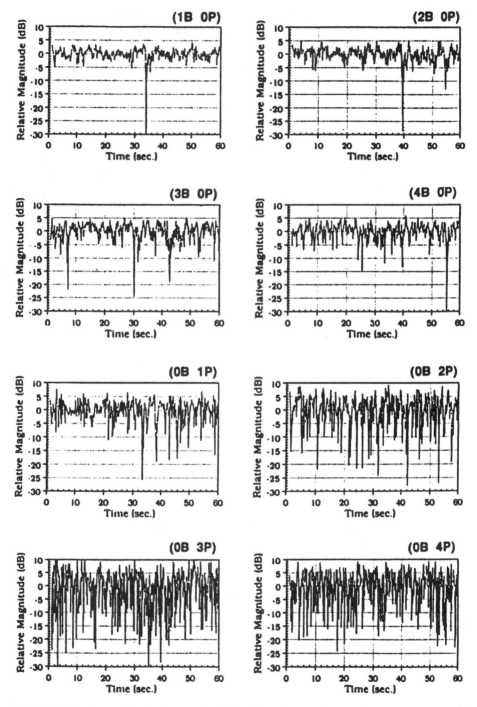

FIGURE 8.4 Temporal variation envelop fading data in an office environment. Base-portable antenna separation is 20 m. Carrier frequency is 1.1 GHz. Labels on the top right corner of each plot, (iB jP), represent motion of i persons around the base antenna and j persons around the portable antenna [23].

the range of several tens to several hundreds of nanoseconds (most often less than 100 ns). As a result, for the same level of intersymbol interference, transmission rates can be much higher in indoor environments.

8.9 INDOOR INFRARED CHANNEL

The majority of currently proposed indoor wireless communication services are based on radio LANs. However, optical LANs are receiving increasing attention for in-building applications because of their many advantages [24].

There are two methods for establishing infrared transmission inside a room: the directive beam configuration (DBC) and the diffuse infrared configuration (DIC) [24]. Under the DBC scheme, two collimated infrared (IR) beams connect a terminal to the network, one for the up-link and one for the down-link. Under the DIC scheme diffused light is reflected and scattered in the room and received with a wide acceptance angle photodetector portable unit located anywhere in the coverage area. Advantages of the DBC are high gains available to directional transmission, resulting in higher attainable transmission rates and lower power requirements. Disadvantages are interruptions in communications caused by shadowing and severe alignment requirements, particularly when the portable unit changes position. The DIC does not require alignment and is suitable for speech transmission or portable computer applications where the moving transceiver changes position frequently. Disadvantages of the DIC are relatively higher power requirements and multipath dispersion caused by walls and furniture. Taking these considerations into account, the DBC has been proposed for wireless access to fixed workstations in a room, while the DIC has a wider range of applications covering speech and data transmission to units with great mobility requirements.

Principles of wireless optical LANs (advantages, disadvantages, system configurations, safety considerations, properties of candidate transceivers, etc.) have been reviewed in [24].

A major concern in designing wireless in-building LANs (for both radio and optics) is multipath dispersion of the signal by the building internal structure, furniture, equipment, and people. Multipath effects are minimal for the DBC since the optical beam is also narrow and directional, and the field of view of the receiver is narrow. For designing DIC systems, on the other hand, multipath dispersion should be a major concern because of wide acceptance angle of the receiver and because, practically, the entire room is illuminated and therefore the chance of reflections and scattering is great.

It can be shown that for indoor *radio* channel cancellations of the received signals occur at radio frequency (RF), with maxima and minima separated by approximately half a wavelength. For an IR wireless transceiver that operates on the intensity modulation-detection principle, on the other hand, the propagation medium can be replaced with an equivalent baseband channel [25]. To derive the baseband model let $x(t)$ denote the normalized message signal ($-1 < x(t) < 1$). The positive signal $A[1 + \mu x(t)]$ then modulates the intensity of the light source, where μ is a positive constant, $0 < \mu < 1$. The intensity of the transmitted signal is thus given by

$$I_T(t) = |f_T(t)|^2 = A[1 + \mu x(t)] \tag{8.8}$$

where $f_T(t)$ is the complex field of the transmitted signal. Denoting the complex field of the signal received through the kth reflection at a given point on the photodetector surface by $f_k(t)$, $k = 0, 1, 2, \ldots, N$,

$$f_k(t) = \sqrt{\alpha_k}\sqrt{A[1 + \mu x(t - t_k)]}e^{-j\omega_0(t-t_k)} \tag{8.9}$$

where $\omega_0 = 2\pi f_0$, and $f_0 \approx 3 \times 10^{14}$ is the frequency of light. The attenuation factor α_k takes into account the inverse square distance-dependent (power) path loss, reflection losses, and the like. The total received signal is

$$f_R(t) = \sum_{k=0}^{N} f_k(t) \tag{8.10}$$

Intensity of the received signal $I_R(t)$ is given by

$$I_R(t) = |f_R(t)|^2 = f_R(t)f_R^*(t) = \sum_{\substack{k=l \\ k \neq l}}^{N}\sum_{l=0}^{N} \beta_{kl}(t)e^{j\omega_0(t_k - t_l)} \tag{8.11}$$

$$\beta_{id}(t) = \sqrt{\alpha_k\alpha_l}\{A[1 + \mu x(t - t_k)]\}^{1/2}\{A[1 + \mu x(t - t_1)]\}^{1/2} \tag{8.12}$$

The N^2 terms of (8.7) can be divided into N terms with $k = 1$ and $N(N - 1)$ terms with $k \neq 1$:

$$I_R(t) = \sum_{k=0}^{N} \alpha_k A[1 + \mu x(t - t_k)] + \sum_{\substack{k=0 \\ k \neq 0}}^{N}\sum_{l=0}^{N} \beta_{kl}(t)e^{j\theta_{kl}} \tag{8.13}$$

where $\theta_{kl} = \omega_0(t_k - t_l) = 2\pi \times 3 \times 10^{14}(t_k - t_l)$; θ_{kl} is critically sensitive to excess path length and changes by 2π when excess path length $c(t_k - t_l)$ changes by a wavelength ($\approx 1\,\mu m$). Since excess path lengths are in the order of centimeters or meters (tens of thousands to millions of wavelengths), θ_{kl} may be modeled as a random variable with a uniform distribution over $[0, 2\pi)$ for any $k\&l$ ($k \neq l$). It should be emphasized that this assertion is independent of the size of the receiving antenna; that is, even if the detector area is infinitesimally small, when the ensemble of receiver sites spread over hundreds of thousands of wavelengths is considered, the geometry of two paths with modulation delays t_k & t_l will lead to a uniform distribution of phase θ_{kl}:

$$l(t) = E[l_R(t)] = \sum_{k=0}^{N} \alpha_k A[1 + \mu x(t - t_k)] + \sum_{\substack{k=0 \\ k \neq 0}}^{N}\sum_{l=0}^{N} \beta_{kl}(t)E[e^{j\theta_{kl}}] \tag{8.14}$$

The mathematical expectation evaluated above is equivalent to spatial integration over the photodetector surface (invoking spatial ergodicity of the process). It should be emphasized again that the photodetector surface spans many thousands of wavelengths. Therefore, the second term in (8.14) containing the rapidly fluctuating θ_{kl} term vanishes in the integration, while the first term is approximately constant over the photodetector surface. The second term in (8.14) is equal to zero since θ_{kl} has a Uniform$[0, 2\pi)$ distribution. Conversion of the intensity variations to an electric signal and removing the direct current (dc) term results in

$$y(t) = \sum_{k=0}^{N} a_k x(t - t_k) \tag{8.15}$$

where a_k is constant. So the entire process can be modeled as baseband transmission, although $x(t)$ can have a very large bandwidth.

Since in indoor environments transmitted IR signals are confined within the same room, the simulation approach based on optical ray tracing can provide an accurate estimate of the channel's impulse response [26]. This is subject to the availability of enough information about source and detector parameters, and about the room in which they operate. Details of a simulation model that estimates impulse response and frequency response of the infrared channel in indoor environments of different shapes and sizes are reported in [26]. An important feature of this model is inclusion of the effect of office furniture and people in the room on the channel's impulse response.

Frequency response measurements at 8 different office environments in a university building were carried out and reported in [25, 27, 28]. Altogether, 160 frequency response profiles were collected (8 rooms × 2 positions per room × 2 perpendicular orientations of the photodetector at each position × 5 rotations in multiples of 45° for each orientation). Results of more extensive experimentation of the IR channel in which over 6000 frequency response profiles were collected at 10 different office environments are reported in [29, 30, 31]. A typical set of the channel's frequency responses are shown in Figure 8.5. Inspection of numerous plots of magnitude of $H(f)$ and many tables presented in these studies showed that the channel is very dynamic with great sensitivity to the position, orientation, and degree of rotation of the photodetector optical receiver. There is great variation between power received in different rooms, different positions within the same room, and different directions of the photodetector at the same position of the same room. The same is true for the dc gain and 3-dB bandwidth of the channel, and for the degree of flatness, and presence of nulls in the frequency response profiles. It should be noted that sensitivity of the results to the direction of the receiver is a property of the receiver's antenna pattern, and not necessarily of the physical environment (the channel).

Major conclusions of these measurements are as follows: (i) Path loss (received power minus transmitted power) is very different in different rooms. Dynamic range (maximum difference) of over 30 dB in relative path-loss values are not uncommon. (ii) Path loss is very sensitive to the position in each room and to the direction of rotation. (iii) There is great variation between path loss within measurements performed at the same position in the room and for the same direction but different degrees of rotation. (iv) For a fixed direction of transmission, the 3-dB bandwidth of the channel W is very sensitive to the direction and orientation of the receiving photodetector. There are also great differences in W for measurements in different rooms, in different positions of the same room, and in data for different orientations of the photodetector for measurements at the same position of the same room. (v) There are three types of behavior in the frequency response data. In some cases the channel is very flat. In others frequency response profiles show relatively rapid roll-off in the magnitude of $H(f)$. This roll-off, however, is monotonic. In the third set of data the frequency response profiles contain one or more distinct "nulls." This means that magnitude of $H(f)$ decreases when f increases up to a point and then increases again.

On the basis of the above a postdetection (baseband) diversity scheme, to be labeled "direction diversity" was proposed for future optical wireless receivers [25]. Under this diversity scheme the output of several photodiodes mounted on the receiver facing different directions are combined using any diversity combining technique. The "direction" diversity under this scheme is a substitute for "space" diversity common in radio systems operating in multipath environments.

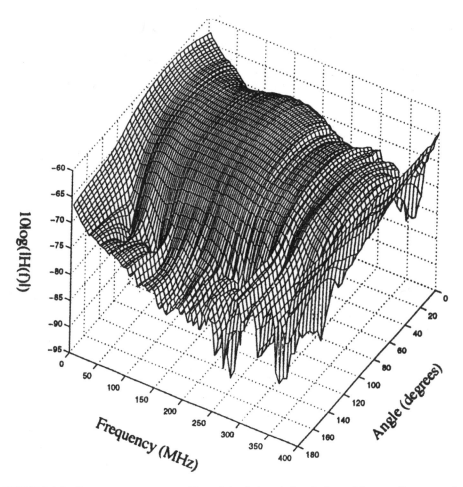

FIGURE 8.5   Frequency response profiles of the indoor infrared channel in an office room for receiver rotation of 180° in 5° increments [29].

8.10   CONCLUSIONS

Wireless indoor communication is an important aspect of picocellular third-generation wireless information networks. In this chapter basic principles of indoor radio and infrared channels were described using impulse response modeling of the channel. The results can be applied in the design of speech and low data rate systems, as well as high bit rate transmission needed for motion video and multimedia applications.

REFERENCES

1. H. Hashemi, "The Indoor Radio Propagation Channel," *Proc. IEEE*, Vol. 81, No. 7, pp. 943–968, July 1993.

2. H. Hashemi, "Principles of Digital Indoor Radio Propagation," Proceedings of the CECC '91 Conference, Calgary, Alberta, Canada, April 8–10, 1991, pp. 48–53.

3. H. Hashemi, "Propagation Modeling for PCS Systems," a 3-hour tutorial presented at the Fourth IEEE International Conference on Universal Personal Communications, ICUPC '95, Tokyo, Japan, November 6, 1995.

4. H. Hashemi, "Indoor Propagation Modeling and Prediction," 28th EuMC & Wireless '98 Workshop Proceedings, the MTT-S European Wireless '98 Conference, Amsterdam, the Netherlands, October 7, 1998, pp. 29–43.

5. H. Hashemi, "Impulse Response Modeling of the Indoor Radio Propagation Channel," *IEEE J. Selected Areas in Commun.*, Vol. 11, No. 7, pp. 967–978, Sept. 1993.

6. H. Hashemi, "Simulation of the Urban Radio Propagation Channel," Proc. of the 1977 National Telecommunication Conference, Los Angeles, Paper 38-1.

7. H. Hashemi, "Simulation of the Urban Radio Propagation Channel," *IEEE Trans. Veh. Tech.*, Vol. VT-28, No. 3, pp. 213–224, Aug. 1979.

8. H. Hashemi, "Simulation of Digital Transmission in Cellular Mobile Radio Environment," presented at the 38th Vehicular Technology Conference, VTC'88, Philadelphia, June 1988, Proc. of the Conference, pp. 530–536.

9. H. Hashemi, "A Simulation Package for Digital Cellular Mobile Radio Applications," paper presented at the IEEE Global Communication Conference, GLOBECOM'89, Dallas, Nov. 1989, Proc. of the Conference, Vol. 1, pp. 112–117.

10. H. Hashemi and S. Best, "Simulation of Urban Radio Propagation Channel in the Presence of Distant Reflectors," Proceedings of the Wireless '91 Conference, Calgary, Alberta, Canada, July 8–10, 1991.

11. H. Hashemi, "Simulation of Urban Radio Propagation in the Presence of Distant Reflectors," Proceedings of the IEEE GLOBECOM '91 Conference, Phoenix, Arizona, Dec. 2–5, 1991, pp. 1940–1946.

12. H. Hashemi, "Principles of Digital Mobile Radio Propagation," Proceedings of the CECC '91 Conference, Calgary, Alberta, Canada, April 8–10, 1991, pp. 19–23.

13. H. Hashemi, B. Azimi Sajjadi, and V. Tabatabaei, "Practical Design Considerations for Digital Mobile Radio Systems Operating in the Mountainous Terrain," *Proc. IEEE Veh. Tech. Conf.*, VTC '93, Secaucus, New Jersey, May 18–20, pp. 53–56, 1993.

14. H. Hashemi, D. Tholl, and G. Morrison, "Statistical Modeling of the Indoor Radio Propagation Channel–Part I," *Proc. IEEE Veh. Tech. Conf.*, VTC '92, Denver, Colorado, pp. 338–342, May 1992.

15. H. Hashemi, D. Lee, and D. Ehman, "Statistical Modeling of the Indoor Radio Propagation Channel–Part II," *Proc. IEEE Veh. Tech. Conf.*, VTC '92, Denver, Colorado, pp. 839–843, May 1992.

16. H. Nikookar and H. Hashemi, "Statistical Modeling of Signal Amplitude Fading of Indoor Radio Propagation Channels," Proceedings of the 2nd International Conference on Universal Personal Communications, Ottawa, Canada, Oct. 1993, pp. 84–88.

17. H. Nikookar and H. Hashemi, "Phase Modeling of the Indoor Radio Propagation Channel," *Proc. 44th IEEE Veh. Tech. Conf.*, VTC '94, Stockholm, Sweden, June 7–10, pp. 1759–1763, 1994.

18. H. Nikookar and H. Hashemi, "Phase Modeling of Indoor Radio Propagation Channels," *IEEE Trans. Veh. Tech.*, Vol. 49, No. 2, pp. 594–606, March 2000.

19. H. Hashemi and D. Tholl, "Statistical Modeling and Simulation of the RMS Delay Spread of Indoor Radio Propagation Channels," *IEEE Trans. Veh. Tech.*, Vol. 43, No. 1, pp. 110–120, Feb. 1994.

20. H. Hashemi and D. Tholl, "Analysis of the RMS Delay Spread of Indoor Radio Propagation Channels," Proceedings of the IEEE International Conference on Communications, ICC'92, Chicago, Illinois, June 14–17, pp. 875–881, 1992.

21. H. Hashemi, D. Tholl, and T. Vlasschaert, "A Study of CW Spatial Fading of the Indoor Radio

Propagation Channel," Proceedings of the Fourth IEEE International Conference on Universal Personal Communication, ICUPC'95, Tokyo, Japan, November 6–9, pp. 201–205, 1995.

22. H. Hashemi, "A Study of Temporal and Spatial Variations of the Indoor Radio Propagation Channel," Proceedings of the IEEE PIMRC '94 Conference, The Hague, The Netherlands, September 19–23, 1994, pp. 127–134, invited paper.

23. H. Hashemi, M. McGuire, T. Vlasschaert, and D. Tholl, "A Study of Temporal Variations of the Indoor Radio Propagation Channel," *IEEE Trans. Veh. Tech.*, Special Issue on Future PCS Technologies, Vol. 43, No. 3, pp. 733–737, August 1994.

24. H. Hashemi, M. Kavehrad, and G. Yun, "Wireless Optical LANs," a tutorial presentation at the Fifth Intern. Conf. on Wireless Comm., Wireless'93, Calgary, Canada, July 12–14, 1993.

25. H. Hashemi, G. Yun, M. Kavehrad, F. Behbahani, and P. Galko, "Indoor Propagation Measurements at Infrared Frequencies for Wireless Local Area Networks Applications," *IEEE Trans. Veh. Tech.*, Vol. 43, No. 3, pp. 562–576, August 1994.

26. M. Abtahi and H. Hashemi, "Simulation of Indoor Propagation Channel at Infrared Frequencies in Furnished Office Environments," Proceedings of the Sixth IEEE International Conference on Personal, Indoor, and Mobile Radio Communications, PIMRC '95, Toronto, Canada, September 27–29, 1995, pp. 306–310.

27. H. Hashemi, G. Yun, M. Kavehrad, F. Behbahani, and P. Galko, "Frequency Response Measurements of the Wireless Indoor Channel at Infrared Optics," Proceedings of the International Zurich Seminar on Digital Communications, IZS '94, Zurich, Switzerland, March 8–11, pp. 273–284, 1994.

28. H. Hashemi, G. Yun, M. Kavehrad, F. Behbahani, and P. Galko, "Frequency Response Measurements of the Wireless Indoor Channel at Infrared Frequencies," Proceedings of the IEEE International Conference on Communications ICC '94, New Orleans, Louisiana, May 1–4, pp. 1511–1515, 1994.

29. M. R. Pakravan, M. Kavehrad, and H. Hashemi, "Measurement of Rotation Effects in an Indoor Infrared Channel," Proceedings of the 48th IEEE Vehicular Technology Conference, VTC '98, Ottawa, Canada, May 18–21, pp. 2100–2103, 1998.

30. M. R. Pakravan, M. Kavehrad, and H. Hashemi, "Effects of Rotation on the Path Loss and the Delay Spread in Indoor Infrared Channel," Proceedings of the IEEE International Conference on Communications, ICC'98, Atlanta, Georgia, June 7–11, pp. 817–821, 1998.

31. M. R. Pakravan, M. Kavehrad, and H. Hashemi, "Indoor Wireless Infrared Channel Characterization by Measurement," *IEEE Trans. Veh. Tech.*, Vol. 50, No. 4, pp. 1053–1073, July 2001.

CHAPTER 9

Propagation Loss Prediction Models

MASAHARU HATA

9.1 INTRODUCTION

Propagation loss prediction models play a very important role in the design of cellular mobile radio communication systems by specifying the key system parameters such as transmission power, frequency reuse, and so on. Several prediction models have been proposed for cellular mobile radio systems operating in the quasi-microwave frequency band.

Some of them were derived in a statistical manner from measurement data, and some were derived analytically based on diffraction effects. Each model uses specific parameters to achieve reasonable prediction accuracy. For example, one relatively long-range prediction model, intended for macrocell systems, uses base and mobile station antenna heights and frequency. On the other hand, a prediction model for short-range estimation that was designed for microcell systems uses building heights, street width, and so on. When the cell size is quite small, for example, for a specific area, deterministic methods such as ray tracing are necessary for accurate prediction. Therefore, it is important for designing mobile systems to select the most appropriate prediction model with the goal of efficient cell coverage.

This chapter summarizes the propagation loss prediction models commonly used in land mobile communication system design and discusses their applicability in various mobile propagation environments.

9.2 EMPIRICAL MODELS

Table 9.1 summarizes the propagation loss prediction models used to design current cellular systems. The Okumura–Hata model [1, 2] is the empirical formula based on field measurements made in a typical mobile propagation environment (see Fig. 9.1). The

Wireless Communications in the 21st Century, Edited by Shafi, Ogose, and Hattori.
ISBN 0-471-155041-X © 2002 by the IEEE.

TABLE 9.1 Propagation Loss Prediction Models for Cellular Mobile Radio Communications

Prediction Model	Okumura–Hata	Lee Ibrahim-Person	Sakagami	Walfisch–Bertoni COST-231 (Walfisch-Ikegami) ITU-R TG8/1 (Xia-Bertoni)
Category		Empirical model		Analytical model
References	[1] 1968, [2] 1980	[3] 1982, [4] 1983	[5] 1991	[6] 1980, [7] 1988, [8] 1991, [9–12] 1992
Parameters				
$T_x - R_x$ distance	Yes			Yes
Frequency	Yes			Yes
Antenna height	Yes			Yes
	(base station and mobile station)			(base station and mobile station)
Structure of buildings, roads	No	No	Yes (average rooftop level, average building separation) (street width, street angle)	
Applicable range				
$T_x - R_x$ distance	1–20 km	1–10 km	0.5–10 km	0.02–5 km
Less than 1 km	No	No	Yes	Yes
Frequency	150–1500 MHz (applicable up to 2.2 GHz)	168–900 MHz (in Ibrahim–Person model)	450–2200 MHz	800–2000 MHz (in COST-231-Walfisch–Ikegami model)
Antenna height	hbase: 30–200 m	Not clear	hbase: 20–100 m	hbase: 4–50 m
below roof tops	No	No	No	Yes
Remarks	Flat urban Simple formula		Applicable for flat urban and suburban areas Data base of buildings and streets are needed	

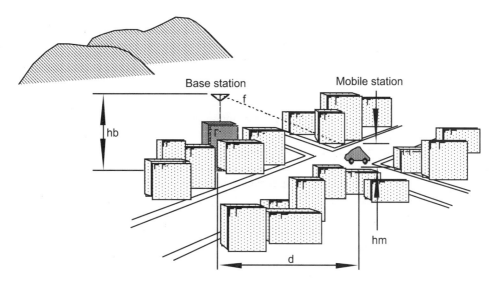

FIGURE 9.1   Parameters used in Okumura–Hata model.

median path loss L is expressed in logarithmic scale as a linear function of d, the distance between the transmitter and receiver:

$$L \text{ (dB)} = A + B\log(d) \tag{9.1}$$

where A and B are functions of frequency, and base and mobile station antenna heights. The formula for urban, quasi-smooth terrain is summarized as follows:

$$
\begin{aligned}
L \text{ (dB)} = {}& 69.55 + 26.16\log(f) - 13.82\log(h_b) - k(h_m) \\
& + [44.9 - 6.55\log(h_b)]\log(d)
\end{aligned}
\tag{9.2}
$$

where $k(h_m)$ is the correction factor for mobile station antenna heights for a small or medium-sized city:

$$k(h_m) = [1.1\log(f) - 0.7]h_m - [1.56\log(f) - 0.8]$$

And for a large city,

$$
\begin{aligned}
k(h_m) &= 8.29[\log(1.54h_m)]^2 - 11 & f \leq 200 \text{ MHz} \\
&= 3.2[\log(11.75h_m)]^2 - 4.97 & f \geq 400 \text{ MHz}
\end{aligned}
$$

For each $k(h_m)$, if $h_m = 1.5$ m, $k(1.5) = 0$ dB.

Applicable range:

f, frequency (MHz)   150–2200 MHz
h_b, base station antenna height (m)   30–200 m
h_m, mobile station antenna height (m)   1–10 m
d, distance from transmission point (km)   1–20 km

Based on the field measurement results, the applicable upper frequency range was extended from 1500 to 2200 MHz [13]. In the formula, L is defined as the loss between isotropic antennas. Therefore, if dipole antennas are used at base and mobile stations, the first term of the right-hand side of Eq. (9.2) should be 65.25 instead of 69.55. The following prediction formulas described in this chapter also consider the path loss between isotropic antennas.

In the Okumura–Hata model, urban and quasi-smooth terrain is the base of the prediction formula. The influence of irregular terrain is defined by terrain correction factors and given by prediction curves [1]. For suburban and open (rural) areas, correction formulas have also been produced based on measurement data [2]. However, how to select correction formulas for an actual application remained uncertain; the environmental definition was not really clear.

Considering that the categorization of urban, suburban, and rural area depends on the degree of urbanization, a correction method using a ground cover factor was proposed by Akeyama et al. [14]. The ground cover factor, α, is defined as the percentage of the area covered by buildings within $500 \times 800\,\mathrm{m}^2$. The deviation from reference median path loss S is shown in Figure 9.2 and expressed by:

$$
S\,(\mathrm{dB}) =
\begin{aligned}
&-19\log(\alpha) + 26 && (5\% \leq \alpha) \\
&-9.75[\log(\alpha)]^2 - 3.74\log(\alpha) + 20 && (1\% < \alpha < 5\%) \\
&20 && (\alpha \leq 1\%)
\end{aligned}
\tag{9.3}
$$

The path loss with the Akeyama et al. correction is then given by:

$$
L\,(\mathrm{dB}) = L\,(\mathrm{dB})\ \text{given by Eq. (9.2)} - S(\mathrm{dB})
\tag{9.4}
$$

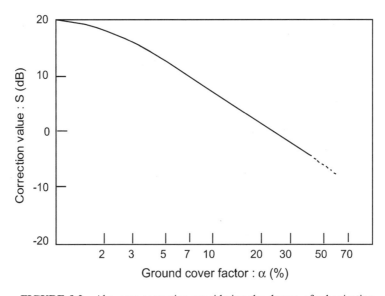

FIGURE 9.2   Akeyama correction considering the degree of urbanization.

This correction makes the Okumura–Hata model applicable to suburban and rural areas. With this correction, an adjustment constant is often added to the formula in actual applications.

Several propagation loss prediction models were proposed after the Okumura–Hata model. All of them also follow Eq. (9.1). In the Lee model [3], the median path loss is given by

$$L \text{ (dB)} = A_1 + \beta \log(d) + F_l \tag{9.5}$$

where A_1 is the path loss at a range of 1 km and β is the slope of the path loss curve. Term F_l is the adjustment factor for actual base station antenna heights, antenna gain, and transmission power with respect to a reference. Values A_1 and β were derived from experiments for urban, suburban, and rural areas. For an actual environment, it is necessary to select A_1 and β by comparing the environment under consideration with the reference environment that most closely resembles it. The influence of terrain is given as the adjustment factor F_l by calculating the effective base station antenna height.

The Ibrahim–Person model [4] uses two parameters, land usage factor (L_a) and degree of urbanization (U_d) to reflect the environmental condition; L_a is defined as the percentage of the $500 \times 500 \, \text{m}^2$, which are covered by buildings, regardless of their height; U_d is defined as the percentage of building site area, within the square, occupied by buildings having four or more floors. The median path-loss formula is given by:

$$
\begin{aligned}
L \text{ (dB)} = &-20 \log(0.7h_b) - 8 \log(h_m) + f/40 + 26 \log(f/40) \\
&- 86 \log[(f + 100)/156] + \{40 + 14.15 \log[(f + 100)/156]\} \log(d) \\
&+ 0.265 L_a - 0.37 H_g + K_u
\end{aligned}
\tag{9.6}
$$

Applicable range:

f, frequency (MHz)   168–900 MHz

h_b, base station antenna height (m)

h_m, mobile station antenna height (m) $\leq 3 \, \text{m}$

d, distance from transmission point (km)   1–10 km

where H_g is the correction term for average ground height, and $K_u = 0.087 U_d - 5.5$ for highly urbanized areas, otherwise $K_u = 0$.

In the models described above, the distance d is more than 1 km and the base station antenna height h_b is higher than the rooftop level of surrounding buildings. This means that the models are suitable for macrocell systems. To reduce the applicable range to less than 1 km, a prediction formula has been proposed by Sakagami et al. [5]. As the model was based on measurement data considering detailed data on the buildings and streets in the prediction area, the formula uses many parameters indicating the layout of buildings and roads:

$$
\begin{aligned}
L \text{ (dB)} = &\, 100 - 7.1 \log(W) + 0.023\theta + 1.4 \log(h_s) + 6.1 \log\langle H \rangle \\
&- [24.37 - 3.7(H/h_{bo})^2] \log(h_b) + [43.42 - 3.1 \log(h_b)] \log(d) \\
&+ 20 \log(f) + \exp\{13[\log(f) - 3.23]\}
\end{aligned}
\tag{9.7}
$$

Applicable range:

f, frequency (MHz)   450–2200 MHz

d, distance from transmission point (km)   0.5–10 km

W, street width (m)   5–50 m

θ, street angle to the base station (deg.)   0–90°

h_s, building height along the street (m)   5–80 m

$\langle H \rangle$, average building height (m)   5–50 m

h_b, base station antenna height above the mobile station ground (m)   20–100 m

h_{b0}, base station antenna height from the base station ground (m)

H, building height near the base station (m) $H \leq h_{b0}$

The Sakagami model may indicate the ultimate empirical formula for short-range path-loss prediction. For smaller area prediction, it seems necessary to consider the actual propagation paths between base and mobile station antennas using precise environment data of the prediction area.

9.3 ANALYTICAL MODELS

To overcome the range limitations of the empirical formulas and try to explain the propagation mechanism, analytical models have been proposed [6–12]. As seen in the Sakagami model, the influences of buildings and streets are significant factors determining the path loss in small-area propagation environments.

Figure 9.3 shows the parameters used in the Walfisch–Bertoni model [7]. The median path loss, L (dB), is expressed as the summation of three independent terms: the free-space

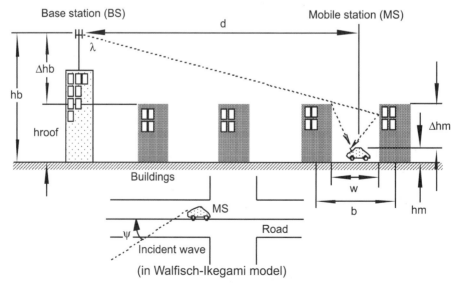

FIGURE 9.3 Parameters used in Walfisch–Bertoni model.

loss (L_{fs}), the diffraction loss from rooftop to street (L_{rts}), and the reduction due to plane wave multiple diffraction past rows of buildings (L_{md}):

$$L \text{ (dB)} = L_{fs} + L_{rts} + L_{md} \tag{9.8}$$

where L_{fs} is a function of wavelength and distance; L_{rts} can be calculated by the geometrical theory of diffraction (GTD) [15], and is a function of average rooftop level, mobile station antenna height, street width, and wavelength; L_{md} has been evaluated in closed forms by Xia and Bertoni [9] and is a function of average rooftop level, average separation distance between the rows of buildings, base station antenna height, street width, wavelength, and distance.

Recently, simplified formulas have been proposed by Xia [12]. For base station antennas above the average rooftop level, the prediction formula is given by:

$$L \text{ (dB)} = -10 \log\left(\frac{\lambda}{4\pi d}\right)^2 - 10 \log\left[\frac{\lambda}{2\pi^2 r}\left(\frac{1}{\theta} - \frac{1}{2\pi + \theta}\right)^2\right]$$
$$- 10 \log\left[(2.35)^2\left(\frac{\Delta h_b}{d}\sqrt{\frac{b}{\lambda}}\right)^{1.8}\right] \tag{9.9}$$

When the base station antenna is near the average rooftop level, the path loss is given by:

$$L \text{ (dB)} = -10 \log\left(\frac{\lambda}{2\sqrt{2}\pi d}\right)^2 - 10 \log\left[\frac{\lambda}{2\pi^2 r}\left(\frac{1}{\theta} - \frac{1}{2\pi + \theta}\right)^2\right]$$
$$- 10 \log\left(\frac{b}{d}\right)^2 \tag{9.10}$$

For base station antennas below the average rooftop level, the path loss is given by:

$$L \text{ (dB)} = -10 \log\left(\frac{\lambda}{2\sqrt{2}\pi d}\right)^2 - 10 \log\left[\frac{\lambda}{2\pi^2 r}\left(\frac{1}{\theta} - \frac{1}{2\pi + \theta}\right)^2\right]$$
$$- 10 \log\left\{\left[\frac{b}{2\pi(d - b)}\right]^2 \frac{\lambda}{\sqrt{\Delta h_b^2 + b^2}}\left(\frac{1}{\phi} - \frac{1}{2\pi + \phi}\right)^2\right\} \tag{9.11}$$

where λ is the wavelength and d is the distance between base and mobile station. The other parameters, Δh_b, Δh_m, b, and w, are defined in Figure 9.3, and r, θ, and ϕ are defined by:

$$r = \sqrt{\Delta h_m^2 + (w/2)^2} \qquad \theta = \tan^{-1}[\Delta h_m/(w/2)] \qquad \phi = -\tan^{-1}(\Delta h_b/b)$$

In the COST-231–Walfisch–Ikegami model [8], the angle of incident wave ψ shown in Figure 9.3 is also used as an additional parameter. The diffraction loss from rooftop to street, L_{rts}, and the multiscreen diffraction loss, L_{md}, in Eq. (9.8) are given by:

$$L_{rts} \text{ (dB)} = -16.9 - 10 \log(w) + 10 \log(f) + 20 \log(\Delta h_m) + L_{ori} \qquad \text{for } h_{roof} > h_m$$
$$0 \qquad \text{for } L_{rts} < 0 \tag{9.12}$$

where

$$L_{\text{ori}} = -10 + 0.354\psi \qquad \text{for } 0 \le \psi < 35°$$
$$2.5 + 0.075(\psi - 35) \qquad \text{for } 35 \le \psi < 55°$$
$$4.0 - 0.114(\psi - 55) \qquad \text{for } 55 \le \psi \le 90°$$

$$L_{\text{md}} \text{ (dB)} = L_{\text{bsh}} + k_a + k_d \log(d) + k_f \log(f) - 9\log(b)$$
$$0 \qquad \text{for } L_{\text{md}} < 0 \tag{9.13}$$

where

$$L_{\text{bsh}} = -18\log(1 + \Delta h_b) \qquad \text{for } h_b > h_{\text{roof}}$$
$$0 \qquad \text{for } h_b < h_{\text{roof}}$$
$$k_a = 54 \qquad \text{for } h_b > h_{\text{roof}}$$
$$54 - 0.8\Delta h_b \qquad \text{for } d \ge 0.5 k_m \text{ and } h_b \le h_{\text{roof}}$$
$$54 - 0.8\,\Delta h_b(d/0.5) \qquad \text{for } d < 0.5 k_m \text{ and } h_b = h_{\text{roof}}$$
$$k_d = 18 \qquad \text{for } h_b > h_{\text{roof}}$$
$$18 - 15\,\Delta h_b / h_{\text{roof}} \text{ for } h_b < h_{\text{roof}}$$
$$k_f = -4 + 0.7(f/925 - 1) \qquad \text{for medium-sized cities and suburban centers with moderate tree density}$$
$$-4 + 1.5(f/925 - 1) \qquad \text{for metropolitan centers}$$

The applicable range in the COST-231–Walfisch–Ikegami model is

f, frequency (MHz)   800–2000 MHz
h_b, base station antenna height (m)   4–50 m
h_m, mobile station antenna height (m)   1–3 m
d, distance from transmission point (km)   0.02–5 km

The Walfisch–Bertoni model supports distances less than 1 km and base station antenna heights lower than rooftop level. This means the model is suitable for microcell systems. However, if the model is applied to a street microcell system whose base station antenna heights are below the average rooftop level, the prediction error may be large. This is because it is necessary to consider not only free line-of-sight propagation but also wave guiding and diffraction at corners. The model does not correspond to this situation.

Figure 9.4 shows the two-path model with street canyon propagation. A calculation of the theoretical signal level for the two-path process showed that it roughly followed a free space power law close to the base station before making a transition to the faster attenuation rate of the inverse fourth-power law. As shown in Figure 9.5, measurement results also showed that close to the base station the propagation followed the free space power law, and beyond this distance the propagation followed the inverse fourth-power law

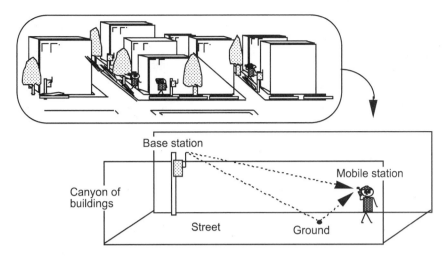

FIGURE 9.4 Geometry of the two-path model for street microcells.

[16, 17]. The distance at which the path-loss law changes is called the breakpoint distance and given by:

$$b_k = \frac{k h_m h_b}{\lambda} \qquad (9.14)$$

where λ is the wavelength, and h_m and h_b are the mobile and base station antenna heights, respectively. The factor k takes values from π to 4π, and depends on the actual

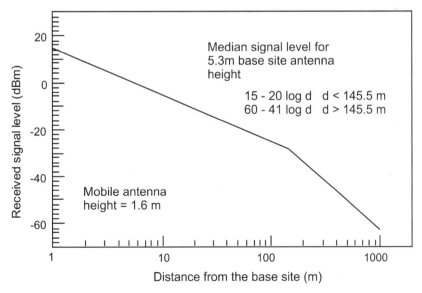

FIGURE 9.5 Microcellular propagation laws that characterized the measurement data taken in the Ginza of Tokyo.

environment such as vehicular traffic, trees, and traffic signals. For the example shown in Figure 9.5, it has been reported that $k = 2\pi$.

Based on this fact and measurement results, a path-loss prediction formula for street microcells was proposed by Ichitsubo et al. [18]. For a line-of-site street, the path loss is given by:

$$L \text{ (dB)} = P(L_1) - 15.5 \log(w_1) + F_{\text{loss}} + 59.9 \tag{9.15}$$

for $L_1 < b_k$

$$P(L_1) = 20 \log(L_1)$$

for $L_1 > b_k$

$$P(L_1) = 43.3 \log(L_1) - 23.3 \log(b_k)$$

For a street with intersections, the path loss is given by:

$$L \text{ (dB)} = P(L_1, L_2) - 20.2 \log(w_1) - 7.8 \log(w_2) + F_{\text{loss}} + 59.8 \tag{9.16}$$

for $L_1, L_2 < b_k$

$$P(L_1, L_2) = 20 \log(L_1) + 20 \log(L_2)$$

for $L_1, L_2 > b_k$

$$P(L_1, L_2) = 32.5 \log(L_1) + 37.5 \log(L_2) - 30 \log(b_k)$$

for $L_1 > b_k, L_2 < b_k$

$$P(L_1, L_2) = 32.5 \log(L_1) + 20 \log(L_2) - 12.5 \log(b_k)$$

for $L_1 < b_k, L_2 > b_k$

$$P(L_1, L_2) = 20 \log(L_1) + 37.5 \log(L_2) - 17.5 \log(b_k)$$

For parallel streets, the path loss is given by:

$$L \text{ (dB)} = P(L_1) + 40.4 \log(L_2) + 18.6 \log[(L_2 + L_3)/L_2] - 15.4 \log(w_1)$$
$$- 19.9 \log(w_2) - 8.5 \log(w_3) + F_{\text{loss}} + 40.6 \tag{9.17}$$

for $L_1 < b_k$

$$P(L_1) = 20 \log(L_1)$$

for $L_1 > b_k$

$$P(L_1) = 2.5 \log(L_1)$$

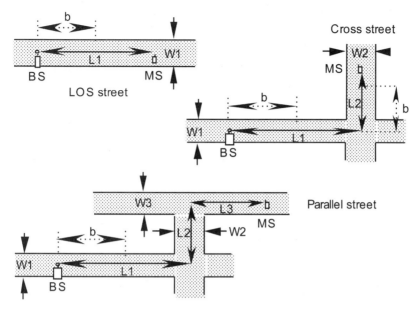

FIGURE 9.6 Parameters used in street microcell path-loss prediction.

The parameters used in the formula are defined in Figure 9.6; F_{loss} represents frequency dependency.

9.4 DETERMINISTIC METHODS

The history of path-loss prediction models shows that information about buildings and streets is necessary and geometrical paths between transmitter and receiver should be considered for accurate prediction. The ray-tracing method is a relatively precise approach for small areas such as indoor picocells and street microcells [19–21].

In the ray-tracing method, rays are launched from the transmitter, and geometrical reflection, transmission, and diffraction are repeated for walls and edges of buildings, as shown in Figure 9.7. The rays arriving at the receiver are tracked as traces, and the field strength at the receiving point is calculated by summing the electric fields of all arrival rays. The field strength of each ray is determined by calculating all reflection, transmission, and diffraction losses in the propagation path. The reflection and transmission losses are usually calculated using Fresnel reflection and transmission coefficients, and diffraction loss is calculated by using the geometrical theory of diffraction (GTD). This means that the prediction accuracy depends on how to find exact ray paths between the transmitter and receiver.

As shown in Figure 9.8, there are two approaches to ray tracing: the imaging method and the launching method. In the imaging method, the reflection and transmission points are determined geometrically by considering the transmitter's equivalent image. The rays reaching the receiving point are located by examining all combinations of reflection, transmission, and all diffraction points between the transmitter and receiver. This method offers good prediction accuracy but needs long computation time when a lot of images

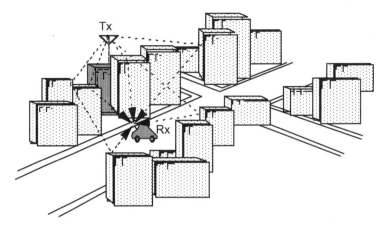

FIGURE 9.7   Reflection, transmission, and diffraction between T_x and R_x.

must be considered. In the launching method, a ray is launched at every angle $\Delta\theta$ from the transmitter, and its path is traced through reflection, transmission, and diffraction. The rays that reach reception area ΔS are considered to have arrived at the reception point. Since ray tracing is performed for each discrete angle $\Delta\theta$, the computation time is shorter than that of the imaging method. The prediction accuracy of this method depends on the chosen values of $\Delta\theta$ and ΔS. A combination of these two methods has been reported to simplify three-dimensional ray tracing [22].

With both methods, there is a trade-off between the prediction accuracy and the computation time. How to reduce the reflection, transmission, and diffraction points, or how to find the paths that most strongly contribute to the receiving level is the key to optimizing this trade-off. The other way to reduce the computation time is to model the buildings and roads in the prediction area. Regarding roads, for example, each intersection and the street between intersections can be transferred to a node and an element component, and the data of street width and building height are stored as the component

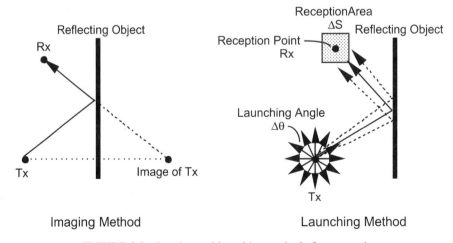

FIGURE 9.8   Imaging and launching methods for ray tracing.

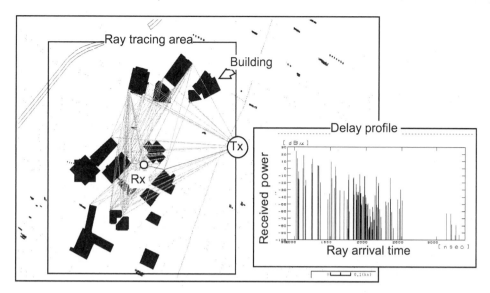

FIGURE 9.9   Example of ray-tracing results for macrocell system.

factors. In a similar manner, the prediction area can be transformed into a plane with pixels, and each pixel holds data on terrain and buildings. By introducing such modeling to the database structure, preprocessing becomes possible before the ray-tracing process, and it also becomes possible to refer to the database quickly during the calculations.

When the ray-tracing method is applied to a wide area, a huge database is required, and computation time exponentially increases with the number of traces. However, the enormous increase in the processing performance of computers and the commercialization of CD-ROM map, including the data of terrain and obstacles, has increased interest in this method for microcell and macrocell system design tools [23, 24]. Figure 9.9 shows an example of a ray-tracing result for a macrocell system.

9.5 SUMMARY

Figure 9.10 shows the applicable areas and simplicity of the major prediction models. The cell size of current cellular systems in urban areas is shrinking to cope with increasing demand. For designing of microcells and indoor picocells, the ray-tracing method is now practical. The Walfisch–Bertoni model is also useful for microcells somewhat larger than street microcells. Macrocells with cell radii larger than 1 km are still needed for suburban and rural areas to realize cost-efficient systems. The Okumura–Hata model with the Akeyama correction and terrain correction is useful in these application areas due to its simplicity.

Each prediction model has its own applicable conditions. In particular, the range of application area is different from each other. Prediction accuracy of each model is assured under the indicated applicable conditions. Therefore, when we design cellular mobile radio systems, it is important to select the prediction model appropriate for the intended cell size. In actual case, it is necessary to use one or several prediction models for determining the path loss [25].

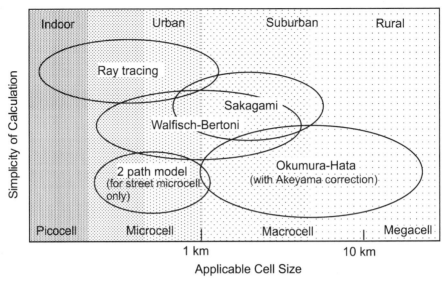

FIGURE 9.10   Application areas and simplicity of prediction models.

The applicable frequency range of propagation loss prediction models proposed so far is up to around 2 GHz. The third-generation mobile communication system, the international mobile telecommunications 2000 (IMT-2000), will be introduced in the 2-GHz frequency band. IMT-2000 will use the bandwidth of more than 5 MHz to support user bit rates of up to 2 Mbits/s. The wideband characteristics of the propagation channel, for example, the characteristics of individual paths in multipath propagation are necessary for such system design. The fourth-generation system will provide multimedia services beyond IMT-2000 by using microwave and millimeter wave frequency bands and so will require even wider bandwidth. In such a situation, microcells and picocells will be introduced to compensate the increased path loss. The space–time equalization technology, which combines adaptive equalizers and adaptive array antennas, may be the breakthrough needed to overcome the increase in path loss and delay spread [26]. The path loss, the direction of arrival (DOA) and the time of arrival (TOA) of each path are essential characteristics in developing these technologies. The ray-tracing method appears to be most effective in these cases and so will become more important in future system designs.

REFERENCES

1. Y. Okumura, E. Ohmori, T. Kawano, and K. Fukuda, "Field Strength and its Variability in VHF and UHF Land Mobile Service," *Rev. Elec. Commun. Lab.*, Vol. 15, pp. 825–873, 1968.

2. M. Hata, "Empirical Formula for Propagation Loss in Land Mobile Radio Services," *IEEE Trans. Veh. Tech.*, Vol. 29, pp. 317–325, Aug. 1980.

3. W. C. Y. Lee, *Mobile Communication Engineering*, McGraw-Hill, New York, 1982.

4. M. F. Ibrahim and J. D. Person, "Signal Strength Prediction in Build-up Areas. Part 1: Median Signal Strength," *IEE Proc.*, Vol. 130, Part F, pp. 377–384, 1983.

5. S. Sakagami and K. Kuboi, "Mobile Propagation Loss Prediction for Arbitrary Urban Environments," *Elec. Commun. Japan*, Part 1, Vol. 74, No. 10, pp. 87–99, 1991.

6. F. Ikegami and S. Yoshida, "Analysis of Multipath Propagation Structure in Urban Mobile Radio Environments," *IEEE Trans. on AP*, Vol. 28, pp. 531–537, 1980.

7. J. Walfisch and H. L. Bertoni, "Theoretical Model of UHF Propagation in Urban Environments," *IEEE Trans. on AP*, Vol. 36, No. 12, pp. 1788–1796, 1988.

8. "COST: Urban Transmission Loss Models for Mobile Radio in the 900- and 1800-MHz Bands," *COST 231 TD (90) 119 Rev. 1*, Florence, Jan. 1991.

9. H. H. Xia and H. L. Bertoni, "Diffraction of Cylindrical and Plane Waves by an Array of Absorbing Half Screens," *IEEE Trans. on AP*, Vol. 40, No. 2, pp. 170–177, Feb. 1992.

10. L. R. Maciel, H. L. Bertoni, and H. H. Xia, "Unified Approach to Prediction of Propagation over Buildings for all Ranges of Base Station Antenna Height," *IEEE Trans. Veh. Tech.*, Vol. 42, No. 1, pp. 41–45, Feb. 1993.

11. "ITU-R Recommendation: Guideline for Evaluation of Radio Transmission Technologies for IMT-2000," *Rec. ITU-R M. 1225*, 1997.

12. H. H. Xia, "A Simplified Analytical Model for Predicting Path Loss in Urban and Suburban Environments," *IEEE Trans. Veh. Tech.*, Vol. 46, No. 4, pp. 1040–1046, Nov. 1997.

13. S. Kozono and A. Taguchi, "Quasi-microwave Propagation Loss in Urban Area", *Trans. IEICE Japan*, Vol. J70-B, No. 10, pp. 1249–1250, Oct. 1987.

14. A. Akeyama, T. Nagatsu, and Y. Ebine, "Mobile Radio Propagation Characteristics and Radio Zone Design Method in Local Cities," *Rev. Elec. Commun. Lab.*, Vol. 30, pp. 308–317, 1982.

15. M. Born and E. Wolf, *Principles of Optics*, Pergamon Press, Oxford, 1974.

16. E. Green, "Measurements and Models for the Radio Characterization of Microcells," IEEE ICCS'90 Proceedings, pp. 1263–1267, Singapore, Nov. 1990.

17. E. Green and M. Hata, "Microcellular Propagation Measurements in an Urban Environment," IEEE PIMRC'91 Proceedings, pp. 324–328, King's College London, Sept. 1991.

18. S. Ichitsubo and T. Imai, "Propagation Loss Prediction in Microcell with Low Base Site Antenna," *Trans IEICE Japan*, Vol. J75-B-II, No. 8, pp. 596–598, Aug. 1992.

19. V. Erceg, A. J. Rustako, Jr., and R. S. Roman, "Diffraction around Corners and its Effects on the Microcell Coverage Area in Urban and Suburban Environments at 900 MHz, 2 GHz, and 6 GHz," *IEEE Trans. Veh. Tech.*, Vol. 43, No. 3, pp. 762–766, Aug. 1994.

20. S. Y. Seidel and T. S. Rappaport, "Site-Specific Propagation for Wireless In-building Personal Communication System Design," *IEEE Trans. Veh. Tech.*, Vol. 43, No. 4, pp. 879–891, Nov. 1994.

21. M. C. Lawton and J. P. McGeehan, "The Application of a Deterministic Ray Launching Algorithm for the Prediction of Radio Channel Characteristics in Small-Cell Environments," *IEEE Trans. Veh. Tech.*, Vol. 43, No. 4, pp. 955–969, Nov. 1994.

22. T. Imai and T. Fujii, "Indoor Microcell Area Prediction System Using Ray-Tracing for Mobile Communication Systems," IEEE PIMRC'96 Proceedings, pp. 24–28, Taipei, Oct. 1996.

23. T. Kurner, D. J. Cichon, and W. Wiesbeck, "Concepts and Results for 3D Digital Terrain-based Wave Propagation Models: An Overview," *IEEE J. Select. Areas Commun.*, Vol. 11, No. 7, pp. 1002–1012, Sept. 1993.

24. T. Imai and T. Fujii, "Propagation Loss in Multiple Diffraction Using Ray-Tracing," IEEE AP-S'97 Proceedings, pp. 2572–2575, Montreal, June 1997.

25. C. Smith, "Propagation Models," *Cellular Business*, pp. 72–76, Oct. 1995.

26. A. J. Paulraj and C. B. Papadias, "Space-Time Processing for Wireless Communications," *IEEE Signal Processing Mag.*, pp. 49–83, Nov. 1997.

CHAPTER 10

Path-Loss Measurements for Wireless Mobile Systems

DONGSOO HAR and HOWARD H. XIA

10.1 OVERVIEW

In the early stage of wireless mobile systems development, measurements were mainly made to characterize radio propagation in macrocellular environments since initial cellular systems deployments typically involve macrocells only. The macrocellular measurements were designed to examine the influence of radio and geometric parameters, such as frequency, transmitting antenna height, and polarization on received signal strength. Some results of these measurements also demonstrate the propagation effects associated with terrain and foliage. These measurements were typically conducted over a distance range of about 10 km for macrocells employing elevated base station antenna radiating power on the order of 10 W. Generally, particular building structures and street configurations have little effects on path-loss within a macrocell served by a base station antenna well above the building clutter. Therefore, measurement results obtained in a typical macrocellular environment can often be applied to similar environments if appropriate adjustment is used to account for slight variations of building and street configuration.

As the demand for wireless mobile services dramatically increases in recent years, microcells and picocells have been deployed in dense areas and inside buildings to increase capacity and to provide indoor coverage. To cope with microcell deployments, measurements have been carried out for small cells with low base station antennas at frequency bands used for cellular and personal communications services. When the base station antenna is about the same height or even below the surrounding buildings, the dependence of radio signals on street orientation and building heights becomes more significant. Therefore microcellular measurements usually involve propagation effects due to street configuration and building structure in addition to terrain profile and morphology. The microcellular measurements were commonly conducted along line-of-sight (LOS) path and/or non-LOS routes. The measurement results for non-LOS routes can be used to study the complex diffraction mechanism associated with path-loss over rooftops or

Wireless Communications in the 21st Century, Edited by Shafi, Ogose, and Hattori.
ISBN 0-471-155041-X © 2002 by the IEEE.

around building corners. Following the review of radio propagation measurements made for microcells, we will compare some empirical models developed based on the measurements with a few well-established theoretical models. To support the introduction of in-building wireless systems or the need to provide indoor coverage by outdoor base stations, propagation inside buildings has recently become an important area for measurements. Due to the complexity associated with the in-building environments, the theoretical in-building modeling also calls for more measurements to validate and/or verify the accuracy of its prediction.

10.2 MACROCELLULAR MEASUREMENTS

Before the commercial introduction of Advanced Mobile Phone System (AMPS) in North America and its deviations around the world in the early 1980s, extensive measurements [1–5] were made to examine fundamental propagation characteristics in macrocellular environments. Young [1] made propagation measurements at four radio frequencies of 150, 450, 900, and 3700 MHz and used the measurement results to address the pros and cons of using these specific frequencies for mobile radio applications. In the early propagation analysis, signal strength variation over distance was often represented as excess path-loss. The excess path-loss is defined as the path-loss exceeding that calculated by the plane earth formula at a certain distance from a transmitter. Egli's study [2] represents such an approach. The excess path-loss is shown to be dependent on frequency and terrain irregularity. Another study done by Black and Reudink [3] demonstrates path-loss dependence on street grid in macrocellular environments. With a base station antenna placed at height of 500 ft (∼150 m) above the street level, Black and Reudink made measurements at distances greater than one mile. It was found in the measurements that path-loss on parallel streets (or cross streets) was usually 10 dB or more greater than that on the corresponding streets perpendicular to the street where base station is located. A standard deviation ranging from 5 to 10 dB was found in [3] for log-normal slow fading on all the streets. Macrocell measurements employing two different mobile antenna heights combined with six locations of base station antennas were later made by Ott and Plitkins [4] at 820 MHz around the urban area of Philadelphia. It was found in the measurements that there was no significant dependence of path-loss on mobile antenna height. In addition, measurements were made by Lee [5] to study signal strength gain due to spatial diversity at base station site. The measurements in [5] demonstrated that spacing between two antennas is the key factor in determining the correlation between received signals and the diversity gain. Extensive measurements for very high frequency (VHF) (200 MHz) and ultrahigh frequency (UHF) (453, 922, 1310, 1430, and 1920 MHz) bands around Tokyo were reported by Okumura et al. [6]. Path-loss along with correction factors accounting for irregular terrain, environment type, mixed land–sea path, and mobile antenna height were given by Okumura et al. [6] based on their measurements made over a distance ranging from 1 to 100 km and for a base station antenna height ranging from 30 to 1000 m.

Effects of foliage or grove/forest on path-loss were discussed by Reudink and Wazowicz [7], and Lagrone [8]. Based on measurements of seasonal fluctuation of path-loss due to foliage at 900 MHz, it was reported in [7] that dense foliage in spring causes more significant radio signal attenuation. Using also the measurement results corresponding to 11 GHz, the excess loss due to foliage is found to be greater at the higher frequency as compared to that at 900 MHz. For more macrocell measurements, readers are referred to [9–13].

10.3 MICROCELLULAR MEASUREMENTS

Due to the presence of direct path between base station and mobile station, path-loss for LOS path is typically less than that for non-LOS paths. Measurement results obtained by Whitteker [14] showed that there is a difference of about 20 dB between the path-loss measured along the LOS path and that measured in the route just turning around the corner, which is significantly more severe than that observed by Black and Reudink [3] in the macrocellular environments. In some dense urban environments having buildings aligned along both sides of streets, the LOS path-loss break point is found to be pushed forward beyond the theoretical location due to waveguide effects, as observed by Rustako et al. [15].

More comprehensive propagation measurements involving LOS path and a variety of non-LOS paths were conducted by Xia et al. in the San Francisco Bay Area [16, 17]. The measurements were made to characterize microcellular radio signal variation in the cellular and personal communications system (PCS) frequency bands (900 and 1900 MHz). In the measurements, the base station antenna was placed at heights h_b of 3.2, 8.7, and 13.4 m while the mobile antenna height h_r was fixed at 1.6 m. The propagation mechanism associated with LOS paths was discussed in details in [16] using the LOS measurement results obtained in various urban, suburban, and rural areas. The Sunset District and the Mission District of San Francisco, which have attached buildings of quasi-uniform height built on a rectangular street grid on flat terrain, were selected as typical low-rise environments for the measurements. Figure 10.1 shows the test routes used in the measurements with a transmitter located in the middle of a block in a street that is part of a rectangular street grid. Measurements were performed for radial distances up to 3 km. The zig-zag measurement results in [17] show that signal strength decreases of 10 to 20 dB as the mobile turned a corner from a perpendicular street into a parallel street. Therefore, the measurement results for the two different segments of the zig-zag path were treated as separate groups. On the parallel streets the propagation path is *transverse* to the rows of buildings. On the perpendicular streets the propagation path has a long *lateral* segment down the street. Signal strength on the *staircase* route showed continuous variation with distance traveled by the mobile, so that measurement results were treated as one group. Path-loss curves were then generated for the zig-zag and staircase groupings. These path-loss curves were later used by Har, Xia, and Bertoni [18] to establish the HXB empirical microcell models. Difference in signal strength between parallel streets and perpendicular streets was also observed by Wagen [19]. Measurements made along LOS streets and neighboring parallel streets in Dallas, Texas, are reported by Xia et al. [20]. Figure 10.2 shows a comparison of signal strengths obtained for the LOS path and the first parallel street. A constant gap of about 35 dB between LOS and parallel street measurements is shown in Figure 10.2 before the LOS break point. The gap becomes narrower after the break point distance at about 200 m and eventually vanishes at a distance of 1 km. Due to the anisotropic propagation characteristics observed in the measurements, cell shape formed by radio signal contours in dense building environments, as demonstrated by Har et al. [18] and Goldsmith and Greenstein [21], is more closely approximated by a diamond rather than a regular hexagon.

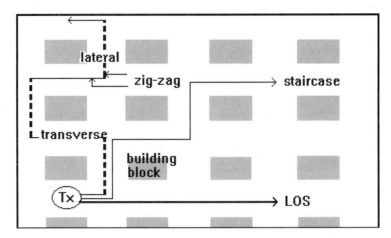

FIGURE 10.1   Measurement test routes (adapted from [17]).

Extensive radio propagation measurements in European cities were performed by many universities and institutions under the COST 231 program. Combining the COST 231 measurements with previously published theoretical model [22], Walfisch–Ikegami models were developed for predicting radio signal propagation in different environments. The COST 231 measurement results were also used to validate the performance of other prediction models [23–25].

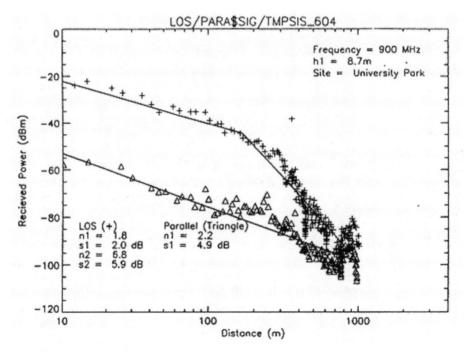

FIGURE 10.2   Received signal strength for LOS and parallel street measurements in University Park, Dallas (adapted from [20]).

Effects of terrain variation on propagation in small urban cells were studied by Lampard and Vu-Dinh [26]. Also, signal strength attenuation due to trees has been estimated [27–29]. Measurements made in a few specific environments such as tunnels or railways are reported [30–34].

Empirical microcell models have been established based on some of the aforementioned microcellular measurements. Here we compare two of these empirical microcell models, that is, the COST 231–Walfisch–Ikegami model [25] and the Har–Xia–Bertoni model [18], with two theoretical path-loss prediction models, the Walfisch–Bertoni model and the Xia–Bertoni model, which have been introduced in Chapter 9. The comparison is made for low-rise environments with relative base station antenna height Δh_b in a range of -5 m $< \Delta h_b < 5$ m. The relative base station antenna height is measured over the average rooftop level of surrounding buildings. In Figure 10.3, we plot the path-loss, excluding the loss due to diffraction at the last rooftop adjacent to the receiver at street level, predicted by these models at a distance of 1 km for frequencies of 900 MHz and 1.9 GHz. It seems that the theoretical models, Walfisch–Bertoni model and Xia–Bertoni model, are slightly more pessimistic compared to the empirical models, COST231–Walfisch–Ikegami model and Har–Xia–Bertoni model. The singularity of Walfisch–Bertoni model and Xia–Bertoni model at $\Delta h_b = 0$ m, that is, base station antenna height is at the rooftop level, results from the unbounded value of multiple diffraction loss at rooftops.

10.4 INDOOR MEASUREMENTS

As more and more voice usage has migrated from wireline networks to wireless networks, in-building radio signal coverage becomes increasingly important. In-building coverage can be provided by indoor wireless systems such as wireless local area network (LAN) or wireless PBX as well as outdoor cellular or PCS systems. Due to difficulties associated with theoretical propagation prediction for complex in-building environments, indoor propagation measurements are more frequently used to assist system design and deployment as compared to outdoor propagation measurements. Most in-building prediction models are site-specific to account for propagation phenomena due to different geometrical properties including internal and external building structures and electromagnetic properties such as permittivity and conductivity.

Path-loss range dependence measurements were made by Lafortune and Lecours [35] at 917 MHz in a university building and by Rappaport and Mcgillen [36] at 1.3 GHz inside a few factories. These measurements showed a disparity in range dependence between LOS paths and non-LOS paths. The path-loss exponent of LOS paths was found to be between 1.5 and 1.8 in both cases while it is about 3 for the non-LOS paths. An even wider path-loss exponent variation was reported by Valenzuela et al. [37], which ranges from 1 to 6. Other in-building measurements were made by Alexander [38], Saleh and Valenzuela [39], and Bultitude [40] along LOS paths such as hallways. These measurements showed similar values of the path-loss exponent. Honcharenko et al. [41] examined the influence of furniture and ceiling fixtures on wavefront spreading along LOS paths in an office environment. Break point associated with the in-building LOS paths that separates distinct path-loss range dependences can be estimated using the concept of first Fresnel zone clearance as in the cases of outdoor LOS measurements [16]. Range dependence of non-LOS paths is mainly determined by floor plan, building structure and wall configuration.

Path loss - (rooftop-to-receiver diffraction loss)

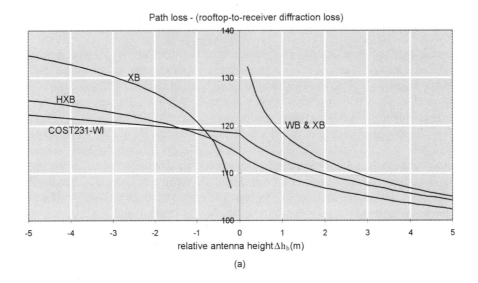

(a)

Path loss - (rooftop-to-receiver diffraction loss)

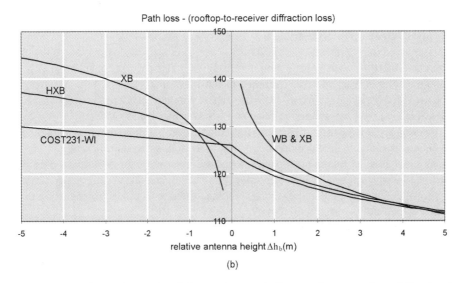

(b)

FIGURE 10.3 Comparison of path-loss values excluding rooftop-to-receiver diffraction loss according to four path-loss models for (a) 0.9 GHz and (b) 1.9 GHz. Relevant parameters are antenna separation $R_k = 1$ km, average building height $h_{BD} = 8$ m, average spacing of building row $d = 50$ m, distance between base station antenna and first building row $r_{lh} = 50$ m (only for XB model).

Measurements showing building penetration loss in the case of an external transmitting antenna have been reported [42–44]. The penetration loss shown in these measurements is associated with attenuation of wall, room, floor, and building. Penetration loss was found to be dependent on architectural details in addition to construction materials. For example, thin wood or plaster walls typically give a loss less than 4 dB while concrete walls generate a loss ranging from 10 to 20 dB. Also, loss due to propagation through a wall generally decreases as window size increases. When an external antenna has a clear LOS path to a building, typical floor losses due to propagation through a single floor at a frequency band of 900 to 1800 MHz were found to be in a range from 4 to 37 dB [25] with a standard deviation of 5 to 15 dB.

Specular processes associated with reflection and transmission at walls are used by Honcharenko et al. [45] to describe wave propagation over a building floor [46]. These processes generally depend on configuration of interior and exterior walls. Because of difficulty in representing surface roughness and electrical property of inhomogeneous wall construction, simplified reflection coefficients [45] with phenomenological roughness factor [47] are developed by Honcharenko et al. [45] based on their measurement results. The simplified reflection coefficients allow reflection calculation based on walls with homogeneous materials and ideally smooth surface. Predictions over a floor were compared with measurements by Lafortune and Lecours [35] for 917 MHz in a university building and by Honcharenko et al. [45] for 852 MHz in an office/laboratory building.

With the antennas located on different floors, signals associated with a direct path from a transmitter to a receiver might experience significant loss by passing through walls and floors. As a result, other alternative paths may become relatively more important. In fact, measurements made by Honcharenko et al. [48] and LeBel and Melangon [49] suggest that rays can exit the building through windows by diffraction and then propagate along the building surface. The rays finally reach the receiver by another diffraction process back into the floor where the receiver is located. Such a complex propagation mechanism can only be analyzed by a site-specific model. In Figure 10.4, we present a typical case observed by Honcharenko et al. [48]. It shows the signal calculated for the *propagation through floors*, the signal exiting and entering back to receiver via *diffracted paths*, and the

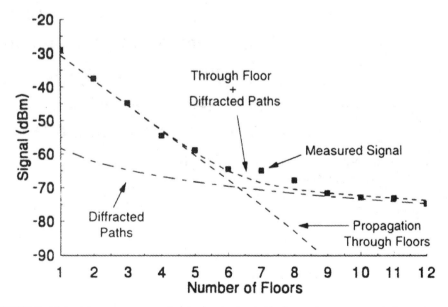

FIGURE 10.4 Measurements and calculated signal level vs. the number of floors between transmitter and receiver (adapted from [48]).

combined signal due to both the *through floors* and the *diffraction* processes. Measurement results show remarkable agreement with the signal level predicted as the combined effects of the above two processes. The above observation confirms that diffraction can play an important role in in-building propagation if the direct path is blocked.

10.5 SUMMARY

Propagation measurements show that path-loss exhibits significantly different characteristics in the outdoor micro- and macrocells and indoor picocells. Therefore, specific considerations have to be taken in wireless mobile systems deployment for each specific operating environment according to its propagation behaviors so as to achieve optimal system design and to exercise best engineering practices.

REFERENCES

1. W. R. Young, Jr., "Comparison of Mobile Radio Transmission at 150, 450, 900, and 3700 Mc," *Bell Sys. Tech. J.*, Vol. 31, pp. 1068–1085, Nov. 1952.

2. J. J. Egli, "Radio Propagation above 40 Mc over Irregular Terrain," in *Proc. IRE*, Vol. 45, No. 10, pp. 1382–1391, Oct. 1957.

3. D. M. Black and D. O. Reudink, "Some Characteristics of Mobile Radio Propagation at 836 MHz in the Philadelphia Area," *IEEE Trans. Veh. Tech.*, Vol. VT-21, pp. 45–51, May 1972.

4. G. D. Ott and A. Plitkins, "Urban Path-Loss Characteristics at 820 MHz," *IEEE Trans. Veh. Tech.*, Vol. VT-27, No. 4 pp. 189–197, Nov. 1978.

5. W. C. Y. Lee, "Effects on Correlation between Two Mobile Radio Base Station Antennas," *IEEE Trans. Commun.*, Vol. COM-21, pp. 1214–1223, Nov. 1973.

6. Y. Okumura, E. Ohmori, T. Kawano, and K. Fukada, "Field Strength and Its Variability in VHF and UHF Land-Mobile Radio Service," *Rev. Elec. Commun. Lab.*, Vol. 16, pp. 825–873, Sept.– Oct. 1968.

7. D. O. Reudink and M. F. Wazowicz, "Some Propagation Experiments Relating Foliage Loss and Diffraction Loss at X-Band and UHF Frequencies," *IEEE Trans. Veh. Tech.*, Vol. 22, No. 4, pp. 114–122, Nov. 1973.

8. A. H. Lagrone, "Propagation of VHF and UHF Electromagnetic Waves over a Grove of Trees in Full Leaf," *IEEE Trans. Antennas Propagat.*, Vol. 25, No. 6, pp. 865–869, 1977.

9. K. K. Kelly, "Flat Suburban Area Propagation at 820 MHz," *IEEE Trans. Veh. Tech.*, Vol. VT-27, pp. 198–204, 1978.

10. A. Atefi and J. D. Parsons, "Urban Radio Propagation in Mobile Radio Frequency Bands," Comms 86, Birmingham (IEE Conf. Publication No. 262, pp. 13–18).

11. M. F. Ibrahim and J. D. Parsons, "Signal Strength Prediction in Built-up Areas. Part 1: Median Signal Strength," *IEE Proc.*, Vol. 130, Part F, No. 5, pp. 377–384, 1983.

12. V. Graziano, "Propagation Correlations at 900 MHz," *IEEE Trans. Veh. Tech.*, Vol. 27, pp. 182–189, 1978.

13. J. F. Aurand and R. E. Post, "A Comparison of Prediction Model for 800 MHz Mobile Radio Propagation," *IEEE Trans. Veh. Tech.*, Vol. 34, No. 4, pp. 149–153, 1985.

14. J. H. Whitteker, "Measurements of Path Loss at 910 MHz for Proposed Microcell Urban Mobile Systems," *IEEE Trans. Veh. Tech.*, Vol. 37, No. 3, 1988.

15. A. J. Rustako, Jr., N. Amitay, G. J. Owens, and R. S. Roman, "Radio Propagation at Microwave Frequencies for Line-Of-Sight Microcellular Mobile and Personal Communications," *IEEE Trans. Veh. Tech.*, Vol. 40, pp. 203–210, 1991.

16. H. H. Xia, H. L. Bertoni, L. R. Maciel, A. Lindsay-Stewart, and R. Rowe, "Radio Propagation Characteristics for Line-of-Sight Microcellular and Personal Communications," *IEEE Trans. Antenna Propagat.*, Vol. 41, No.10, pp. 1439–1447, Oct. 1993.

17. H. H. Xia, H. L. Bertoni, L. R. Maciel, R. Rowe, and A. Lindsay-Stewart, "Microcellular Propagation Characteristics for Personal Communications in Urban and Suburban Environments," *IEEE Trans. Veh. Tech.*, Vol. 43, No. 3, pp. 743–752, August 1994.

18. D. Har, H. H. Xia, and H. L. Bertoni, "Path Loss Prediction Model for Microcells," *IEEE Trans. Veh. Tech.*, Vol. VT-48, pp. 1453–1462, Sept. 1999.

19. J. F. Wagen, "Signal Strength Measurements at 881 MHz for Urban Microcells in Downtown Tampa," Proc. Globecom Conf., Phoenix, AZ, 1991, pp. 1313–1317.

20. H. H. Xia, S. Kim, and H. L. Bertoni, "Microcellular Propagation Measurements in Dallas City," IEEE Vehicular Technology Conf., Secaucus, NJ, 1993.

21. A. J. Goldsmith and L. J. Greenstein, "A Measurement-based Model for Predicting Coverage Areas of Urban Microcells," *IEEE J. Selected Areas Commun.*, Vol. 11, No. 7, Sept. 1993.

22. J. Walfisch and H. L. Bertoni, "A Theoretical Model of UHF Propagation in Urban Environments," *IEEE Trans. Antennas Propagat.*, Vol. 36, pp. 1788–1796, 1988.

23. D. J. Cichon, T. C. Becker, and M. Döttling, "Ray Optical Prediction of Outdoor and Indoor Coverage in Urban Macro- and Microcells," Proc. IEEE Vehicular Technology Conf., Atlanta, GA, 1996, pp. 41–45.

24. T. Kürner, R. Fauß, and A. Wäsch, "A Hybrid Propagation Modelling Approach for DCS 1800 Macrocells," Proc. IEEE Vehicular Technology Conf., Atlanta, GA, 1996, pp. 1628–1632.

25. COST 231 Final Report, in *Propagation Prediction Models*, pp. 17–21.

26. G. Lampard and T. Vu-Dinh, "The Effect of Terrain on Radio Propagation in Urban Microcells," *IEEE Trans. Veh. Tech.*, Vol. 42, No. 3, Aug. 1993.

27. R. Leppänen, J. Lähteenmäki, and S. Tallqvist, "Radiowave Propagation at 900 and 1800 MHz Bands in Wooded Environments," COST 231 TD(92) 112, Helsinki 1992; paper available from Jaakko Lähteenmäki, VTT Information Technology, P. O. Box 1202, SF02044 VTT, Finland.

28. W. J. Vogel and J. Goldhirsh, "Tree Attenuation at 869 MHz derived from Remotely Piloted Aircraft Measurements," *IEEE Trans. Antennas Propagat.*, Vol. 34, pp. 1460–1464, 1986.

29. E. K. Tameh and A. R. Nix, "The Use of Measurement Data to Analyse the Performance of Rooftop Diffraction and Foliage Loss Algorithms in a 3-D Integrated Urban/Rural Propagation Model," Proc. IEEE Vehicular Technology Conf., Ottawa, Canada, 1998, pp. 303–307.

30. P. Aikio, R. Gruber, and P. Vainikainen, "Wideband Radio Channel Measurements for Train Tunnels," Proc. IEEE Vehicular Technology Conf., Ottawa, Canada, 1998, pp. 460–464.

31. T. Klemenschits, "Narrow Band Measurements in Road Tunnels," COST 231 TD(92) 14, Vienna, Austria, 1992.

32. CSELT-SIRTI (Italy), "Propagation Measurements inside Tunnels," COST 231 TD(91) 83, Leidschendam, 1991.

33. M. Göller, "Radio Channel Measurements on Lines of German Railway in 900 MHz Frequency Band," COST 231 TD (92) 20, Vienna, Austria, 1992.

34. M. Uhlirz, "GSM in High-Speed Trains—Experimental Results," COST 231 TD(94) 45, Lisbon, Portugal, 1994.

35. J. F. Lafortune and M. Lecours, "Measurement and Modeling of Propagation Losses in a Building at 900 MHz," *IEEE Trans. Veh. Tech.*, Vol. 39, pp. 101–108, May 1990.

36. T. S. Rappaport and C. D. Mcgillem, "UHF Fading in Factories," *IEEE J. Sel. Areas Commun.*, Vol. 7, No. 1, Jan. 1989.

37. R. A. Valenzuela, D. Chizhik, and J. Ling, "Measured and Predicted Correlation between Local Average Power and Small Scale Fading in Indoor Wireless Communication Channels," IEEE Vehicular Technology Conf., May 1998, pp. 2104–2108.

38. S. E. Alexander, "Radio Propagation within Buildings at 900 MHz," *Electron. Lett.*, Vol. 19, p. 860, Sept. 1983.

39. A. A. M. Saleh and R. A. Valenzuela, "A Statistical Model for Indoor Multipath Propagation," *IEEE J. Select. Areas Commun.*, Vol. SAC-5, pp. 138–146, Feb. 1987.

40. R. J. C. Bultitude, "Measurement, Characterization and Modeling of Indoor 800/900 MHz Radio Channels," *IEEE Commun. Mag.*, Vol. 25, pp. 5–12, June 1987.

41. W. Honcharenko, H. H. Xia, S. Kim, and H. L. Bertoni, "Measurements of Fundamental Propagation Characteristics Inside Buildings in the 900 and 1900 MHz Bands," Proc. IEEE Vehicular Technology Conf., Sycaucus, NJ, 1993, pp. 879–882.

42. H. H. Hoffman and D. C. Cox, "Attenuation of 900 MHz Radio Waves Propagating into a Metal Building," *IEEE Trans. Antennas Propagat.*, Vol. 30, No. 4, pp. 808–811, 1982.

43. Y. L. C. De Jong, M. H. A. J. Herben, and A. Mawira, "Transmission of UHF Radiowaves through Buildings in Urban Microcell Environments," *IEE Electron. Lett.*, Vol. 35, No. 9, pp. 743–745, April 1999.

44. H. W. Arnold, R. R. Murray, and D. C. Cox, "815 MHz Radio Attenuation Measured within Two Commercial Buildings," *IEEE Trans. Antennas Propagat.*, Vol. 37, pp. 1335–1339, 1989.

45. W. Honcharenko, H. L. Bertoni, J. Dailing, J. Qian, and H. D. Yee, "Mechanism Governing Propagation on Single Floors in Modern Office Buildings," *IEEE Trans. Veh. Tech.*, Vol. 41, pp. 496–504, 1992.

46. H. L. Bertoni, W. Honcharenko, L. R. Maciel, and H. H. Xia, "UHF Propagation Prediction for Wireless Personal Communications," *Proc. IEEE*, Vol. 82, No. 9, Sept. 1994.

47. O. Landron, M. J. Feuerstein, and T. S. Rappaport, "In Situ Microwave Reflection Coefficient Measurements for Smooth and Rough Exterior Wall Surfaces," Proc. IEEE Vehicular Technology Conf., Secaucus, NJ, 1993, pp. 77–80.

48. W. Honcharenko, H. L. Bertoni, and J. Dailing, "Mechanism Governing Propagation between Different Floors in Building," *IEEE Trans. Antennas Propagat.*, Vol. 42, pp. 787–790, 1993.

49. J. LeBel and P. Melangon, "The Development of a Comprehensive Indoor Propagation Model," Int. Symp. on Personal, Indoor and Mobile Radio Communications, London, UK, 1991.

PART 4

TECHNOLOGIES

CHAPTER 11

Coding and Modulation for Power-Constrained Wireless Channels

EZIO BIGLIERI, GIUSEPPE CAIRE and GIORGIO TARICCO

11.1 INTRODUCTION

This chapter provides a *tour d'horizon* of the issues arising in digital wireless transmission when the availability of a limited-power/energy source contrasts with quality-of-service requirements. Specifically, we focus our attention on the trade-offs involved at the physical layer, and, in particular, discuss coding/modulation (C/M) from the point of view of power/energy efficiency.

C/M choices are strongly affected by the channel model. We first examine the Gaussian channel because this has shaped the discipline of C/M. Since this channel is characterized by a constant signal-to-noise ratio (SNR), C/M over it has focused on reducing the average power for a given transmission reliability. Over wireless channels, due to fading and interference, the signal-to-disturbance ratio becomes a random variable, which brings into play optimum power allocation. This consists of choosing, based on channel measurements, the minimum transmit power that can compensate for the channel effects and hence guarantee a given quality of service.

Here we examine three wireless channel models, namely the flat independent fading channel (where the signal attenuation is constant over one symbol interval, and changes independently from symbol to symbol), the block-fading channel (where the signal attenuation is constant over an N-symbol block, and changes independently from block to block), and a channel operating in an interference-limited mode. This last model takes into consideration the fact that in a multiuser environment a central concern is overcoming interference, which may limit the transmission reliability more than noise. For each of them we discuss criteria for C/M choice and power allocation strategies.

Wireless Communications in the 21ˢᵗ Century, Edited by Shafi, Ogose, and Hattori.
ISBN 0-471-155041-X © 2002 by the IEEE.

11.2 DESIGNING A C/M SCHEME: THE GAUSSIAN CHANNEL PERSPECTIVE

The most general statement about the selection of a digital C/M scheme is that it should be done by making the best possible use of the resources available for transmission, namely bandwidth, power, and complexity, in order to achieve the reliability required. In summary, C/M selection should be based on four factors: the bit error probability $P_b(e)$, the bandwidth efficiency $\mathcal{R}/\mathcal{B}$ ($\mathcal{R}$ the bit rate, $\mathcal{B}$ the user signal bandwidth), the signal-to-noise ratio $\mathcal{E}_b/N_0$ ($\mathcal{E}_b$ is the received energy per bit and $N_0/2$ is the two-sided power spectral density of the thermal noise) necessary to achieve $P_b(e)$, and the complexity of the transmit/receive scheme. The first factor tells us how reliable the transmission is, the second measures the efficiency in bandwidth expenditure, the third measures how efficiently the modulation scheme makes use of the available power, and the fourth measures the cost of the equipment. It should be observed that complexity also refers to processing power, and hence power source life. However, we shall not delve any further on the latter point.

The ideal system achieves a small $P_b(e)$ with a high $\mathcal{R}/\mathcal{B}$ and a low $\mathcal{E}_b/N_0$. Now, information theory places bounds on the values of these parameters that can be achieved by any C/M scheme. Moreover, complexity considerations push the operating point of a real-life system further away from the theoretical limits.

Let us take a closer look at the factors affecting the performance of a digital modulation scheme. Consider first the bandwidth of the modulated signal $x(t)$, whose duration we denote by T. This signal can take on any of a multiplicity of time functions as its value. Assume we can represent every possible realization of $x(t)$ by a linear combination of v orthonormal signals $\psi_1(t), \ldots, \psi_v(t)$. Then the coefficients $x_1, x_2, \ldots, x_v$, in the linear combination

$$x(t) = \sum_{i=1}^{v} x_i \psi_i(t),$$

interpreted as the coordinates of a point in the v-dimensional Euclidean space, provide a *signal-space representation* of $s(t)$. If all this is done in such a way as to minimize v, then the *Shannon bandwidth* [35] of the modulated signal is $\mathcal{B} = v/2T$. This can be expressed in hertz, although it is often more appropriate to measure it in *dimensions per second*. The Shannon bandwidth is the minimum amount of bandwidth that the signal *needs*, in contrast to more common definitions of bandwidth of a set of modulated signals (e.g., the 3-dB bandwidth, or the equivalent-noise bandwidth, or the bandwidth containing 99% of the signal power, etc.—see, e.g., [1, Chapter 5]). Any of these, which can be called *Fourier bandwidth* of the modulated signal, expresses the amount of bandwidth that the signal actually *uses*. For a thorough discussion about this point, see [35]: Here it suffices to mention the fact that a spread-spectrum signal can be defined as one whose Fourier bandwidth is much larger than its Shannon bandwidth.

The information rate of the source, $\mathcal{R}$, is the rate in bits/second that can be accepted by the modulator. It is related to the number of waveforms used by the memoryless modulator, M, and to the duration of these waveforms, T, by the equality

$$\mathcal{R} = \frac{\log_2 M}{T} \tag{11.1}$$

The average power expended by the modulator is $\mathcal{P} = \mathcal{E}/T$, where $\mathcal{E}$ is the average energy of the modulator signals. Each signal carries $\log_2 M$ information bits. Thus, defining $\mathcal{E}_b$ as the average energy expended by the modulator to transmit one bit, so that $\mathcal{E} = \mathcal{E}_b \log_2 M$, we have

$$\mathcal{P} = \mathcal{E}_b \frac{\log_2 M}{T} = \mathcal{E}_b \mathcal{R} \tag{11.2}$$

The *signal-to-noise ratio* η is the ratio between the average signal power and the average noise power over the signal (Shannon) bandwidth, namely, $N_0 \mathcal{B}$. We have, under the assumption that the channel introduces no attenuation,

$$\eta = \frac{\mathcal{P}}{N_0 \mathcal{B}} = \frac{\mathcal{E}_b}{N_0} \frac{\mathcal{R}}{\mathcal{B}} \tag{11.3}$$

Expression (11.3) shows that the signal-to-noise ratio η is the product of two quantities, namely, $\mathcal{E}_b/N_0$, the energy per bit divided by twice the power spectral density, and $\mathcal{R}/\mathcal{B}$, the *bandwidth* (or *spectral*) *efficiency* of a modulation scheme. The latter is usually measured in bits/second/hertz, and tells us how many bits per second can be transmitted in a given bandwidth $\mathcal{B}$ (another way of expressing this is by saying that $\mathcal{R}/\mathcal{B}$ tells how many bits are transmitted in a signal dimension pair). For example, if a system transmits data at a rate of 9600 bits/s in a 4800-Hz-wide bandwidth, then its spectral efficiency is 2 bits/s/Hz (or 2 bits/dimension pair). The higher the bandwidth efficiency, the more efficient the use of the available bandwidth made by the modulation scheme. Table 11.1 summarizes the spectral efficiencies achieved by some wireless systems and standards [42].

We can define the *asymptotic power efficiency* γ of a modulation scheme as follows. For high signal-to-noise ratios the error probability can be closely approximated by a complementary error function whose argument is $d_{\min}/2\sqrt{N_0}$, $d_{\min}$ being the minimum Euclidean distance between any two elements of the modulator signal set. The parameter γ expresses how efficiently a modulation scheme makes use of the available signal energy to generate a given minimum Euclidean distance. We have

$$\gamma = \frac{d_{\min}^2}{4 \mathcal{E}_b} \tag{11.4}$$

TABLE 11.1 Bandwidth Efficiency of Some Wireless Systems

System	$\mathcal{R}/\mathcal{B}$ (bit/s/Hz)
CT2	0.72
GSM	1.35
IS-54	1.62
PDC	1.68

TABLE 11.2 Maximum Bandwidth- and Power Efficiency of Some M-ary Modulation Schemes: PAM, PSK, QAM, and Orthogonal FSK

Modulation	$\mathcal{R}/\mathcal{B}$	γ
PAM	$2 \log_2 M$	$\dfrac{3 \log_2 M}{M^2 - 1}$
PSK	$\log_2 M$	$\sin^2 \dfrac{\pi}{M} \cdot \log_2 M$
QAM	$\log_2 M$	$\dfrac{3 \log_2 M}{2\, M - 1}$
FSK	$2 \dfrac{\log_2 M}{M}$	$\dfrac{1}{2} \log_2 M$

which provides an approximation to error probability that is asymptotically tight for large signal-to-noise ratios:

$$P(e) \approx \frac{M - 1}{2} \operatorname{erfc}\left(\sqrt{\gamma \frac{\mathcal{E}_b}{N_0}} \right) \tag{11.5}$$

Thus we may say that, at least for high signal-to-noise ratios, a modulation scheme is better than another if its asymptotic power efficiency is greater (at low signal-to-noise ratios the situation is much more complicated than this, but the asymptotic power efficiency still plays a role). Some pairs of values of $\mathcal{R}/\mathcal{B}$ and γ that can be achieved by practical modulation schemes without memory are summarized in Table 11.2.

The performance of a channel when coding is used on it can be expressed by its capacity; for the additive white Gaussian noise (AWGN) channel it takes the value

$$C(\eta) = \mathcal{B} \log_2(1 + \eta) \quad \text{bits/s} \tag{11.6}$$

The "Shannon capacity bound" states that there exists a C/M scheme that achieves, over the AWGN channel, arbitrarily low error probabilities provided that

$$\mathcal{R} < C(\eta) \tag{11.7}$$

Now, in a power-limited environment, the desired system performance should be achieved with the smallest possible power. One solution is the use of standard error control codes, which increase the power efficiency by adding extra bits to the transmitted symbol sequence. This procedure requires the modulator to operate at a higher data rate and, hence, requires a larger bandwidth. In a bandwidth-limited environment, increased efficiency in both power and frequency utilization can be obtained by choosing an integrated C/M solution, where higher order modulation schemes (e.g., 8-PSK instead of 4-PSK) are combined with low-complexity coding schemes. The *trellis-coded modulation* [4] solution combines the choice of a modulation scheme with that of a convolutional code, while the receiver, instead of performing demodulation and decoding in two separate steps, combines the two operations into one. The redundancy necessary to power savings is obtained by a factor-of-2 expansion of the signal constellation size. Table 11.3 summarizes

TABLE 11.3   Asymptotic Coding Gains of TCM (in dB)

Number of States	Coding Gain (8-PSK)	Coding Gain (16-QAM)
4	3.0	4.4
8	3.6	5.3
16	4.1	6.1
32	4.6	6.1
64	4.8	6.8
128	5.0	7.4
256	5.4	7.4

some of the energy savings ("coding gains") in decibels that can be obtained by doubling the constellation size and using TCM. These are considered for coded 8-PSK (relative to uncoded 4-PSK) and for coded 16-QAM (relative to uncoded 8-PSK). These gains can actually be achieved only for high signal-to-noise ratios, and decrease as the latter decrease. The complexity of the resulting decoder/demodulator is proportional to the number of states.

11.3 WIRELESS CHANNEL: A NEW PERSPECTIVE

The consideration of wireless channels—where nonlinearities, Doppler shifts, fading, shadowing, and interference from other users make the simple AWGN channel model far from realistic—forces one to revisit the Gaussian channel paradigms described in the previous section. One early example of this is the design of modulation schemes: Starting from standard phase-shift keying, which has an excellent performance over the Gaussian channel, variations of its basic scheme were derived to cope with nonlinear effects, tight bandwidth constraints, and the like, and resulting in effective modulations like GMSK and $\pi/4$-QPSK [1, Chapter 6]. Another, more recent, example is the study of "fading codes," that is, C/M schemes that are specifically optimized for a Rayleigh channel. These will be examined in Section 11.4.

11.3.1 Fading

Signal fading occurs as a process of random fluctuations of the signal level due to the relative movement of the signal source and the receiver and to the presence of obstacles along the signal path. A most important feature of fading is its time-varying nature. A simple fading channel model is represented in Figure 11.1.

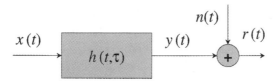

FIGURE 11.1   Block diagram of the time-varying fading channel with impulse response $h(t, \tau)$ and additive Gaussian noise $n(t)$. *Here $y(t) = \int_{-\infty}^{\infty} h(t, \tau)x(t - \tau)\, d\tau$.*

Three sources of signal attenuation, modeled through the time-varying channel impulse response $h(t, \tau)$, can be distinguished (see [51] for a tutorial addressing fading manifestations and the degradations it causes):

- *Propagation Path Loss*   It depends on the distance between the transmitter and the receiver, whose randomness is only due to the position of the mobile unit. Its changes are very limited for practical systems over time intervals of interest. Path loss can be compensated for by varying the transmitted power as a function of the distance between a base station and a mobile unit.
- *Shadowing*   It depends on the presence of obstacles in the signal path and on the relative position of the mobile unit with respect to the base station. It is assumed to be a slow process and is commonly modeled as having lognormal statistics.
- *Multipath Propagation*   It causes rapid fluctuations of the received signal even though the mobile unit moves over very short distances. It is often referred to as *short-term* or *fast* fading. The signal fluctuations due to fast fading are strongly frequency dependent. This particularly affects wideband communication systems.

11.3.2   Impact of Decoding Delay

Another relevant factor in the choice of a C/M scheme is the decoding delay that one should allow: In fact, recently proposed, extremely powerful codes (the "turbo codes" of [2]) suffer from a considerable decoding delay, and hence their application might be useful for data transmission but not for real-time speech. For real-time speech transmission, which imposes a strict decoding delay constraint, channel variations with time may be rather slow with respect to the maximum allowed delay.

11.3.3   Unequal Error Protection

In some analog source coding applications, like speech or video compression, the sensitivity of the source decoder to errors in the coded symbols is typically not uniform: The quality of the reconstructed analog signal is rather insensitive to errors affecting certain classes of bits, while it degrades sharply when errors affect other classes. This happens, for example, when analog source coding is based on some form of hierarchical coding, where a relatively small number of bits carries the "fundamental information" and a larger number of bits carries the "details" like in the case of the MPEG2 standard.

Assuming that the source encoder produces frames of binary coded symbols, each frame can be partitioned into classes of symbols of different "importance" (i.e., of different sensitivity). Then, it is apparent that the best coding strategy aims at achieving lower bit error rate (BER) levels for the important classes while admitting higher BER levels for the unimportant ones. This feature is referred to as "unequal error protection" (UEP). (See, e.g., [7–9] for the inclusion of UEP in turbo coding.)

A conceptually similar solution to the problem of avoiding that degradations of the channel have a catastrophic effect on the transmission quality is "multiresolution modulation" (see, e.g., [16]). This generates a hierarchical protection scheme by using a signal constellation consisting of clusters of points spaced at different distances. The minimum distance between two clusters is higher than the minimum distance within a cluster. The most significant bits are assigned to clusters, and the least significant bits to signals in a cluster.

11.4 FLAT INDEPENDENT FADING CHANNEL

This simplest fading channel model assumes that the duration of a modulated symbol is much greater than the delay spread caused by multipath propagation. If this occurs, then all frequency components in the transmitted signal are affected by the same random attenuation and phase shift, and the channel is frequency flat. If in addition the channel varies very slowly with respect to the symbol duration, then the fading level remains approximately constant during the transmission of one symbol (if this does not occur the fading process is called *fast*.)

The assumption of nonselectivity allows the fading to be modeled as a process affecting the transmitted signal in a multiplicative form. The assumption of slow fading reduces this process to a random variable during each symbol interval. In conclusion, if $x(t)$ denotes the complex envelope of the modulated signal transmitted during an interval of length T, then the complex envelope of the signal received at the output of a channel affected by slow, flat fading and additive white Gaussian noise can be expressed in the form

$$r(t) = Re^{j\Theta}x(t) + n(t) \tag{11.8}$$

where $n(t)$ is a complex Gaussian noise, and $Re^{j\Theta}$ is a Gaussian random variable, with R having a Rice or Rayleigh probability density function (pdf).

If we can further assume that fading is so slow that we can estimate the phase Θ with sufficient accuracy, and hence compensate for it, then coherent detection is feasible, and model (11.8) can be further simplified to

$$r(t) = Rx(t) + n(t) \tag{11.9}$$

In the following we will assume for simplicity that $\mathbb{E}[R^2] = \mathbb{E}[|n(t)|^2] = 1$, so that the signal power and the SNR coincide.

It should be immediately apparent that with this simple model of fading channel the only difference with respect to an AWGN channel resides in the fact that R, instead of being a constant attenuation, is now a random variable, whose value affects the amplitude, and hence the power, of the received signal. A key role here is played by the channel state information (CSI), that is, the fade level, which may be known at the transmitter, at the receiver, or both. Knowledge of CSI allows the transmitter to adapt $x(t)$, and the receiver to adapt its detection strategy, to the fade level.

Figure 11.2 compares the error probability over the Gaussian channel with that over the Rayleigh fading channel without power control (a binary, equal-energy uncoded modulation scheme is assumed, which makes CSI at the receiver irrelevant). This simple example shows how considerable the loss in energy efficiency is. Moreover, in the power-limited environment we are considering in this chapter, the simple device of increasing the transmitted energy to compensate for the effect of fading is not directly applicable. A solution is consequently the use of coding, which can compensate for a substantial portion of this loss.

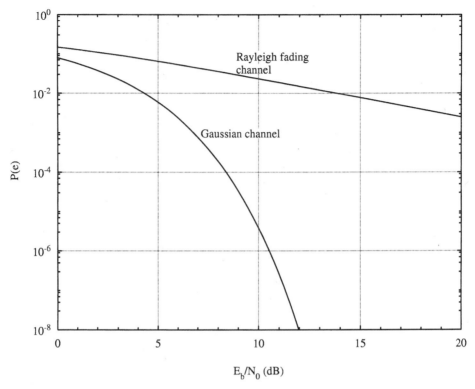

FIGURE 11.2 Error probabilities of binary transmission over the Gaussian channel and over a Rayleigh fading channel.

Under suitable ergodicity assumptions, the capacity of the fading channel with CSI at the receiver is obtained by taking the expectation of the random variable $C(R^2\eta)$ over the probability distribution of R. This is always lower than the capacity of an AWGN channel with the same average power. Focusing on binary input channels, if we compare the capacities of a Rayleigh fading channel with and without CSI at the receiver and that of a Gaussian channel (see Fig. 11.3), we see that the loss in capacity due to fading is much smaller than the loss in terms of error probability, so that coding can actually be highly beneficial in the fading channel. The expressions for the capacities of Figure 11.3 are the following. For perfect CSI:

$$C = \mathbb{E}_R[C(R)] \tag{11.10}$$

where $C(R)$ is the capacity of the channel when the fading is R:

$$C(R) = 1 + \int_y p(y|R, 1) \log_2 \frac{p(y|R, 1)}{p(y|R, 1) + p(y|R, -1)} dy \tag{11.11}$$

In the absence of CSI:

$$C = 1 + \int_y p(y|1) \log_2 \frac{p(y|1)}{p(y|1) + p(y|-1)} dy \tag{11.12}$$

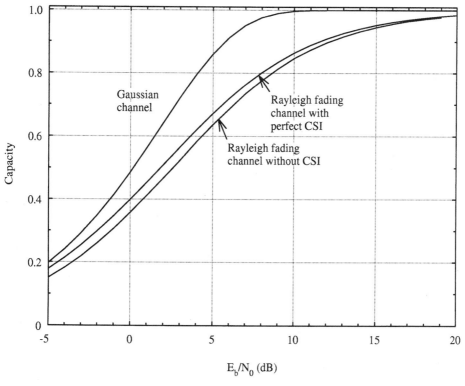

FIGURE 11.3   Capacity of binary transmission over the Gaussian channel and over a Rayleigh fading channel with and without channel-state information at the receiver.

11.4.1   Optimum Codes for the Slow, Flat Rayleigh Fading Channel

Analysis of coding for the slow, flat Rayleigh fading channel proves that Hamming distance (also called "code diversity" in this context) plays the central role here. Assume transmission of a coded sequence $\mathbf{x} = (x_1, x_2, \ldots, x_n)$ where the components of $\mathbf{x}$ are signals selected from a constellation. We do not distinguish here among block or convolutional codes (with soft decoding) or block- or trellis-coded modulation. We also assume that, thanks to perfect (i.e., infinite-depth) interleaving, the fading random variables affecting the various signals x_k are independent. Finally, it is assumed that the detection is coherent, that is, that the phase shift due to fading can be estimated and hence removed. We write, for the components of the received sequence $(r_1, r_2, \ldots, r_n)$:

$$r_k = R_k x_k + n_k \tag{11.13}$$

where the R_k are independent, and, under the assumption that the noise is white, the RV's n_k are also independent.

We can compute the pairwise error probability of this transmission scheme: This is defined as the probability that, when $\mathbf{x}$ is transmitted, the receiver prefers $\hat{\mathbf{x}}$ to $\mathbf{x}$ (see, e.g., [1]). In this case, the pairwise error probability can be bounded above by

$$P\{\mathbf{x} \to \hat{\mathbf{x}}\} \leq \prod_{k \in \mathfrak{K}} \frac{1}{|x_k - \hat{x}_k|^2 / 4N_0} = \frac{1}{[\delta^2(\mathbf{x}, \hat{\mathbf{x}})/4N_0]^{H(\mathbf{x}, \hat{\mathbf{x}})}} \tag{11.14}$$

where

$$\delta^2(\mathbf{x}, \ \hat{\mathbf{x}}) \triangleq \left[\prod_{k \in \Re} |\mathbf{x}_k - \hat{\mathbf{x}}_k|^2 \right]^{1/H(\mathbf{x}, \ \hat{\mathbf{x}})}$$

is the geometric mean of the nonzero squared Euclidean distances between the components of $\mathbf{x}$, $\hat{\mathbf{x}}$, and $H(\mathbf{x}, \ \hat{\mathbf{x}})$ is the Hamming distance between $\mathbf{x}$ and $\hat{\mathbf{x}}$, that is, the number of components in which $\mathbf{x}$ and $\hat{\mathbf{x}}$ differ. The latter result shows the important fact that the error probability is (approximately) inversely proportional to the *product* of the squared Euclidean distances between the components of $\mathbf{x}$, $\hat{\mathbf{x}}$ that differ, and, to a more relevant extent, to a power of the signal-to-noise ratio whose exponent is the Hamming distance between $\mathbf{x}$ and $\hat{\mathbf{x}}$, also called the code *diversity*. This result holds under the assumption that perfect CSI is available at the receiver.

Thus, optimum C/M schemes should maximize their Hamming distance: See, for example [21] and references therein. Now, if the channel model is uncertain or not stationary enough in time to design a C/M scheme closely matched to it, then the best proposition may be that of a "robust" solution, that is, a solution that provides suboptimum (but close to optimum) performance on a wide variety of channel models. As we shall see later, the use of antenna diversity with maximal-ratio combining provides good performance on a wide variety of fading environments. Another robust solution is offered by bit-interleaved coded modulation [11].

We may also think of space, or time, or frequency diversity as a special case of coding. In fact, the various diversity schemes may be seen as implementations of the simple repetition code, whose Hamming distance turns out to be equal to the number of diversity branches [45].

11.4.2 Power Allocation Strategies

Another strategy, which can be used in conjunction with coding and diversity, is based on the simple observation that the increase of signal power reflected in Figure 11.2 is based on an average fading level, and consequently power increase may be allocated more efficiently on a symbol-by-symbol basis, provided that CSI is available at the transmitter.

Consider the simplest such strategy. The flat, independent fading channel with coherent detection yields the received signal (11.9). Assume that the CSI R is known at the transmitter front-end, that is, the transmitter knows the value of R during the transmission (this assumption obviously requires that R is changing very slowly). Under these conditions, let the average transmitted power be S_0 and the transmitted signal

$$x(t) = \sqrt{S(R)}s(t) \tag{11.15}$$

where $s(t)$ has unit power. One possible optimization criterion (constant error probability over each symbol) requires that

$$S(R) = S_0 \frac{R^{-2}}{\mathbb{E}[R^{-2}]} \tag{11.16}$$

This way, the channel is transformed into an equivalent additive white Gaussian noise channel. The received SNR turns out to be

$$\eta = \frac{S_0}{\mathbb{E}[R^{-2}]} \leq \eta_0 (= S_0)$$

by Jensen's inequality, denoting by η_0 the average SNR at the receiver without power control, which implies that error performance is bounded by that of the AWGN channel.

This technique ("channel inversion") is simple to implement, since the encoder and decoder are designed for the AWGN channel, independently of the fading statistics: For instance, it is common in spread-spectrum systems with near-far interference imbalances. However, it may suffer from a large capacity penalty. For example, with Rayleigh fading, $\mathbb{E}[R^{-2}]$ diverges and the capacity is zero.

To avoid divergence of the average power (or an inordinately large value thereof) a possible strategy is the following. Choose

$$S(R) = S_0 \frac{\min(R^{-2},\ R_0^{-2})}{\mathbb{E}[\min(R^{-2},\ R_0^{-2})]} \tag{11.17}$$

that is, invert the channel only if the power expenditure is not too large; otherwise, compensate only for a part of the channel attenuation. By appropriately choosing the value of the threshold R_0 we trade off a decrease of the average received power value for an increase of error probability.

11.4.3 Capacity Analysis

The channel capacity with channel inversion is the capacity of an AWGN channel with SNR $S_0/\mathbb{E}[R^{-2}]$:

$$C(S_0) = \mathcal{B} \log \left[1 + \frac{S_0}{\mathbb{E}[R^{-2}]} \right]$$

With the truncated inversion policy that only compensates for fading above a certain threshold fade depth R_0, the capacity is found in Eq. (12) of [19].

More effective power control schemes are described in [18, 19]. Suppose that we allow the transmit power S to vary with R, subject to the average power constraint

$$\mathbb{E}[S(R)] \leq S_0 \tag{11.18}$$

Using standard methods from the calculus of variations, it can easily be shown that the capacity is maximized when

$$\frac{S(R)}{S_0} = \begin{cases} \dfrac{1}{R_0^2} - \dfrac{1}{R^2} & R \geq R_0 \\ 0 & R < R_0 \end{cases}$$

for some threshold value R_0. If R is below this threshold during a symbol interval, then no data is transmitted over that interval.

The resulting channel capacity is given by

$$C(S) = \mathbb{E}\{\log[1 + R^2 S(R)]\}$$

Now, from Jensen's inequality applied to the function $\log(1 + t)$, we have

$$C(S) \leq \log\{1 + \mathbb{E}[R^2 S(R)]\}$$

Since $\mathbb{E}[R^2 S(R)]$ may be larger than S_0, the capacity of a Rayleigh-fading channel with power control may be higher than the capacity of a Gaussian channel with the same average transmitted power S_0. This turns out to be true for very low values of S_0, as shown in Figure 11.4. Optimal power allocation is also described as "water-filling" in time as it resembles the "water-filling" in frequency used to calculate the capacity of a colored Gaussian channel [15].

11.4.4 Using Multiple Antennas

As briefly mentioned before, multiple antennas can be used as an alternative to coding, or in conjunction with it, to provide diversity. Consider first receive-antenna diversity. The standard approach to it is based on the fact that, since each antenna generates its own channel, the probability that the signal will be simultaneously faded on all channels can be

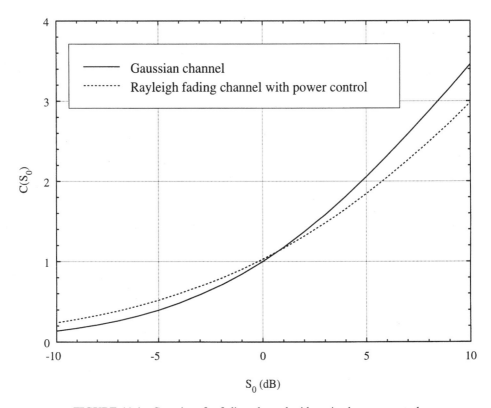

FIGURE 11.4 Capacity of a fading channel with optimal power control.

made small, and hence the detector performance improves. Another perspective, which was investigated by the authors in [56–58], is based upon the observation that, under fairly general conditions, a channel affected by fading can be turned into an AWGN channel by increasing the number of antenna-diversity branches and using maximum-ratio combining (which requires knowledge of CSI at the receiver—see, e.g., [1, Chapter 13]). Consequently, it can be expected (and it was indeed verified by analyses and simulations) that a coded modulation scheme designed to be optimal for the AWGN channel will perform asymptotically well also on a fading channel with diversity, at the only cost of an increased receiver complexity.

Since changes in the physical channel affect the reception very little, an upside of this solution is its robustness, which allows us to argue that the use of "Gaussian" codes (i.e., codes designed by maximizing their minimum Euclidean distance) along with diversity reception provides indeed a solution to the problem of designing robust coding schemes for the mobile radio channel. In fact, [56] proves that, as the number of diversity branches increases, the performance of Gaussian codes comes closer and closer to that of the best codes matched to the fading channel. The assumption of branch independence, although important, is not critical: in effect, [56] also shows that branch correlations as large as 0.5 degrade system BER only slightly. Moreover, the complexity introduced by diversity can be traded for delay: in some cases diversity makes interleaving less necessary, so that a lower interleaving depth (and consequently a lower overall delay) can be compensated by an increase of the diversity order.

When differential, rather than coherent, detection is used [57], a BER floor occurs that can be reduced by introducing diversity. As for the effect of co-channel interference, even its BER floor is reduced as the diversity order is increased (although for its elimination multiuser detectors should be employed).

Recent work (see, e.g., [28, 29, 54]) has explored the ultimate performance limits in a fading environment of systems in which multiple antennas are used at both transmitter and receiver side. It has been shown that, in a system with t transmit and r receive antennas and a slow fading channel modeled by an $t \times r$ matrix with random i.i.d. complex Gaussian entries (the "independent Rayleigh fading" assumption), the average channel capacity with perfect CSI at the receiver is about $m \triangleq \min(t, r)$ times larger than that of a single-antenna system for the same transmitted power and bandwidth. The capacity increases by about m bits/second/hertz for every 3-dB increase in signal-to-noise ratio (SNR). The effect of correlation among antennas is studied in [46].

A further performance improvement can be under the assumption that CSI is available at the transmitter as well. Obtaining transmitter CSI from multiple transmitting antennas is particularly challenging, because the transmitter should achieve instantaneous information about the fading channel. On the other hand, if transmit CSI is missing, the C/M scheme employed should guarantee good performance with the majority of possible channel realizations.

Codes specifically designed for a multiple-antenna system use degrees of freedom in both space and time and are usually called *space–time codes*. Specific designs of space–time codes have been proposed, for example, in [48–50, 52, 53]).

11.5 BLOCK-FADING CHANNEL

Different fading channel models prompt different coding strategies. Thus, it is essential to use a model that closely represents the actual propagations effects. In the frequent situation of a channel coherence time much longer than one symbol interval, the "block-fading channel" model proves exceedingly useful. It applies to a channel in which several adjacent symbols (referred to in the sequel as a *block*) are affected by the same fading value, as, for example, in an indoor wireless data network or a personal communication system with mobile terminals moving at walking speed. Here the channel gain, albeit random, varies so slowly with time that it can be assumed as constant along a block. More generally, fading blocks can be thought of as separated in time (e.g., in a time division system [39]), as separated in frequency (e.g., in a multicarrier system), or as separated both in time and in frequency (e.g., with slow time–frequency hopping [10, 25, 26]). On a block-fading channel, even when very long code words are transmitted, perfect interleaving may not be achieved because of delay limitations. In particular, following [12, 39], we assume that a code word of length $n = MN$ spans a number M of fading blocks. As explained in [39], M can be regarded as a measure of the interleaving delay of the system, so that systems subject to a strict delay constraint are characterized by a fixed (and usually small) value of M.*

The analysis developed for the flat, slow Rayleigh fading channel holds, *mutatis mutandis*, for the block-fading channel: It suffices in this case to interpret the variables x_k as *blocks of symbols*, rather than symbols. In this situation, the relevant criterion becomes the *Hamming block distance*, that is, the number of *blocks* in which two code words differ. An application of singleton bound shows that the maximum block-Hamming distance achievable on an M-block fading channel is limited by

$$D \leq 1 + \left\lfloor M\left(1 - \frac{\mathcal{R}}{\log_2 |\mathcal{S}|}\right) \right\rfloor$$

where $|\mathcal{S}|$ is the size of the signal set $\mathcal{S}$ and $\mathcal{R}$ is the code rate, expressed in bits/symbol. Note how binary signal sets ($|\mathcal{S}| = 2$) are not effective in this case: For example, a code of rate $\frac{1}{2}$ over any binary alphabet (e.g., 2PSK, 2FSK, or 4PSK with Gray mapping) can achieve at most diversity $1 + \lfloor M/2 \rfloor$. In the GSM full-rate case, corresponding to $M = 8$, we get maximum diversity 5, and actually it turns out that the 16-state binary convolutional code selected in the GSM full-rate standard achieves diversity 5, and hence is optimal from this point of view [25]. A larger code diversity can be obtained by increasing the size of the signal set. For example, if we use quaternary PAM, the maximum achievable diversity with a rate $\frac{1}{2}$ code is $1 + \lfloor 3M/4 \rfloor$, that for $M = 8$ equals 7. In [25] a computer search for binary convolutional codes maximizing code diversity when used with 2PAM and 4PAM is presented.

11.5.1 Coding with Finite-Depth Interleaving

Average channel capacity can be thought of as the *long-term* average of mutual information between the transmitter and the receiver. A nonzero average rate is achievable even if the channel may be very poor during some intervals, provided that code words span a time duration much larger than the deep fade duration (e.g., via interleaving). As an alternative, if the transmitter has knowledge of the channel state (channel state feedback), the transmitter can be turned off during deep fades, thus saving energy, and used at a

---

*$M = 2$ in the IS-54 standard. $M = 4$ in the half-rate GSM standard, and $M = 8$ in its full-rate version. In all systems the number of channel symbols in each block exceeds 100 [42].

higher rate and power when the channel is good. This, in essence, is the capacity-achieving coding scheme of [18] (see also [12], where this point is discussed for block-fading channels, and [61], which describes a retransmission protocol that interrupts the communication when the channel quality is bad).

Unfortunately, some applications are delay sensitive (a typical example is real-time voice) and basically constant rate. In these cases, both arbitrary large interleaving depths and variable-rate coding are not allowed. Then, a block-fading channel model with $N > 1$ should be considered.

11.5.2   Using Multiple Antennas on the Block-Fading Channel

Marzetta and Hochwald in [33] derive the capacity of a block-fading channel with multiple antennas and no delay constraints in the absence of CSIT and CSIR. References [27–29, 54] investigate the outage probability of a single-block fading channel with CSIR. References [13] and [41] study the performance of a single-block fading channel with deterministic frequency-selective fading and when the number of transmit/receive antennas approaches infinity (with and without CSIT), respectively.

In [3] the outage probability at a given *fixed* code rate is minimized. Given an M-block fading channel with t transmit and r receive antennas (hereafter referred to as the M-block $t \times r$ fading channel), [3] finds the transmission scheme that minimizes the outage probability under the constraint that the average transmitted power (to be defined properly) shall not exceed a given threshold. The optimal transmission scheme consists of a standard ("Gaussian") code for the AWGN channel, followed by a suitable beam former (described by a $t \times t$ matrix) that may vary from block to block. These beam-forming matrices can be explicitly evaluated once the fading-gain matrices are known. With this approach, the problems of coding and beam forming are decoupled, and no special space-coding design is needed to minimize the outage probability: This stands in contrast to the case of no CSIT, where specific space-code constructions prove to be useful [52].

11.6   INTERFERENCE-LIMITED CHANNEL

The design of C/M schemes is further complicated when a multiuser environment is taken into account. An important feature of the mobile radio channel is the coexistence of several user signals over the same channel, as a consequence of the frequency reuse scheme adopted [31]: we have here a *multiuser* communication channel.

Figure 11.5 illustrates schematically the multiuser fading channel. Independent transmitted signals experience different attenuations, represented by time-varying filters with impulse responses $h_i(t, \tau)$. The channel outputs are given by

$$r(t) = \sum_{i=1}^{M} \int h_i(t, \tau) x_i(t - \tau) \, d\tau + n(t) \tag{11.19}$$

where $n(t)$ is a Gaussian random process representing noise, and $h_i(t, \tau)$ take into account all the propagation effects that characterize the signal fading process.

The main problem here, and in general in communication systems that share channel resources, is the presence of multiple access interference (MAI). This is generated by the

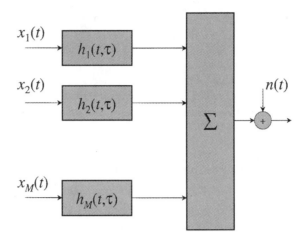

FIGURE 11.5   Block diagram of a multiuser fading channel. M-user signals are combined together after passing through different time-varying channels described by the time-varying impulse responses $h_i(t, \tau)$. Gaussian noise $n(t)$ is added at the receiver.

fact that every user receives, besides the signal that is specifically directed to it, also some power from transmission to other users. This is not true only when code division multiple access (CDMA) is used, but also, for example, with space-division multiple access, in which intelligent antennas are directed toward the intended user. Multiuser detection was born in the context of terrestrial cellular communication, and hence implicitly assumed a MAI-limited environment where thermal noise is negligible with respect to MAI (high-SNR condition). For this reason coding was seldom considered, and hence most multiuser detection schemes known from the literature are concerned with symbol-by-symbol decisions. (For recent results on coding for multiuser channels, see [6, 17, 24, 38].)

The channel models examined in the previous sections assumed implicitly that the transmission was noise limited. Now, in a multiuser environment, a central concern is overcoming interference. In an interference-limited environment, energy can be conserved while guaranteeing quality of service by using an interference-sensitive power management scheme, that is, by modulating the transmitted power according to the interference. Specifically, power levels can be selected as a function of the interference in order to provide a certain data transmission rate. Now, it should be kept in mind that if the power transmitted by a user is reduced, that user will be more vulnerable to interference. Thus, a constraint on the transmission quality must be introduced.

The optimization problem here takes the form [43, 44]

$$\min_{S(i)} \int_0^\infty S(i) p_I(i) \, di \tag{11.20}$$

where I is the random variable, taking values i, denoting the interference power level, $p_I(\cdot)$ its probability density function, and $S(i)$ the transmitted power depending on the interference level, so that the signal-to-interference ratio at the receiver is

$$\eta_I = S(i)/i$$

With $e(i, S)$ denoting the probability of error, the constraint under which (11.20) should be solved is

$$\int_0^\infty e[i, S(i)] p_I(i) \, di \leq p_0 \tag{11.21}$$

The appropriate choice for $e(i, S)$ depends on several conditions: the network topology, the protocol employed, the modulation scheme. For example, in [43, 44], it is assumed that e has the general form

$$e(i, S) = \frac{1}{1 + \beta\eta_I} = \frac{i}{i + \beta S}$$

(where the actual value of β is determined by physical factors) and that the interference is uniformly distributed, that is,

$$p_I(i) = \frac{1}{i_{\max}}, \qquad 0 \leq i \leq i_{\max}$$

Based on these assumptions, a closed form for $S(i)$ can be obtained [43, 44]. From the latter, it can be seen, for example, that, under the assumption of a battery life inversely proportional to $\mathbb{E}[S]$, the choice $e(i, S) = 10^{-2}$ achieves a 13% increase in the useful life of the battery.

In general, while the optimal $S(i)$ depends explicitly on the expression for the error probability $e(S, i)$, the form of the optimal power management function $S(i)$ is independent of the distribution of the interference power. In fact, direct calculation shows that the Fréchet differential [30] of the Lagrangian functional

$$\int_0^\infty [S(i) + \lambda e(i, S)] p_I(i) \, di$$

vanishes for

$$\frac{\partial e}{\partial S} = -\frac{1}{\lambda} \tag{11.22}$$

irrespective of $p_I(i)$.

11.7 CONCLUSIONS

We have examined some of the trade-offs involved at the physical layer in digital wireless transmission when the availability of a limited-energy source is taken into account. Specifically, we have described the choice of coding/modulation schemes. Since the optimization of the latter is critically dependent on the channel model, we have examined four of these, namely, the Gaussian channel (which, although a poor model for many of the wireless channels, has shaped the discipline of coding and modulation) the independent-fading channel, the block-fading channel, and a channel where the transmission quality is limited by interference rather than by noise. Code selection, power allocation strategies, and information-theoretic bounds are discussed for each model.

REFERENCES

1. S. Benedetto and E. Biglieri, *Principles of Digital Transmission with Wireless Applications*, Kluwer/Plenum, New York, 1999.

2. C. Berrou and A. Glavieux, "Near Optimum Error Correcting Coding and Decoding: Turbo-Codes," *IEEE Trans. Commun.*, Vol. 44, No. 10, pp. 1261–1271, October 1996.

3. E. Biglieri, G. Caire, and G. Taricco, "Limiting Performance of Block-Fading Channels with Multiple Antennas," *IEEE Trans. Inform. Theory*, Vol. 47, No. 4, pp. 1273–1289, May 2001.

4. E. Biglieri, D. Divsalar, P. J. McLane, and M. K. Simon, *Introduction to Trellis-Coded Modulation with Applications*, MacMillan, New York, 1991.

5. E. Biglieri, J. Proakis, and S. Shamai (Shitz), "Fading Channels: Information-Theoretic and Communication Aspects," *IEEE Trans. Inform. Theory, 50th Anniversary Issue*, Vol. 44, No. 6, pp. 2619–2692, October 1998.

6. G. D. Boudreau, D. Falconer, and S. Mahmoud, "A Comparison of Trellis Coded versus Convolutionally Coded Spread-Spectrum Multiple-Access Systems," *IEEE J. Select. Areas Commun.*, Vol. 8, No. 4, pp. 628–639, May 1990.

7. F. Burkert, G. Caire, J. Hagenauer, T. Hindelang, and G. Lechner, "'Turbo' Decoding with Unequal Error Protection Applied to GSM Speech Coding," *Proc. IEEE GLOBECOM 1996*, London, UK, pp. 2044–2048, 18–22 Nov. 1996.

8. G. Caire and E. Biglieri, "Parallel Concatenated Codes with Unequal Error Protection," *IEEE Trans. Commun.*, Vol. 46, No. 5, pp. 565–567, May 1998.

9. G. Caire and G. Lechner, "Turbo Codes with Unequal Error Protection," *IEE Electronics Lett.*, Vol. 32, No. 7, pp. 629–631, 28 March 1996.

10. G. Caire, R. Knopp, and P. Humblet, "System Capacity of F-TDMA Cellular Systems," *IEEE Trans. Comm.*, Vol. 46, No. 12, pp. 1649–1661, Dec. 1998.

11. G. Caire, G. Taricco, and E. Biglieri, "Bit-Interleaved Coded Modulation," *IEEE Trans. Inform. Theory*, Vol. 44, No. 3, pp. 927–946, May 1998.

12. G. Caire, G. Taricco, and E. Biglieri, "Optimal Power Control for the Fading Channel," *IEEE Trans. Inform. Theory*, Vol. 45, No. 5, pp. 1468–1489, July 1999.

13. C.-N. Chuah, D. Tse, and J. M. Kahn, "Capacity of Multi-Antenna Array Systems in Indoor Wireless Environment," GLOBECOM'98, Sydney, Australia, Nov. 8–12, 1998.

14. R. Clarke, "A Statistical Theory of Mobile Radio Reception," *Bell Syst. Tech. J.*, Vol. 47, pp. 957–1000, 1968.

15. T. M. Cover and J. A. Thomas, *Elements of Information Theory*, Wiley, New York, 1991.

16. K. Fazel, "Matched Combined Channel Coding and Modulation to the Hierarchical TV Source Coding Scheme," In R. De Gaudenzi and M. Luise (Eds.), *Audio and Video Digital Radio Broadcasting Systems and Techniques*, Elsevier Science, Amsterdam, The Netherlands, 1994, pp. 265–276.

17. J. R. Foerster and L. B. Milstein, "Coding for a Coherent DS-CDMA System Employing an MMSE Receiver in a Rayleigh Fading Channel," *Proc. 1998 IEEE International Symposium on Information Theory (ISIT'98)*, Cambridge, MA, p. 281, 16–21 Aug. 1998.

18. A. J. Goldsmith, *Design and Performance of High-Speed Communication Systems over Time-Varying Radio Channels*, Ph.D. Thesis, University of California at Berkeley, 1994.

19. A. J. Goldsmith and P. P. Varaiya, "Capacity of Fading Channels with Channel Side Information," *IEEE Trans. Inform. Theory*, Vol. 43, No. 6, pp. 1986–1992, Nov. 1997.

20. S. Hanly and D. Tse, "Multiaccess Fading Channels. Part II: Delay-Limited Capacities," Technical Report UCB/ERL M96/69, Electronics Research Laboratory, College of Engineering, University of California, Berkeley, 1996.

21. S. H. Jamali and T. Le-Ngoc, *Coded-Modulation Techniques for Fading Channels*, Kluwer Academic Publishers, New York, 1994.

22. G. Kaplan and S. Shamai (Shitz), "Error Probabilities for the Block-Fading Gaussian Channel," A.E.Ü., Vol. 49, No. 4, pp. 192–205, 1995.

23. G. Kaplan, S. Shamai (Shitz), and Y. Kofman, "On the Design and Selection of Convolutional Codes for an Uninterleaved, Bursty Rician Channel," *IEEE Trans. Commun.*, Vol. 43, No. 12, pp. 2914–2921, Dec. 1995.

24. C. Y. Keung and R. S.-K. Cheng, "Coded CDMA Systems with and without MMSE Multiuser Receiver," *Proc. 3rd Asia Pacific Conference on Communications (APCC '97)*, Sydney, NSW, Australia, pp. 767–771, 7–10 Dec. 1997.

25. R. Knopp, Coding and Multiple Access over Fading Channels, Ph.D. Thesis, Ecole Polytechnique Fédérale de Lausanne, Switzerland, 1997.

26. R. Knopp and P. A. Humblet, "Information Capacity and Power-Control in Single-Cell Multiuser Communications," *Proc. of the IEEE International Conference on Communications (ICC'95)*, pp. 331–335, Seattle, Wa., June 18–22, 1995.

27. G. J. Foschini, "Layered Space-Time Architecture for Wireless Communication in a Fading Environment when Using Multi-Element Antennas," *Bell Labs Tech. J.*, Vol. 1, No. 2, pp. 41–59, Autumn 1996.

28. G. J. Foschini and M. J. Gans, "On Limits of Wireless Communications in a Fading Environment when Using Multiple Antennas," *Wireless Personal Commun.*, Vol. 6, No. 3, pp. 311–335, March 1998.

29. G. J. Foschini and R. A. Valenzuela, "Initial Estimation of Communication Efficiency of Indoor Wireless Channels," *Wireless Networks*, Vol. 3, No. 2, pp. 141–154, 1997.

30. D. G. Luenberger, *Optimization by Vector Space Methods*, Wiley, New York, 1969.

31. V. H. MacDonald, "The Cellular Concept," *Bell Sys. Tech. J.*, Vol. 58, pp. 15–41, Jan. 1979.

32. E. Malkamäki and H. Leib, "Coded Diversity on Block-Fading Channels," *IEEE Trans. Inform. Theory*, Vol. 45, No. 2, pp. 771–781, March 1999.

33. T. L. Marzetta and B. M. Hochwald, "Capacity of a Mobile Multiple-Antenna Communication Link in Rayleigh Flat Fading," *IEEE Trans. Inform. Theory*, Vol. 45, No. 1, pp. 139–157, January 1999.

34. R. McEliece and W. Stark, "Channels with Block Interference," *IEEE Trans. Inform. Theory*, Vol. 30, No. 1, pp. 44–53, Jan. 1984.

35. J. L. Massey, "Towards an Information Theory of Spread-Spectrum Systems," In S. G. Glisic and P. Leppänen, Eds., *Code Division Multiple Access Communications*, Kluwer Academic, Boston, 1995, pp. 29–46.

36. R. McEliece and W. Stark, "Channels with Block Interference," *IEEE Trans. Inform. Theory*, Vol. 30, No. 1, pp. 44–53, Jan. 1984.

37. A. Narula, M. D. Trott, and G. Wornell, "Performance Limits of Coded Diversity Methods for Transmitter Antenna Arrays," *IEEE Trans. Inform. Theory*, Vol. 45, No. 7, pp. 2418–2433, Nov. 1999.

38. I. Opperman and B. Vucetic, "Capacity of a Coded Direct Sequence Spread Spectrum System over Fading Satellite Channels Using an Adaptive LMS-MMSE Receiver," *IEICE Trans. Fundamentals*, Vol. E79-A, No. 12, pp. 2043–2049, Dec. 1996.

39. L. Ozarow, S. Shamai, and A. D. Wyner, "Information Theoretic Considerations for Cellular Mobile Radio," *IEEE Trans. Veh. Tech.*, Vol. 43, No. 2, pp. 359–378, May 1994.

40. J. Proakis, *Digital Communications*, 3rd ed. McGraw-Hill, New York, 1995.

41. G. Raleigh and J. Cioffi, "Spatio-Temporal Coding for Wireless Communication," *IEEE Trans. Commun.*, Vol. 46, No. 3, pp. 357–366, March 1998.

42. T. S. Rappaport, *Wireless Communications. Principles and Practice*, Prentice Hall, Upper Saddle River, NJ, 1996.

43. J. M. Rulnick and N. Bambos, "Mobile Power Management for Maximum Battery Life in Wireless Communication Networks," *Proc. INFOCOM'96*, San Francisco, CA, March 26–28, 1996, pp. 443–450.

44. J. M. Rulnick and N. Bambos, "Performance Evaluation of Power-Managed Mobile Communication Devices," *IEEE Intern. Conf. Commun. (ICC'96)*, Dallas, TX, June 23–27, 1996.

45. N. Seshadri and C.-E. W. Sundberg, "Coded Modulations for Fading Channels—An Overview," *European Trans. Telecomm.*, Vol. ET-4, No. 3, pp. 309–324, May–June 1993.

46. D. Shiu, G. Foschini, M. J. Gans, and J. M. Kahn, "Fading Correlation and its Effect on the Capacity of Multi-Element Antenna Systems," *IEEE Trans. Commun.*, January 1998.

47. D. Shiu and J. M. Kahn, "Power Allocation Strategies for Wireless Systems with Multiple Transmit Antennas," *IEEE Trans. Commun.*, June 1998.

48. D. Shiu and J. M. Kahn, "Design of High-Throughput Codes for Multiple-Antenna Wireless Systems," *IEEE Trans. Inform. Theory*, January 1999.

49. D. Shiu and J. M. Kahn, "Layered Space-Time Codes for Wireless Communications Using Multiple Transmit Antennas," *IEEE Int. Conf. Telecomm. (ICC'99)*, Vancouver, BC, June 6–10, 1999.

50. A. S. Stefanov and T. M. Duman, "Turbo Coded Modulation for Systems with Transmit and Receive Antenna Diversity: System Model, Decoding Approaches, and Practical Considerations," *IEEE J. Selected Areas Commun.*, Vol. 19, No. 5, pp. 958–968, May 2001.

51. B. Sklar, "Rayleigh Fading Channels in Mobile Digital Communication Systems. Part I: Characterization," *IEEE Commun. Mag.*, Vol. 35, No. 7, pp. 90–100, July 1997.

52. V. Tarokh, N. Seshadri, and A. R. Calderbank, "Space-Time Codes for High Data Rate Wireless Communication: Performance Criterion and Code Construction," *IEEE Trans. Inform. Theory*, Vol. 44, No. 2, pp. 744–765, March 1998.

53. "Space-Time Block Codes from Orthogonal Designs," *IEEE Trans. Inform. Theory*, Vol. 45, No. 5, pp. 1456–1467, July 1999.

54. I. E. Telatar, "Capacity of Multi-Antenna Gaussian Channels," *European Trans. Telecomm.*, Vol. 10, No. 6, pp. 585–595, Nov./Dec. 1999.

55. D. Tse and V. Hanly, "Multi-Access Fading Channels—Part I: Polymatroid Structure, Optimal Resource Allocation and Throughput Capacities," *IEEE Trans. Inform. Theory*, Vol. 44, No. 7, pp. 2796–2815, November 1998.

56. J. Ventura-Traveset, G. Caire, E. Biglieri, and G. Taricco, "Impact of Diversity Reception on Fading Channels with Coded Modulation—Part I: Coherent Detection," *IEEE Trans. Commun.*, Vol. 45, No. 5, pp. 563–572, May 1997.

57. J. Ventura-Traveset, G. Caire, E. Biglieri, and G. Taricco, "Impact of Diversity Reception on Fading Channels with Coded Modulation—Part II: Differential Block Detection," *IEEE Trans. Commun.*, Vol. 45, No. 6, pp. 676–686, June 1997.

58. J. Ventura-Traveset, G. Caire, E. Biglieri, and G. Taricco, "Impact of Diversity Reception on Fading Channels with Coded Modulation—Part III: Co-channel Interference," *IEEE Trans. Commun.*, Vol. 45, No. 7, pp. 809–818, July 1997.

59. J. Wolfowitz, *Coding Theorems of Information Theory*, 2nd ed., Springer, New York, 1964.

60. E. Zehavi, "8-PSK Trellis Codes for a Rayleigh Channel," *IEEE Trans. Commun.*, Vol. 40, No. 5, pp. 873–884, May 1992.

61. M. Zorzi and R. R. Rao, "Error Control and Energy Consumption in Communications for Nomadic Computing," *IEEE Trans. Computers (Special Issue on Mobile Computing)*, Vol. 46, pp. 279–289, Mar. 1997.

CHAPTER 12

Modulation and Demodulation Techniques for Wireless Communication Systems

SEIICHI SAMPEI

12.1 INTRODUCTION

When developments of digital wireless communication technologies were started in the late 1970s, applicable modulation schemes were limited only to the constant envelope modulation schemes, such as Gaussian-filtered minimum shift keying (GMSK) [1, 2] due to their high transmit power efficiency. However, since the late 1980s, development of high spectral efficient modulation scheme, such as $\pi/4$-shift quaternary phase shift keying (QPSK) ($\pi/4$-QPSK) [3] and quadrature amplitude modulation (QAM) [4, 5], have been started because much higher system capacity is recognized as the most important issue to satisfy rapidly growing demand for wireless communication services. For such developments, however, precise fading compensation is mandatory. Therefore, various types of pilot signal-aided fading compensation techniques, that is, pilot tone-aided [4], pilot symbol-aided [5], and pilot code-aided schemes [6], have also been developed. As a result, wireless engineers at present can freely select any modulation scheme for wireless communication systems considering various constraint conditions specific for each system. In other words, wireless engineers are requested to fully understand what the features of each modulation/demodulation scheme are and how to effectively combine modulation/demodulation and antifading techniques considering propagation path conditions for the service areas, required transmission quality, and so on.

Thus, this chapter will first discuss classification of modulation schemes from the viewpoint of spectral efficiency, power efficiency, and transmission quality. Then, this chapter will explain basic features of each modulation and demodulation scheme followed by pilot signal-aided fading compensation techniques. When we employ a pilot signal-aided fading compensation technique, we can obtain bit error rate (BER) performances of almost the same as that for theoretical ones if we appropriately select its parameters.

Wireless Communications in the 21st Century, Edited by Shafi, Ogose, and Hattori.
ISBN 0-471-155041-X © 2002 by the IEEE.

Therefore, we will explain only theoretical BER performances in the explanation of modulation and demodulation schemes.

Then, this chapter will also discuss two important and advanced modulation schemes, that is, orthogonal frequency division multiplexing (OFDM) and adaptive modulation.

12.2 OUTLINE OF MODULATION AND DEMODULATION TECHNIQUES

For the design of wireless multimedia communication systems, the most important issue is how to appropriately select modulation and demodulation schemes considering (1) high spectral efficiency, (2) high power efficiency, and (3) high fading immunity.

Figure 12.1 shows classification of the modulation schemes based on these requirements. During the late 1970s and early 1980s, constant envelope modulation schemes were mainly studied aiming at achieving high power efficient terminals using a C-class amplifier. As a result, GMSK [1, 2], phase-locked loop (PLL) QPSK [7], four-level frequency modulation (FM) [8], and tamed frequency modulation (FM) [9] were developed. Among them, GMSK is widely applied to the practical systems, such as global systems for mobile communications (GSM) [10] and digital European cordless telecommunications (DECT) [11].

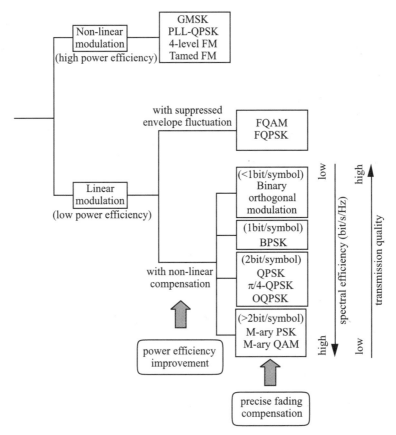

FIGURE 12.1 Classification of modulation schemes.

In the mid-1980s, when capacity limitation of analog cellular systems was becoming more and more a serious problem in each country, development of linear modulations having 2 bits/s/Hz transmission ability were started [3]. When we apply a linear modulation to a wireless communication system, however, we have to achieve high spectral efficiency and high power efficiency at the same time. For this purpose, there are two strategies. The first one is the suppression of envelope fluctuation by introducing cross correlation between in-phase channel (I-ch) and quadrature channel (Q-ch) of the baseband signal, and the second one is the application of nonlinear compensator to the transmitter amplifier. Feher's QPSK (FQPSK) and FQAM [12] are categorized in the former strategy. Because phase for FQPSK is smoothly changed while keeping its envelope almost constant, their spectra are very compact even if a nonlinear amplifier amplifies the signal.

As for the latter strategy, there are many techniques, such as predistortion type [3, 13, 14], Cartisian loop [15], feed-forward type [16], and linear amplification with nonlinear components (LINC) [17] techniques. Owing to these nonlinear compensation techniques, $\pi/4$-QPSK is applied to the Japanese [18] and North American [19] digital cellular and personal communication systems (PCS).

Since the late 1980s, there has been development of modulation schemes with much higher spectral efficiency (> 2 bits/s/Hz), such as 16QAM [4, 5]. In this case, however, very precise fading compensation techniques are required because information is included in both amplitude and phase components in the modulated signal. This can be solved by various types of pilot signal-aided fading compensation techniques, which will be explained later. As a result, 16QAM or multicarrier 16QAM have been applied to the multichannel access (MCA) system and extended specialized mobile radio (ESMR) system [20].

Thanks to these extensive studies, wireless communication system engineers at present have the freedom to apply any modulation scheme to wireless multimedia communication systems. Moreover, we can apply different types of modulation schemes to each part of the transmitted signal. For example, 16QAM is applied to the traffic channel, QPSK is applied to the associated control channel, and binary orthogonal modulation is applied to very important part of the signal. Or, we can adaptively control the modulation scheme according to the traffic and propagation path conditions. This technique is called adaptive modulation [21].

12.3 GMSK

Minimum shift keying (MSK) is an FM scheme with its modulation index of 0.5 and its signal state diagram as shown in Figure 12.2. Let us assume that the transmitted bit for time $t = nT_b$ (T_b is a bit duration) is d_n. When $d_n = 1$, signal phase is shifted with an angle frequency of $0.5\pi/T_b$, whereas it is rotated with an angle frequency of $-0.5\pi/T_b$ in the case of $d_n = 0$. This means that phase position for any bit timing is 0, $\pi/2$, $-\pi/2$, or π, which is just the same as those for QPSK. Therefore, both the coherent detection and differential detection schemes can be applied to MSK with almost the same configurations as those for QPSK. Moreover, frequency discriminator detection is applicable to MSK because MSK is an FM.

GMSK is a modification of MSK. Figure 12.3 shows a configuration of the GMSK modulator [1]. As shown in this figure, a pre-Gaussian filter is inserted before the MSK modulator. In the case of MSK, BER performance for coherent detection is the same as

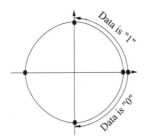

FIGURE 12.2   Signal state diagram of MSK.

that for differentially encoded BPSK with coherent detection. On the other hand, BER for GMSK is degraded due to intersymbol interference by the premodulation Gaussian filter. The BER performance for GMSK with coherent detection under additive white Gaussian noise (AWGN) conditions is given by

$$P_{GMSK}(\gamma) \cong \text{erfc}(\sqrt{\beta\gamma}) \tag{12.1}$$

where γ is energy per bit to noise spectral density (E_b/N_0), β is a degradation factor due to premodulation Gaussian filter, and $\beta = 1$ corresponds to the performance for MSK. When the BER is calculated under flat Rayleigh fading conditions, it is given by

$$P_{GMSK}^{Ray}(\gamma_0) = \int_0^\infty P_{GMSK}(\gamma) \frac{\gamma}{\gamma_0} \exp\left(\frac{\gamma}{\gamma_0}\right) d\gamma$$

$$= 1 - \frac{1}{\sqrt{1 + 1/(\beta\gamma_0)}} \tag{12.2}$$

where γ_0 is the average E_b/N_0.

Table 12.1 summarizes the relationship between the bandwidth of the premodulation Gaussian filter normalized by the bit rate (B_bT_b), 99.99% bandwidth normalized by a bit duration, and β.

When the received signal is multiplied by the 1-bit-duration delayed received signal (1-bit differential detection), the shifted amount of the modulated phase is detected, thereby the transmitted data sequence is regenerated. In this case, however, BER performance is severely degraded due to intersymbol interference, especially when B_bT_b is small.

To mitigate such intersymbol interference, 2-bit differential detection was developed [2]. In the 2-bit differential detection circuit, the phase difference between $\phi(t)$ and its 2-bit delayed phase $\phi(t - 2T_b)$ is detected. In the case of MSK, $\phi(t) - \phi(t - 2T_b)$ is π when the consecutive bits are the same, and $\phi(t) - \phi(t - 2T_b)$ is 0 when the consecutive bits are

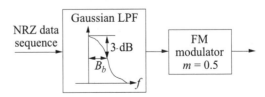

FIGURE 12.3   Configuration of GMSK modulator.

TABLE 12.1   Relationship between $B_b T_b$, 99% Normalized Bandwidth and β

$B_b T_b$	99.99% Bandwidth	β
0.2	1.22	0.76
0.25	1.37	0.84
0.30	1.41	0.89
0.40	1.80	0.94
0.50	2.08	0.97

different. Therefore, when the source bit sequence is differentially encoded, the transmitted data sequence can be regenerated by detecting polarity of $\cos[\phi(t) - \phi(t - 2T_b)]$. Because intersymbol interference to the $2T_b$-separated bit is much smaller than that to the T_b-separated bit, the BER performance for 2-bit differential detection is much better than that for 1-bit differential detection [2].

Frequency discrimination detection is another option for the demodulation scheme. However, when $B_b T_b$ is very small, BER is severely degraded. One of its solution is to apply a decision feedback equalizer (DFE) to the frequency discriminator output in the receiver [22].

Figure 12.4 shows theoretical BER performances of GMSK (a) with coherent detection, (b) with 2-bit differential detection, and (c) with frequency discriminator with DFE, under AWGN and flat Rayleigh fading conditions.

12.4   QPSK

QPSK is a modulation scheme that transmits 2-bit information using four states of the phase. There are two types of phase encoding schemes, absolute phase encoding and differential phase encoding. Figure 12.5 shows a signal state diagram of QPSK with (a) absolute phase encoding and (b) differential phase encoding. In the case of absolute phase encoding, each phase position represents 2-bit information as shown in Figure 12.5(a), whereas the amount of shifted phase represents 2-bit information in the case of differential encoding as shown in Figure 12.5(b).

Figure 12.6 shows a configuration of the QPSK modulator. The transmitted serial data sequence is converted to a 2-bit parallel data sequence at the serial to parallel (S/P) converter, where a symbol duration of the S/P converter output (T_s) is twice as long as T_b. According to the parallel data logic, a baseband signal is generated at the baseband signal generator, where a Gray encoder is employed for absolute phase encoding or a differential encoder is employed for differential encoding, as shown in Figure 12.6. After the baseband signal is filtered by a root Nyquist low-pass filter (LPF), a carrier with its frequency of f_c is modulated by the filtered baseband signal and transmitted.

In the receiver, we can apply both absolute phase coherent detection and differential phase coherent detection. In the case of QPSK with absolute phase coherent detection (CQPSK), we have to regenerate its carrier component with no phase ambiguity using a

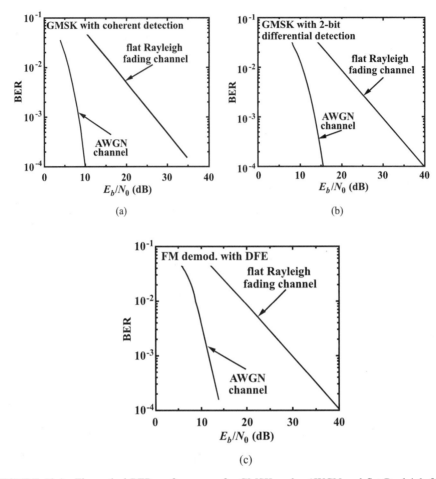

FIGURE 12.4 Theoretical BER performances for GMSK under AWGN and flat Rayleigh fading conditions: (a) coherent detection, (b) two-bit differential detection, and (c) frequency discriminator with DFE.

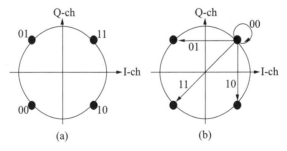

FIGURE 12.5 Signal state diagram of QPSK: (a) absolute phase encoding and (b) differential encoding.

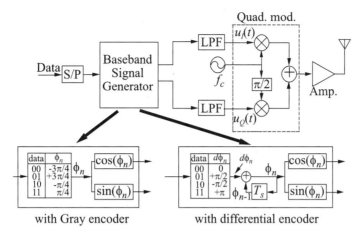

FIGURE 12.6 Configuration of the QPSK modulator.

pilot signal-aided fading compensation technique. The BER performance for Gray-encoded QPSK with coherent detection under AWGN conditions is given by

$$P_{\text{CQPSK}}(\gamma) = \tfrac{1}{2}\,\text{erfc}(\sqrt{\gamma}) \tag{12.3}$$

and the BER performance under flat Rayleigh fading conditions is given by

$$P_{\text{CQPSK}}^{\text{Ray}}(\gamma_0) = \frac{1}{2}\left[1 - \frac{1}{1 + 1/\gamma_0}\right] \tag{12.4}$$

When the regenerated carrier includes phase ambiguity, we have to employ a differential encoder in the transmitter and a differential decoder in the receiver. This scheme is called differentially encoded QPSK (DEQPSK). In this case, the BER becomes twice as large as that for the absolute phase QPSK with coherent detection under both AWGN and flat Rayleigh fading conditions, because a symbol error results in the detection errors of consecutive two-phase transitions.

Another option for the QPSK detection scheme is differential detection. The BER for QPSK with differential detection (DQPSK) under AWGN conditions is approximated by [23]

$$P_{\text{DQPSK}}(\gamma) = \frac{1}{2}\,\text{erfc}\left(2\sin\frac{\pi}{8}\sqrt{\gamma}\right) \tag{12.5}$$

and that under flat Rayleigh fading conditions is given by

$$P_{\text{DQPSK}}^{\text{Ray}}(\gamma_0) = \frac{1}{2}\left[1 - \frac{2\gamma_0}{\sqrt{2(1 + 2\gamma_0)^2 - 4\gamma_0^2}}\right] \tag{12.6}$$

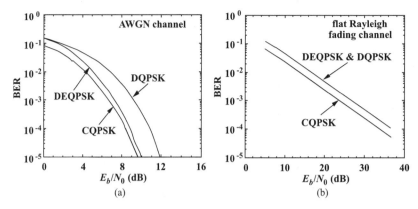

FIGURE 12.7   Theoretical BER performances of CQPSK, DEQPSK, and DQPSK: (a) under AWGN conditions and (b) under flat Rayleigh fading conditions.

Figure 12.7 shows theoretical BER performances for CQPSK, DEQPSK, and DQPSK (a) under AWGN conditions and (b) under flat Rayleigh fading conditions. As shown in this figure, CQPSK gives the best BER performances among these three schemes in both conditions.

12.5 $\pi/4$-QPSK

$\pi/4$-QPSK is a modified QPSK, and both absolute phase encoding and differential phase encoding are applicable to $\pi/4$-QPSK.

Figure 12.8 shows a signal state diagram for absolute phase-encoded $\pi/4$-QPSK. Its difference from that for QPSK is that in-phase and quadrature-phase (I–Q) axes of 0 and $\pi/2$ and those of $-\pi/4$ and $\pi/4$ are alternately changed every T_s second [24]. When $t = 2mT_s$ ($m = 0, 1, 2, \ldots$), four phases represented by a black circle in Figure 12.8(a) are used to transmit 2-bit information as shown in Figure 12.8(b). On the other hand, when $t = (2m + 1)T_s$, I–Q axes are rotated by $-\pi/4$ rad, and four phases represented by a white circle in Figure 12.8(a) are used as shown in Figure 12.8(c). Therefore, when the I–Q axes for the timing $t = (2m + 1)T_s$ is rotated by $\pi/4$ rad, the receiver configuration that is just the same as that for CQPSK can be applied to the $\pi/4$-QPSK with absolute phase coherent detection scheme. Because the effect of this I–Q axes, rotation can be perfectly canceled out when frame synchronization is perfectly taken at the receiver, the BER performance of the absolute phase-encoded $\pi/4$-QPSK is the same as that for the absolute phase-encoded QPSK.

Differential encoding is another option for $\pi/4$-QPSK. Figure 12.9 shows a signal state diagram of differentially encoded $\pi/4$-QPSK. In this diagram, phase shift of $\pi/4$, $3\pi/4$, $-3\pi/4$, and $-\pi/4$ correspond to 2-bit data of 00, 01, 11, and 10, respectively. Therefore, when phase of the received signal is rotated by $-\pi/4$ at every symbol timing, the relationship between 2-bit information and the amount of the shifted phase after $-\pi/4$ rotation is the same as that for DEQPSK [25]. Because such a $-\pi/4$ phase shift does not

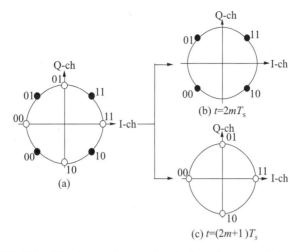

FIGURE 12.8   Signal state diagram for absolute phase-encoded $\pi/4$-QPSK.

cause any degradation at all, the BER performances for differentially encoded $\pi/4$-QPSK are the same as those for DEQPSK.

When the source data sequence is differentially encoded, we can also employ differential detection. Its BER performance is the same as that for DQPSK.

12.6   *M*-ARY QAM

In the *M*-ary QAM system, a symbol is generated according to $\log_2 M$-bit of the source data. Figure 12.10 shows signal state diagrams of 16QAM, 64QAM, and 256QAM. Because coherent detection is essential for the *M*-ary QAM, a pilot signal-assisted fading compensation technique would be necessary for demodulation.

Configuration of *M*-ary QAM modulator is the same as that for CQPSK except for the symbol generation logic. In the case of *M*-ary QAM, after a serial data sequence is converted to $\log_2 M$-bit of parallel data, they are divided into two groups, for example, (a_{4n-3}, a_{4n-2}) and (a_{4n-1}, a_{4n}) in the case of 16QAM. Then, the former group is assigned to the in-phase channel and the latter group is assigned to the quadrature channel.

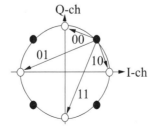

FIGURE 12.9   Signal state diagram of differentially encoded $\pi/4$-QPSK.

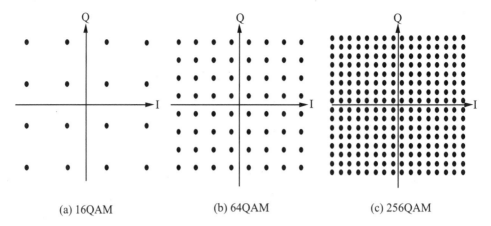

FIGURE 12.10   Signal state diagram: (a) 16QAM, (b) 64QAM, and (c) 256QAM.

The BER for Gray-encoded 16QAM, 64QAM, and 256QAM under AWGN conditions are given by

$$P_{16QAM}(\gamma) = \frac{3}{8}\mathrm{erfc}\left(\sqrt{\frac{2}{5}\gamma}\right) - \frac{9}{64}\mathrm{erfc}^2\left(\frac{2}{5}\gamma\right) \tag{12.7}$$

$$P_{64QAM}(\gamma) = \frac{7}{24}\mathrm{erfc}\left(\sqrt{\frac{1}{7}\gamma}\right) - \frac{49}{384}\mathrm{erfc}^2\left(\frac{1}{7}\gamma\right) \tag{12.8}$$

$$P_{256QAM}(\gamma) = \frac{15}{64}\mathrm{erfc}\left(\sqrt{\frac{4}{85}\gamma}\right) - \frac{225}{2048}\mathrm{erfc}^2\left(\frac{4}{85}\gamma\right) \tag{12.9}$$

When M-ary QAM is transmitted over a flat Rayleigh fading channel, the BER for each modulation is given by

$$P_{16QAM}^{Ray}(\gamma_0) = \frac{3}{8}\left[1 - \frac{1}{\sqrt{1 + 5/(2\gamma_0)}}\right] \tag{12.10}$$

$$P_{64QAM}^{Ray}(\gamma_0) = \frac{7}{24}\left[1 - \frac{1}{\sqrt{1 + 7/\gamma_0}}\right] \tag{12.11}$$

$$P_{256QAM}^{Ray}(\gamma_0) = \frac{15}{64}\left[1 - \frac{1}{\sqrt{1 + 85/(4\gamma_0)}}\right] \tag{12.12}$$

Figure 12.11 shows theoretical BER performances of Gray-encoded QPSK, 16QAM, 64QAM, and 256QAM under AWGN and flat Rayleigh fading conditions.

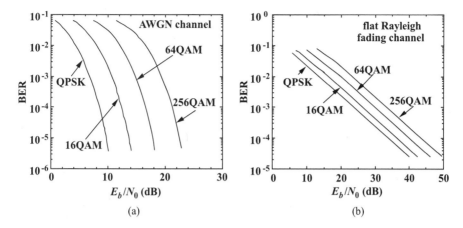

FIGURE 12.11 Theoretical BER performances of Gray-encoded QPSK, 16QAM, 64QAM, and 256QAM: (a) under AWGN conditions and (b) under flat Rayleigh fading conditions.

12.7 PILOT SIGNAL-AIDED FADING COMPENSATION TECHNIQUES

There are three types of pilot signal-aided fading compensation technique [26].

Pilot tone-aided techniques in which one or more tone signal(s) and the information signal are multiplexed in the frequency domain. They can also be called frequency division multiplexing type (FDM-type) techniques.

Pilot symbol-aided techniques in which a known pilot symbol sequence and the information sequence are multiplexed in the time domain. They can also be called time division multiplexing type (TDM-type) techniques.

Pilot code-aided techniques in which a spread spectrum signal using a code orthogonal to traffic channel(s) is multiplexed with the traffic channel(s). They can also be called code division multiplexing type (CDM-type) techniques.

Figure 12.12 shows classification of the pilot signal-aided calibration techniques. When we transmit a carrier component simultaneously, we can exactly regenerate a reference carrier signal with no phase ambiguity. The simplest way to transmit such a carrier component is just to transmit a carrier component. This technique is called the tone calibration technique (TCT) [27]. One disadvantage of the TCT is its difficulty in suppressing mutual interference between the carrier and modulated signal components. Dual tone calibration technique (DTCT) in which a tone is located on each side of the signal spectrum [28] can easily satisfy orthogonality between the tone signals and the modulated signal. However, it is more sensitive to the adjacent channel interference and imbalance between bandpass filters to pick up tone signals.

The transparent tone-in band (TTIB) scheme is a technique that combines advantages of both TCT and DTCT [4]. In this scheme, the baseband signal is split into two components using a pair of quadrature mirror filters and each spectrum is shifted to the other way to create a gap at the center of the spectrum to transmit a carrier component.

In the case of pilot symbol-aided techniques, fading variation is regularly measured at each pilot symbol timing, and fading variation at each information symbol is estimated by interpolating the measured values [5].

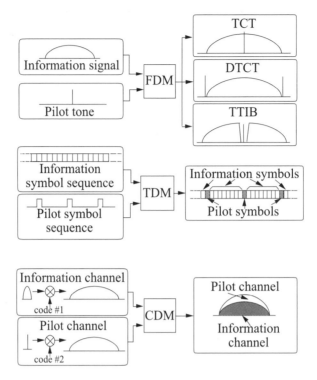

FIGURE 12.12   Classification of pilot signal-aided fading compensation techniques.

Pilot code-aided techniques are mainly applied to the CDMA system to measure instantaneous delay profile necessary for Rake diversity as well as to achieve soft handover [6, 29]. Of course, it is applicable to time division multiple access (TDMA) systems if the assigned bandwidth for each user is sufficiently wide.

Among these three types of the pilot signal-aided techniques, FDM and TDM type of techniques are effective only to compensate for flat Rayleigh fading because these techniques can estimate only channel variation at the center of the assigned bandwidth. On the other hand, the code division multiplexing type (CDM-type) can exactly estimate channel variation of the entire frequency band because it can measure delay profile of the channel. Performance degradation due to these pilot signal-aided fading compensation and their parameter optimization are detailed in Ref. [26].

12.8   ORTHOGONAL FREQUENCY DIVISION MULTIPLEXING

Multicarrier modulation that transmits high bit rate data over a large number of narrowband subcarriers is one of the effective antifrequency selective fading techniques. Let us assume that symbol duration of the multicarrier-modulated signal is T_s and its subcarrier spacing is Δf. When frequency separation of arbitrary two subcarrier is m/T_s (m is an arbitrary integer), they are orthogonal to each other. Therefore, a multicarrier modulation with its subcarrier spacing of $\Delta f = 1/T_s$ is called an orthogonal multicarrier system or orthogonal frequency division multiplexing (OFDM). When we apply OFDM to

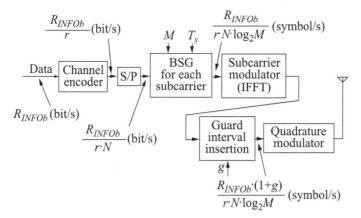

FIGURE 12.13   Configuration of OFDM transmitter.

a wireless communication system, however, it is necessary to introduce a guard interval insertion technique to avoid intersymbol interference (ISI) from the preceding OFDM symbols [30].

Figure 12.13 shows a configuration of the OFDM transmitter that includes a guard interval insertion section. After an information data sequence is channel encoded, the encoded sequence is converted to N-parallel data sequence followed by the N-channel baseband signal generator (BSG), where symbol rate for the BSG output is $1/\log_2 M$ times the rate of BSG input, and M is the number of the modulation level. Of course, we can employ both Gray encoding and differential encoding at this stage. The generated baseband signal is then fed to the subcarrier modulator. Usually, this subcarrier modulation process is carried out using an inverse fast Fourier transform (IFFT) device.

When a guard interval is set to $T_{\mathrm{GI}} = gT_s$ $(0 < g < 1)$, a symbol duration for the OFDM symbol is given by $T_{\mathrm{OFDM}} = T_s + T_{\mathrm{GI}} = (1 + g)T_s$. Therefore, information bit rate for the OFDM system is given by

$$R_{\mathrm{INFOb}} = \frac{N \cdot r \cdot \log_2 M}{T_{\mathrm{OFDM}}} \quad \text{(bits/s)} \tag{12.13}$$

where r is channel coding rate. Baseband signal after guard interval insertion is then given by

$$s(t) = \sum_{l=-\infty}^{\infty} \sum_{k=0}^{N-1} w(t - lT_{\mathrm{OFDM}}) c_{ik} \exp[\, j2\pi \, \Delta f (t - T_{\mathrm{GI}} - lT_{\mathrm{OFDM}})] \tag{12.14}$$

where c_{ik} is the lth complex symbol in the kth subcarrier, and $w(t)$ is the time-windowing function for each OFDM symbol. An example of $w(t)$ is given by

$$w(t) = \begin{cases} 1; & -T_{\mathrm{GI}} \leq t \leq T_s \\ 0; & \text{otherwise} \end{cases} \tag{12.15}$$

The OFDM baseband signal is then fed to the quadrature modulator.

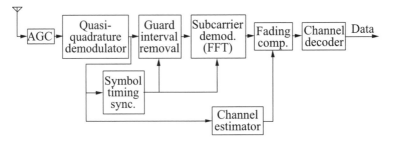

FIGURE 12.14 Configuration of OFDM receiver.

Figure 12.14 shows a configuration of the OFDM receiver. After the received signal is amplified to a proper level by the automatic gain controller (AGC), the quasi-quadrature demodulator generates the received OFDM baseband signal. Using the generated baseband signal, OFDM symbol timing is regenerated for detection of the FFT windowing timing, and the guard interval is removed using the detected FFT windowing timing. The guard interval-removed signal is fed to the FFT to demodulate each subcarrier, and fading distortion for each subcarrier is compensated for at the fading compensator using a pilot signal-aided technique. The fading compensated signal is then fed to the channel decoder to regenerate the transmitted data sequence [31, 32].

Because the BER for each subcarrier is the same as that for the single-carrier transmission systems, the BER for OFDM is the same as that for single-carrier modulation scheme under both AWGN and flat Rayleigh fading conditions. Even under frequency-selective fading conditions, if the maximum delay time of the delayed paths is less than the guard interval, the BER performance for uncoded OFDM is almost the same as the performance under flat Rayleigh fading conditions because guard interval can prevent intersymbol interference to the FFT-windowed period.

When the transmitted data sequence for OFDM is encoded and interleaved in the frequency domain, the BER is, in most cases, better than that for single-carrier systems under frequency-selective fading because frequency diversity effect is usually obtained due to narrower coherence bandwidth than the OFDM signal bandwidth.

The most important OFDM parameters are T_{OFDM}, T_{GI}, N, and M. First of all, T_{GI} should be longer than the longest delay time of the delayed paths. From this viewpoint, T_{GI} should be as long as possible. However, longer T_{GI} degrades transmission efficiency of the OFDM system. Therefore, T_{GI} is usually selected to be longer than the longest delay time for the critical delayed paths in the service area, and T_{OFDM} is selected to satisfy that T_{GI} is 10–20% of the T_{OFDM} [33]. However, such a design method requires a large number of N when the longest delay time of the critical delayed paths is very long and the required information bit rate is very high. Therefore, T_{OFDM}, T_{GI}, N, and M should be finally adjusted to compromise all the constraint conditions.

12.9 ADAPTIVE MODULATION

The adaptive modulation is a promising modulation scheme for next-generation wireless data communication systems. In the adaptive modulation systems, the modulation parameters, such as the modulation level [21], symbol rate [21], or coding rate of the channel coding scheme [34] are adaptively controlled according to the propagation path characteristics or time-varying system traffic load.

There are three types of the adaptive modulation schemes:

Preassigned adaptive modulation that preliminarily selects modulation parameters for each base stations. An example is the FLEX pager system.

Slow adaptive modulation that controls the modulation parameters according to the average path loss due to distance and shadowing, or the traffic load in each base station. Examples are general packet radio service (GPRS) [35], enhanced data rates for global evolution (EDGE) [36], and IEEE802.11a [37].

Fast adaptive modulation that controls the modulation parameters according to the fast variation of propagation path characteristics.

Among them, preassigned and slow adaptive modulation schemes are relatively easier to implement because variations of the average path loss and traffic are sufficiently slow. On the other hand, implementation of the fast adaptive modulation is quite difficult and requires cutting-edge technologies because multipath fading variation is very fast. Therefore, we will discuss only the fast adaptive modulation systems in the following.

12.9.1 Principle of the Fast Adaptive Modulation

Figure 12.15 shows the concept of fast adaptive modulation systems. One of the key issues for this system is how to accurately estimate propagation path characteristics (delay profile) for the next transmission time slot. It can be solved when time division duplex (TDD) is employed as the duplex scheme because delay profiles for the uplink and downlink for the TDD systems are highly correlated provided that the time interval between the transmission and reception time slots are sufficiently short.

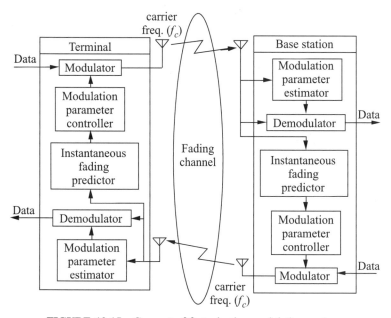

FIGURE 12.15   Concept of fast adaptive modulation systems.

Because different modulation parameters are employed in each slot, the used modulation parameters for each reception time slot should be estimated before demodulation. In the receiver, we also have to estimate delay profile for the next transmission time slot using reciprocity of the uplink and downlink channels in the TDD systems. Using the estimated delay profile, we will estimate instantaneous carrier power to noise spectral density (C/N_0), delay spread for the next transmission time slot, and select a best combination of the modulation parameters according to these values. Using the selected modulation parameters, we will transmit data for the next transmission time slot.

12.9.2 Estimation for Modulation Parameter and Instantaneous Delay Profile

Figure 12.16 shows the basic concept of the modulation parameter and instantaneous fading prediction. The most important key techniques for the fast adaptive modulation system is how to accurately estimate the used modulation parameter in the receiver and channel conditions for the next transmission time slot. For this purpose, two important code words are embedded in the midamble of each slot, that is, channel estimation word and modulation parameter estimation word.

One of the simple and effective ways to encode the modulation parameters is to employ a Walsh code. To simplify discussion, we assume that there are four modulation parameter candidates, QPSK, 16QAM, 64QAM, and 256QAM in Figure 12,16. In this case, the used modulation parameters are estimated by taking correlation between the received code word and all the code word candidates, and searching a code word having the maximum correlation value [21].

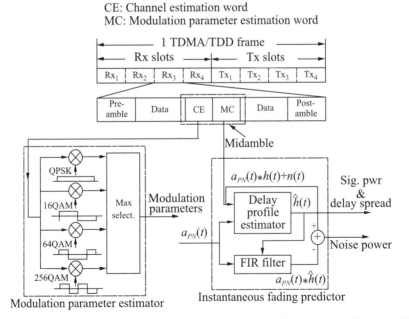

FIGURE 12.16 Basic concept of the modulation parameter and instantaneous fading prediction.

Channel estimation code word consists of an *M*-sequence. By taking correlation between the received channel estimation code word and its original code word, we can measure a complex delay profile for each slot. When we take convolution between the channel estimation code word and the measured delay profile, we can obtain a replica of the noiseless received signal. Therefore, we can estimate noise sequence by subtracting the replica from the received signal, thereby we can estimate noise power. Next, we will estimate delay profile for the next transmission time slot by extrapolating delay profile sequence of the reception time slot. Using the extrapolated profile, we can estimate received signal power and delay spread for the next transmission time slot [21].

12.9.3 Modulation Parameter Selection Chart

According to the estimated C/N_0 and delay spread for the next transmission time slot, the transmitter selects an optimum combination of the modulation parameters. An example is shown in Figure 12.17 [34]. In this example, selectable modulation parameters are modulation scheme (QPSK, 16QAM, 64QAM), symbol rate (full-rate, $\frac{1}{2}$-rate, $\frac{1}{4}$-rate), and channel coding rate ($r = \frac{7}{8}, \frac{3}{4}, \frac{2}{3}, \frac{1}{2}$). Although there are 36 combinations for selection of these parameters, 8 combinations of the modulation parameters are prepared in this example. As shown in Figure 12.17, when C/N_0 is high and delay spread is small, modulation parameters that achieve higher bit rates are selected. On the other hand, when C/N_0 is low or delay spread is large, modulation parameters that achieve lower bit rates are selected. Especially when delay spread is very large, lower symbol rate is selected because reduction of the symbol rate is the most effective way to improve delay spread immunity. To further improve BER performances, this chart also includes nontransmission mode (Dummy in Fig. 12.17) that transmits no information when the channel condition is too bad.

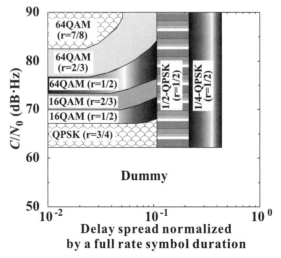

FIGURE 12.17 Example of modulation parameter selection chart.

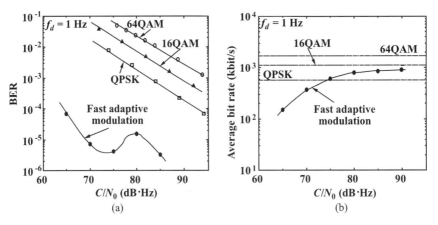

FIGURE 12.18   BER and average bit rate performances of the fixed-rate QPSK, 16QAM, 64QAM, and the fast adaptive modulation using a chart shown in Figure 12.17 under flat Rayleigh fading conditions (Laboratory experiments).

12.9.4   BER Performance of the Fast Adaptive Modulation

Figure 12.18 shows laboratory experimental results of BER and average bit rate performances of the fixed rate QPSK, 16QAM, 64QAM, and the fast adaptive modulation that employs the chart shown in Figure 12.17. Propagation path characteristics for the experiments are flat Rayleigh fading with its maximum Doppler frequency of 1 Hz. As shown in this figure, BER performance of the fast adaptive modulation is drastically improved due to selection of modulation parameters with lower bit rates including nontransmission mode under bad channel conditions. Therefore, although the average bit rate at high C/N_0 is almost the same as that for 16QAM, it is getting lower with the C/N_0 decreases.

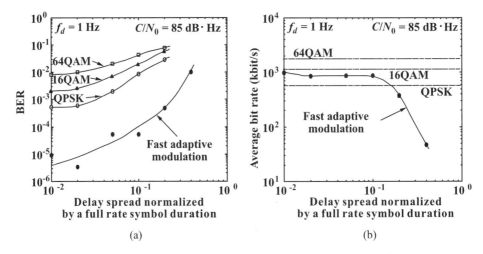

FIGURE 12.19   BER vs. delay spread and average bit rate vs. delay spread performances of the fixed rate QPSK, 16QAM, 64QAM, and the fast adaptive modulation using a chart shown in Figure 12.17 (Laboratory experiments).

Another advantage of the fast adaptive modulation is that it is robust to delay spread because the modulation parameters are also selected according to the delay spread. Figure 12.19 shows BER vs. delay spread and average bit rate vs. delay spread performances of the fixed rate QPSK, 16QAM, 64QAM, and the fast adaptive modulation. Propagation path characteristics for this experiment are 2-ray Rayleigh fading channel with its maximum Doppler frequency of 1 Hz and average C/N_0 of 85 dB·Hz. In this figure, delay spread is normalized by a symbol duration for the full rate transmission. As shown in this figure, when the normalized delay spread is lower than 0.1, BER for the fast adaptive modulation is 1 to 2 orders of magnitude lower than the fixed rate modulation schemes without reducing so much the average bit rate.

12.10 SUMMARY

This chapter has explained modulation and demodulation techniques including fading compensation techniques for wireless communication systems. Moreover, this chapter has focused on OFDM and adaptive modulation techniques including their specific techniques necessary for its application to high-bit-rate indoor/outdoor wireless communication systems. Because these techniques have a potential to flexibly satisfy various and diverged requirements of wireless multimedia communication services, they will, we believe, act an important role in the development of next-generation high-bit-rate wireless data communication systems.

REFERENCES

1. K. Murota and K. Hirade, "GMSK Modulation for Digital Mobile Radio Telephony," *IEEE Trans. Commun.*, Vol. COM-29, No. 7, pp. 1044–1050, July 1980.

2. K. Kinoshita, M. Hata, and H. Nagabuchi, "Evaluation of 16 kbit/s Digital Voice Transmission," *IEEE Trans. Commun.*, Vol. VT-33, 4, pp. 321–326, Nov. 1984.

3. Y. Akaiwa and Y. Nagata, "Highly Efficient Digital Mobile Communications with a Linear Modulation Method," *IEEE J. Selected Areas Commun.*, Vol. 5, No. 5, pp. 890–895, June 1987.

4. A. Bateman, "Feedforward Transparent Tone-in-Band: Its Implementation and Applications," *IEEE Trans. Veh. Tech.*, Vol. 39, No. 3, pp. 235–243, August 1990.

5. S. Sampei and T. Sunaga, "Rayleigh Fading Compensation for QAM in Land Mobile Radio Communications," *IEEE Trans. Veh. Tech.*, Vol. 42, No. 2, pp. 137–147, May 1993.

6. A. Salmasi and K. S. Gilhousen, "On the System Design Aspects of Code Division Multiple Access (CDMA) Applied to Digital Cellular and Personal Communication Networks," 41st IEEE Veh. Tech. Conf., St. Louis, MO, pp. 57–62, May 1991.

7. K. Honma, E. Murata, and Y. Riko, "On a Method of Constant Envelope Modulation for Digital Mobile Communications," IEEE ICC'80, pp. 24.1.1–24.1.5, June 1980.

8. Y. Akaiwa, I. Takase, S. Kojima, M. Ikoma, and N. Saegsa, "Performance of Baseband Bandlimited Multi-Level FM with Discriminator Detection for Digital Mobile Telephony," *Trans. IECE Japan*, Vol. E64, pp. 463–469, July 1981.

9. F. D. Jager and C. B. Dekker, "Tamed Frequency Modulation, a Novel Method to Achieve Spectrum Economy in Digital Transmission," *IEEE Trans. Commun.*, Vol. COM-26, No. 5, pp. 534–542, May 1978.

10. ETSI, "GSM Technical Specification," GSM 05.01–05.05, April 1998.

11. ETSI, "Digital Enhanced Cordless Telecommunications (DECT) Common Interface (CI)," Document EN 300 175-1–175-9, June 1999.

12. K. Feher, *Wireless Digital Communications, Modulation & Spread Spectrum Applications*, Prentice-Hall, Upper Saddle River, NJ, 1995.

13. S. P. Stapleton and F. C. Costescu, "An Adaptive Predistorter for a Power Amplifier Based on Adjacent Channel Emission," *IEEE Trans. Veh. Tech.*, Vol. VT-41, No. 1, pp. 49–56, Feb. 1992.

14. J. K. Cavers, "Amplifier Lineralization Using a Digital Predistorter with Fast Adaptation and Low Memory Requirements," *IEEE Trans. Veh. Tech.*, Vol. 39, No. 4, pp. 374–382, Nov. 1990.

15. M. A. Briffa and M. Faulkner, "Dynamically Biased Cartesian Feedback Linearization," 43rd IEEE Veh. Tech. Conf., Secaucus, New Jersey, pp. 672–675, May 1993.

16. S. Narahashi and T. Nojima, "Extremely Low-Distortion Multi-Carrier Amplifier—Self-Adjusting Feed-Forward (SAFF) Amplifier," IEEE ICC'91, Denver, Colorado, pp. 1485–1490, June 1991.

17. D. C. Cox, "Linear Amplification with Nonlinear Components," *IEEE Trans. Commun.*, Vol. COM-22, pp. 1942–1945, Dec. 1974.

18. S. Sampei, *Application of Digital Wireless Technologies to Global Wireless Communications*, Prentice Hall, Upper Saddle River, NJ, 1997, Chapter 11.

19. EIA, "Dual-Mode Subscriber Equipment Compatibility Specification," EIA specification IS-54, EIA project number 2215, Washington, D.C., May 1990.

20. S. Sampei, *Application of Digital Wireless Technologies to Global Wireless Communications*, Prentice-Hall, Upper Saddle River, NJ, 1997, Chapter 13.

21. T. Ue, S. Sampei, N. Morinaga, and K. Hamaguchi, "Symbol Rate and Modulation Level-Controlled Adaptive Modulation/TDMA/TDD System for High-Bit-Rate Wireless Data Transmission," *IEEE Trans. Veh. Tech.*, Vol. 47, No. 4, pp. 1134–1147, Nov. 1998.

22. K. Ohno and F. Adachi, "Performance Analysis of GMSK Frequency Detection with Decision Feedback Equalization in Digital Land Mobile Radio," *Proc. IEE*, part-F, Vol. 135, pp. 199–207, June 1988.

23. J. G. Proakis, *Digital Communications*, 2nd ed., McGraw-Hill, New York, 1989.

24. C. L. Liu and K. Feher, "4-QPSK Modems for Satellite Sound/Data Broadcast System," *IEEE Trans. Broadcast. Tech.*, Vol. 37, No. 1, pp. 1–8, Jan. 1991.

25. Y. Matsumoto, S. Kubota, and S. Kato, "A New Burst Coherent Demodulator for Microcellular TDMA/TDD Systems," *IEICE Trans. Commun.*, Vol. E77-B, No. 7, pp. 927–933, July 1994.

26. S. Sampei, *Application of Digital Wireless Technologies to Global Wireless Communications*, Prentice Hall, Upper Saddle River, NJ, 1997, Chapter 4.

27. F. Davarian, "Mobile Digital Communication via Tone Calibration," *IEEE Trans. Veh. Tech.*, Vol. VT-36, No. 2, pp. 55–62, May 1987.

28. M. K. Simon, "Dual-Pilot Tone Calibration Technique," *IEEE Trans. Veh. Tech.*, Vol. VT-35, No. 2, pp. 63–70, May 1986.

29. S. Abeta, S. Sampei, and N. Morinaga, "DS/CDMA Coherent Detection System with a Suppressed Pilot Channel," GLOBECOM'94, San Francisco, CA, pp. 1622–1626, Nov. 1994.

30. K. Fazel, S. Kaiser, P. Robertson, and M. J. Ruf, "A Concept of Digital Terrestrial Television Broadcasting," In *Wireless Personal Commun.* (Kluwer), Vol. 2, No. 1&2, pp. 9–27, 1995.

31. M. J. F.-G. Garcia, J. M. Paez-Borrallo, and S. Zazo, "Novel Pilot Patterns for Channel Estimation in OFDM Mobile Systems over Frequency Selective Fading Channels," PIMRC'99, Osaka, Japan, pp. 363–367, Sept. 1999.

32. N. Maeda, S. Sampei, and N. Morinaga, "Performance of a Sub-carrier Transmission Power

Controlled OFDM System for High Quality Data Transmission," PIMRC'99, Osaka, Japan, pp. 368–372, Sept. 1999.

33. S. Hara, K. Fukui, M. Okada, and N. Morinaga, "Multicarrier Modulation Technique for Broadband Indoor Wireless Communications," PIMRC'93, Yokohama, Japan, pp. 132–136, Sept. 1993.

34. S. Sampei, N. Morinaga, and K. Hamaguchi, "Experimental Results of a Multi-mode Adaptive Modulation/TDMA/TDD System for High Quality and High Bit Rate Wireless Multimedia Communication Systems," IEEE VTC'98, Ottawa, Canada, pp. 934–938, May 1998.

35. M. Oliver and C. Ferrer, "Overview and Capacity of the GPRS (General Packet Radio Service)," PIMRC'98, Boston, MA, pp. 106–110, Sept. 1998.

36. A. Furuskar, S. Mazur, F. Muller, and H. Olofsoon, "EDGE: Enhanced Data Rate for GSM and TDMA/136 Evolution," *IEEE Personal Commun.*, Vol. 6, No. 3, pp. 56–66, June 1999.

37. IEEE Standard for Telecommunications and Information Exchange between Systems -LAN/ MAN Specific Requirements—Part 11: Wireless Medium Access Control (MAC) and Physical Layer (PHY) Specifications: High Speed Physical Layer in the 5 GHz Band, 1999.

CHAPTER 13

Fundamentals of Multiple Access Techniques

FUMIYUKI ADACHI

13.1 INTRODUCTION

In the 21st century, major services provided by wireless communications systems will shift from simple voice communication to multimedia communication. The most important issues in wireless multiple access techniques are how flexible transmissions of various data rates can be made, and how efficiently the limited frequency resources can be utilized by as many users as possible.

Any public wireless communications system comprises one or more base stations with which users communicate. The most popular wireless system is a cellular mobile communications system, where a wide geographical area is covered with a group of base stations to serve users (area covered by each base station is called a cell). If there is a sufficient number of communication channels, that is, the number of channels equals that of the users, no sophisticated multiple access techniques are necessary. However, this is obviously frequency inefficient. With multiple access technique, the number of channels can be far fewer than that of the users and one of the channels is used when the user is in communication. In such a wireless communications system, the reverse (mobile station-to-base station) link is the multipoint-to-point communication and a multiple access technique is necessary. On the other hand, the forward (base station-to-mobile station) link is the point-to-multipoint communication and a multiplexing technique is used since no coordination among different users is necessary.

The channel structure of the wireless system and the flow of the mobile station state are illustrated in Figure 13.1. In mobile communications, two types of channels are allocated to each base station: One is the control channel for paging (a mobile) and accessing (a base station) and the other is the communication channel. In general, there are three states in the mobile station: the channel acquisition state, the communication state, and the idle state. A mobile station is in the idle state. Each mobile user who wants to initiate communication must first acquire a communication channel. A mobile user accesses a base station through the control channel to acquire the communication channel. Once the communication

Wireless Communications in the 21st Century, Edited by Shafi, Ogose, and Hattori.
ISBN 0-471-155041-X © 2002 by the IEEE.

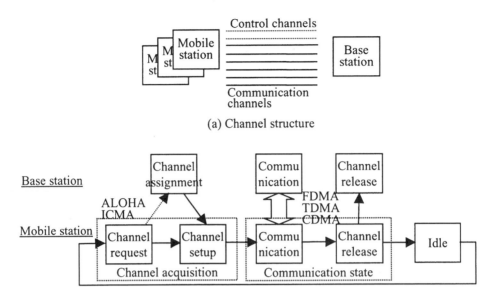

(a) Channel structure

(b) Flow of mobile station state

FIGURE 13.1   Channel structure and mobile station state.

channel is assigned by the base station, that channel is used until the communication is terminated, and then the channel released so that it can be used by another mobile station. Finally, the mobile station returns to the idle state. Different types of multiple access techniques are used in the channel acquisition state and communication state.

In this chapter, fundamentals of multiple access techniques are introduced. The multiple access techniques can be classified into two types: demand-assign-based multiple access and random multiple access techniques. The demand-assign-based multiple access or random multiple access technique is used in the communication state depending on the type of traffic, while the random multiple access technique is used in the channel acquisition state.

The cellular concept is based on frequency reuse; the same frequency is used at different spatially separated cells. In this situation, an important performance measure of wireless systems is the number of available channels per cell, which is defined here as the link capacity, for a given bandwidth. In this chapter, simple capacity equations are also given to aid in establishing a good understanding of different multiple access techniques.

13.2   MULTIPLE ACCESS TECHNIQUES

There are a number of multiple access techniques. They are classified in Figure 13.2. The type that is used depends on the type of traffic. If the data traffic is continuous and a very short transmission delay is required, for example, voice communication, demand-assign-based multiple access is applied. The family of demand-assign-based multiple access includes frequency division multiple access (FDMA), time division multiple access (TDMA), and code division multiple access (CDMA).

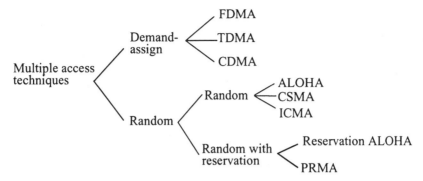

FIGURE 13.2 Multiple access techniques.

In the wireless system for voice communication, as shown in Figure 13.1, the control channel is based on random multiple access and the communication channel is based on FDMA, TDMA, or CDMA. However, demand-assign-based multiple access incurs a disadvantage in that once the channel is assigned, the channel is left idle if a user has nothing to transmit, while other users may have data waiting to be transmitted. This problem is especially acute when data generation is random and has a high peak-to-average rate. In this case, random multiple access is most efficient, in which one communication channel is shared by many users and users transmit their packets in a random or partially coordinated way. The most popular examples are ALOHA and carrier sense multiple access (CSMA). However, if the data occurs in a random manner, but the data length is quite long, then random multiple access combined with a reservation protocol should be used.

13.3 DEMAND-ASSIGN-BASED MULTIPLE ACCESS

With demand-assign-based multiple access, the channels are divided in a static fashion, and each user is allocated one or more channels by a base station during its communication irrespective of whether or not transmitted data is generated. Channels are configured in three ways using the available bandwidth: frequency, time, and code division multiplex techniques. By using the demand-assign scheme, the situation in which multiple users try to access the same channel at the same time can be avoided. How FDMA, TDMA, and CDMA utilize the frequency–time space is compared in Figure 13.3. The FDMA and TDMA channels are orthogonal (no interchannel interference). However, the CDMA channels are quasi-orthogonal and interchannel interference is produced since different users are located at different geographical locations and perfect time synchronization among the locations is not possible. This reduces the link capacity compared to FDMA and TDMA systems in the single cell case (a single base station). However, the situation completely changes in a multiple cell case. This is discussed in Sec. 5.3.3.4.

13.3.1 FDMA

The channels are configured by dividing the bandwidth into a number of frequency bands (channels) with guard bands between them. Each user is allocated a particular channel for its use. The transceiver block diagram is illustrated in Figure 13.4. In the forward link, the

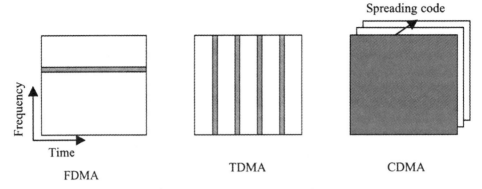

FIGURE 13.3 FDMA, TDMA, and CDMA.

base station broadcasts to the active users in a frequency division multiplex (FDM) format. In the reverse link, each active user transmits to the base station using its own assigned carrier frequency. The reverse link channels are generally not time synchronized. The base station is equipped with a number of radio frequency (RF) transceivers while a user terminal has one transceiver. In order to improve the frequency utilization, the bandwidth of each channel must be narrow. This is achieved by employing a low-bit-rate voice coding technique and a spectrum-efficient linear multilevel modulation technique, for example, Nyquist-filtered phase shift keying (PSK) and quadrature amplitude modulation (QAM) [1]. In each channel, time is divided into frames, the time length of which is related to the

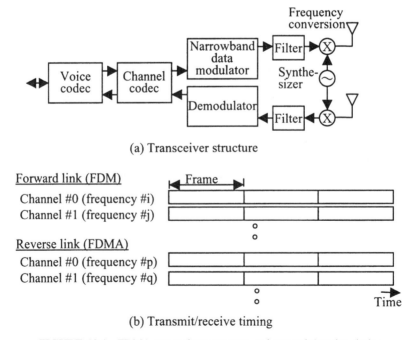

FIGURE 13.4 FDMA transceiver structure and transmit/receive timing.

voice coder and decoder (codec) design, in most cases in the range of 10 to 20 ms. The information data (or voice coder output data) to be transmitted is channel coded for error correction at the receiver and then narrowband data modulated. The transceiver requires a stable carrier frequency synthesizer and a channel selection filter with a sharp roll-off shape. For a higher rate of communication, multiple channels are assigned to a user. However, achieving multirate transmission is relatively difficult compared to TDMA and CDMA, since multiple channels are not necessarily always contiguous in frequency.

For two-way communication, a duplexing technique is necessary. In the case of frequency division duplex (FDD), one band from a pair of separated frequency bands is assigned to the forward link channel and the other is assigned to the reverse link channel. Frame timing synchronization among different channels is not required (however, all the forward link channels from a base station are synchronous). At a mobile transmitter, the frame timing of each reverse link channel is synchronized to the frame timing of the received signal on the corresponding forward link from a base station. In time division duplex (TDD), a single-frequency band is used for both the forward link channel and the corresponding reverse link channel. The FDD and TDD techniques are described in Sec. 13.3.5.

In a wireless system, the propagation channel is characterized as the sum of multiple propagation paths with different time delays produced by obstacles, for example, buildings and terrain structures; this results in a frequency- selective fading channel. The frequency transfer function of the channel is not constant but varies in time and frequency. However, if the communication channel bandwidth is sufficiently narrow, the frequency transfer function can be approximately constant over the bandwidth of each channel (this is generally the case in FDMA channels). In other words, if the data rate is sufficiently slow, the time-delay difference in different multipaths can be much shorter than the data duration. In this situation, the propagation channel is called a frequency-nonselective channel. Thus, the major impairments to communication quality or bit error rate (BER) performance are the amplitude and phase variations in time in the received signal due to fading (this fading is called time-selective fading).

13.3.2 TDMA

In TDMA, time is divided into frames as in the FDMA system; however, each frame is further divided into a fixed number of time slots. Each slot position within a frame is allocated to a different channel and this allocation is fixed. A user is allocated one or more channels (i.e., one or more slots per frame), depending on the requested transmission data rate. In any practical system, TDMA is combined with FDMA; the given frequency bandwidth is divided into a number of frequency bands, each band being used based on TDMA. Figure 13.5 illustrates a simplified structure of the base station transceiver (the duplex technique is FDD). In the forward link, the base station broadcasts to the active users in a time division multiplex (TDM) format. In the reverse link, each active user transmits to the base station only in its own assigned slots.

When users are distributed over an area and communicate with a base station, the difference in propagation times of different users cannot be neglected. All users are synchronized to the frame timing of the received signal on the forward link from the base station with which they are communicating. However, due to different propagation time delays, the regenerated frame timing for different users is different. Based on the regenerated frame timing, each user transmits its burst signal to the base station.

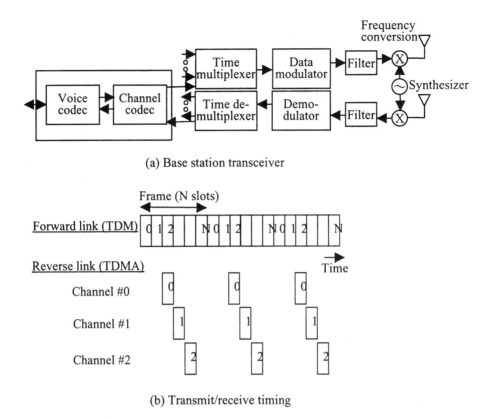

(a) Base station transceiver

(b) Transmit/receive timing

FIGURE 13.5   TDMA transceiver structure and transmit/receive timing.

Accordingly, a round trip delay of $\tau = 2d/c$ is produced, where d is the distance between a user and the base station and c is the speed of light. This is illustrated in Figure 13.6. In a TDMA system, if the guard time is shorter than the maximum round-trip delay, the transmitted burst signals may overlap at the base station receiver, and, thus the signals are not received correctly. So, the guard time determines the communication range of the

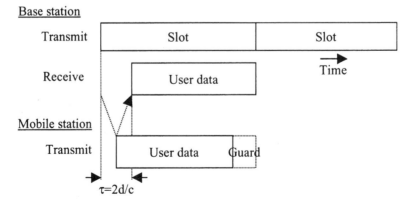

FIGURE 13.6   Transmit and receive timing difference due to propagation time.

TDMA system. Increasing the guard time to avoid the overlap decreases the spectrum efficiency. In order to decrease the guard time, the current cellular mobile communication systems adopt a timing alignment procedure. The base station transmits the timing control command to each user to advance or retard its transmission timing according to the changes in distance between the base station and the user.

The number of radio frequency (RF) transceivers at the base station is reduced by a factor of N compared to the FDMA system, where N is the number of TDMA channels on a RF carrier. Compared to the FDMA system, multirate transmission of various rates is easily accomplished by simply using multiple slots up to N slots within one RF carrier, that is, changing the transmission rate from one to N times the lowest rate is flexibly accomplished. Beyond this maximum rate, multiple carriers must be used as in the FDMA system. In the TDMA system, the RF modulation rate must be N times higher than the FDMA system. Therefore, the bandwidth of each RF carrier and the peak transmit power must be N times the FDMA system, while the average transmit power per channel is the same as in the FDMA system. A high data rate may cause severe intersymbol interference (ISI) since its modulation bandwidth is close to or wider than the channel coherence bandwidth (the propagation channel in this situation is called a frequency-selective channel). An adaptive equalization technique [2] is required to remove the ISI. A training sequence for adaptive equalization is inserted in the slot. A typical slot structure of the reverse link is illustrated in Figure 13.7.

13.3.3 CDMA

In the CDMA system, users share the same wide frequency bandwidth. Channels are configured by different spreading codes defined either by time or by frequency. Direct sequence (DS) and frequency hopping (FH) CDMA are the most popular techniques.

DS-CDMA In any practical system, DS-CDMA is combined with FDMA. The whole frequency bandwidth allocated to the system is divided into the number of frequency bands; each frequency band is used by a DS-CDMA carrier. Channels are directly spread by different spreading code sequences that have a much higher rate than the original information data rate, but they use the same carrier frequency. DS-CDMA has a number of advantages over the FDMA and TDMA systems. One advantage is that basically no management of carrier frequencies and slots is required as in the FDMA and TDMA systems. Another is related to the higher flexibility in multirate data transmissions required by multimedia communications. Data transmission of a wide range of rates is easily accomplished by using multiple spreading codes in parallel [3] or simply changing the spreading factor of a single code while keeping the chip rate the same.

The transceiver structure is illustrated in Figure 13.8. In the forward link, the base station broadcasts to the active users in a code division multiplex (CDM) format. Since all

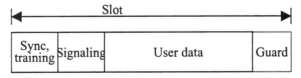

FIGURE 13.7 Time slot structure of the TDMA reverse link.

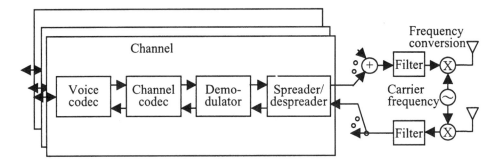

(a) Base station transceiver structure

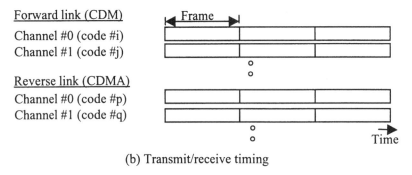

(b) Transmit/receive timing

FIGURE 13.8   DS-CDMA transceiver structure and transmit/receive timing.

forward link channels are time synchronized, orthogonal spreading code sequences can be used (hence, no interchannel interference is produced). In the reverse link, each active user transmits to the base station using its own assigned spreading code sequence. Since the reverse link channels are generally not time synchronized, perfect orthogonalization among different channels is not possible, and hence quasi-orthogonal spreading code sequences such as pseudo-noise (PN) sequences are used.

The information data (or voice coder output data) to be transmitted is first channel coded for error correction at the receiver, and then the carrier is narrowband data modulated. The data-modulated carrier is then multiplied by a spreading code sequence. This process is called spreading modulation. The spreading sequence is, in most cases, a binary sequence comprising a number of chips that can be $+1$ or -1. As an example, generating a spread signal using a spreading code sequence with the spreading factor (SF) of four is illustrated in Figure 13.9, where data modulation and spreading modulation are assumed to be both binary phase shift keying (BPSK). At the receiver, code synchronization must be accomplished first, and then the received spread signal is correlated (or despread) with a synchronously generated replica of the spreading code sequence to recover the data modulated signal. Demodulation and succeeding channel decoding are applied to recover the transmitted data.

The method for achieving multiple access in the reverse link is illustrated in Figure 13.10. Multiple users transmit their data of the bit rate W bits/second (bps) at the same

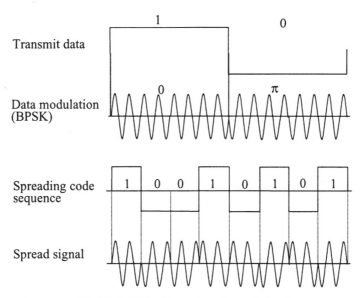

FIGURE 13.9 Generation of spread signal.

time using different spreading code sequences. When the channel coding with rate μ (<1) is used, the data-modulated carrier having the bandwidth of $\mu^{-1} W$ hertz is spread over a bandwidth of B hertz, generally $B \gg W$, where BPSK data modulation is assumed. The spreading factor (SF) of the spreading code is SF $= B/(\mu^{-1} W)$ and B/W is called the processing gain, G. The receiver can distinguish the users if the spreading code sequences have a sufficiently low cross-correlation property to each other. Correlating the received spread signal with a code sequence of the desired user de-spreads the spread signal of this user only, while the other users' spread signals remain spread over the bandwidth of B hertz. Since the power of the de-spread signal having bandwidth $\mu^{-1} W$ hertz is much larger than the other interfering users' power levels, the desired user's signal can be demodulated and channel decoded to recover the transmitted data.

The spreading code design is important because the cross correlation among the different spreading code sequences generates interchannel interference and limits the link capacity (defined as the number of supportable channels). To completely avoid inter-channel interference, orthogonal spreading code sequences such as binary Walsh functions can be used [4]. When 2^K-chip Walsh functions are used, 2^K channels can be orthogonally multiplexed on the same frequency band, which is 2^K times wider than the bandwidth before spreading. In this case, SF is constant and is 2^K. For multirate transmission, different spreading factors are necessary while keeping the chip rate (or the spreading bandwidth) the same. The orthogonal spreading code family known as orthogonal multispreading factor codes or orthogonal variable spreading factor (OVSF) codes can be used [5, 6]. Since orthogonality can only be maintained if the all-code sequences are time synchronized, it makes OVSF codes particularly suitable to the forward link. For the reverse link, since different users are geographically distributed, exact time synchronization is not possible, and hence PN sequences are generally used instead of orthogonal sequences. However, even in the reverse link, when a user wants to transmit different rates of data simultaneously, the OVSF codes can be used to code multiplex different data rates. After code multiplexing, the resultant signal is spread using a PN code sequence.

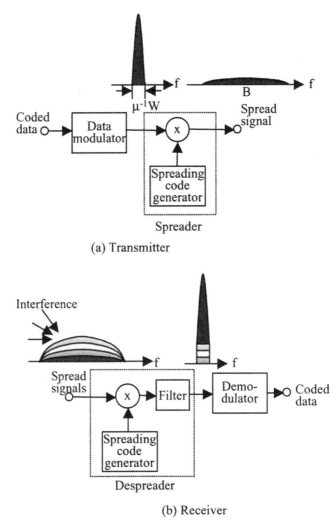

(a) Transmitter

(b) Receiver

FIGURE 13.10 Spreading and de-spreading process of DS-CDMA.

In the reverse link, interchannel interference is produced due to nonzero cross correlation between different PN code sequences. This results in the well-known near-far problem; a user closer to a base station produces a larger degree of interference to other users who are far from the base station. This problem is further worsened by the existence of fast fading and shadowing. For this reason, fast transmit power control (TPC) is indispensable on the reverse link. It regulates the difference in the instantaneous received powers. A simplified model of a closed-loop TPC is illustrated in Figure 13.11. This type of fast TPC can also be applied to the forward link channels to increase the forward link transmit power to a user close to the cell edge while decreasing the transmit power to a user near the base station. At the receiver, the received signal quality represented by the instantaneous received signal power or signal-to-interference plus noise ratio (SIR) is measured for comparison with a predetermined target value. Then, the TPC command is generated and is sent to the transmitter side. If the measured quality is less than the target

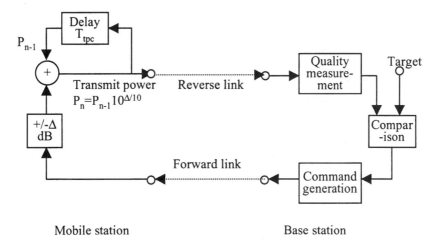

FIGURE 13.11   Closed-loop transmit power control (TPC).

value, the TPC command raises the transmitter power by the step size of Δ dB, otherwise it lowers it by Δ dB. The TPC step size, Δ, and the command interval, T_{tpc}, are important design parameters. Too large a step size raises or lowers the transmit power excessively, while too small a step size cannot track the rapid variations in the received signal power due to multipath fading. The command interval must be short enough to track the multipath fading, and approximately 1 ms is used in cellular mobile communications systems.

The frequency bandwidth of DS-CDMA is in general much wider than TDMA. When the spreading bandwidth is equal to or wider than the coherence bandwidth of the propagation channel, the propagation channel can be seen as a frequency-selective fading channel. Since the spreading code sequences have a good autocorrelation property, the de-spreading process of the receiver can resolve the frequency-selective fading channel into several frequency-nonselective fading channels having time-delay differences with the minimum of one chip duration. At the de-spreader output, multiple replicas of the transmitted data-modulated signal are obtained, each propagated along the resolved frequency-nonselective fading channel. Therefore, they are only subjected to time-selective fading as in the FDMA channels. They are time aligned and coherently combined based on the maximal-ratio combining method. This is known as the Rake receiver [7]. A Rake receiver in DS-CDMA corresponds to an equalizer in TDMA systems. Using the Rake receiver, the DS-CDMA system can reap the benefits from wideband spreading.

FH-CDMA   FH-CDMA is considerably different from DS-CDMA; channels are formed by sequences defined in the frequency domain. The transmitter and receiver structures are illustrated in Figure 13.12. The carrier frequency of the data-modulated carrier is not constant but changes periodically. The frequency-hopping pattern is generated from the code sequence. By changing the carrier frequency, FH-CDMA randomizes the adverse effects of multipath fading and co-channel interference and reduces them with the aid of channel coding and decoding. At the receiver, using a locally generated frequency-hopping pattern, the received signal is translated down to a data-modulated carrier for demodulation and succeeding channel decoding to recover the transmitted data.

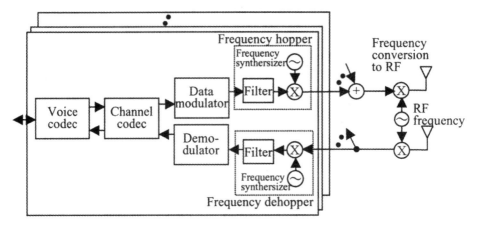

(a) Base station transceiver structure

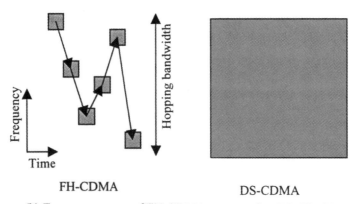

(b) Frequency usage of FH-CDMA compared to DS-CDMA

FIGURE 13.12   FH-CDMA transceiver structure and frequency usage.

The frequency occupation of an FH-CDMA system differs from a DS-CDMA system. When transmitting, DS-CDMA occupies the whole frequency band, while FH-CDMA uses only a small part of the bandwidth, however the location of the frequency part changes over time. When the hopping rate of the carrier is higher than the narrowband modulation rate, FH-CDMA is called fast FH (FFH)-CDMA. In this case, the carrier frequency changes over the data symbol period (if BPSK data modulation is used, the symbol rate is the same as the data bit rate) and the same data symbol is transmitted at different frequencies. On the other hand, when the hopping rate is slower than the data symbol rate, it is called slow FH (SFH)-CDMA. In this case, many data symbols are transmitted at the same carrier frequency. In the DS-CDMA system, the signal bandwidth is expanded (spread) by a factor of SF compared to the original bandwidth, where SF denotes the spreading factor and is equal to the number of chips per data symbol. However, the bandwidth of the FH-CDMA system depends on the range of frequency hopping.

Similarly to the DS-CDMA system, the orthogonality of the different channels can be maintained in the forward link since all channels are time synchronized. The same frequency-hopping pattern can be used for all channels, however, with a certain amount of frequency difference between each other. In the reverse link, orthogonality is possible only for the SFH-CDMA system. In the FFH-CDMA system, some transmit signal bursts may collide or be overlapped by other users' transmitted signal bursts, and hence the design of hopping patterns is important similarly to the design of the spreading code sequences in DS-CDMA.

13.3.4 Comparison of FDMA, TDMA, and CDMA

A detailed capacity comparison among FDMA, TDMA, and CDMA is quite difficult, if not impossible, because the impact of the propagation channel on the transmission performance is different for each case due to their different bandwidths, different modulation/demodulation schemes, different channel coding schemes, and the like. Here, to acquire a basic understanding of these three multiple access techniques, a simple capacity comparison is made assuming terrestrial mobile communications. We consider the case of a single cell (single base station) and then extend the scope to the cellular case (multiple base stations).

Single Cell Case   The information data rate and the channel coding rate are denoted as W bits/second and μ (<1), respectively. No overhead (frame header, guard band, etc.) is assumed. The whole frequency bandwidth allocated to the system is assumed to be B herz. Since the number of N-channel TDMA carriers is one Nth of that of the FDMA system, the number of theoretically available channels, C, is the same for both FDMA and TDMA. Assuming BPSK data modulation, C is given by

$$C_{\text{FDMA}} = C_{\text{TDMA}} = \mu G \tag{13.1}$$

where $G = B/W$. In the case of DS-CDMA, the number of available channels in the forward link is the same as above if orthogonal spreading is used. However, the reverse link channels cannot be perfectly time synchronized due to geographical distribution of users, and thus interchannel interference is produced. Assuming binary PN spreading sequences without spectrum shaping of chip waveform and assuming perfect transmit power control, C is given by [8]

$$C_{\text{CDMA}} \approx \frac{1.5G}{E_b/\eta_0} \left(1 - \frac{N_0}{\eta_0}\right) \tag{13.2}$$

where G is the processing gain and E_b/η_0 is the signal energy per information bit-to-total noise power spectrum density ratio required to achieve the required communication quality, which is often defined as the bit error rate (BER). In Eq. (13.2), $\eta_0 = I_0 + N_0$, where I_0 is the interference power spectrum density and N_0 is the background noise power spectrum density.

What is interpreted from Eqs. (13.1) and (13.2) is discussed below. Equation (13.1) indicates clear dependence of the FDMA and TDMA capacities on the coding rate μ. In DS-CDMA, μ does not appear explicitly in the capacity equation; however, channel coding plays an important role. The use of lower rate coding can decrease the value of required E_b/η_0, thereby increasing the capacity. On the contrary, the FDMA and TDMA capacities decrease as a lower coding rate is used; therefore, a high-rate channel coding is necessary to maximize the FDMA and TDMA capacities. As understood from Eq. (13.1), the FDMA and TDMA capacities are, basically, not directly related to the required BER. This is because the channels are orthogonal, and changing the received power (or transmit power) can control the communication quality. On the other hand, the DS-CDMA capacity is limited by the interference and depends on the values of required E_b/η_0. Lowering the value of required E_b/η_0 increases the capacity. Furthermore, the capacity is affected by η_0/N_0. For the given value of required E_b/η_0, the maximum capacity is obtained when $\eta_0/N_0 = \infty$. Since $\eta_0/N_0 = E_b/N_0(E_b/\eta_0)^{-1}$, this condition corresponds to the case where $E_b/N_0 = \infty$ and the channel is interference limited. However, if η_0/N_0 decreases to 10 dB, the effect of background noise cannot be neglected and the capacity decreases to 90% of its maximum. Assuming $G = 128$ and $E_b/\eta_0 = 5$ dB, $C_{\text{CDMA}} = 55.6$ when $\eta_0/N_0 = 10$ dB, while $C_{\text{FDMA}} = C_{\text{TDMA}} = 64$ if $\mu = \frac{1}{2}$.

Another important property of CDMA that is distinct from FDMA and TDMA is that its capacity can be directly increased by introducing voice- or data-activated transmission and directive antennas. If data is not always present, the transmitter can only be activated when data is present and the amount of interference is decreased. This contributes to a capacity increase in proportion to the voice or data activity ratio α (this is called the statistical multiplexing effect). This is also true for packet data transmission. Furthermore, if N_a directional antennas are used rather than an omni direction antenna, each covering a different direction with the angle width of $2\pi/N_a$ radians (see Figure 13.13), the capacity increases by roughly N_a times. Equation (13.2) is modified to

$$C_{\text{CDMA}} \approx \frac{1.5G}{E_b/\eta_0}\left(1 - \frac{N_0}{\eta_0}\right)\frac{N_a}{\alpha} \qquad (13.3)$$

In the case of FDMA and TDMA, introducing voice- or data-activated transmission is also possible. However, the implementation is not easy because users must hop from one

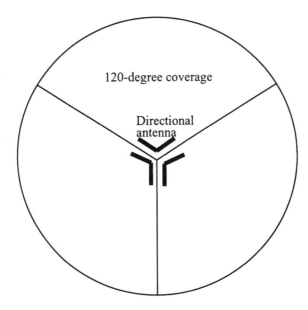

FIGURE 13.13 Use of directional antennas with $N_a = 3$, each covering $120°$.

channel to the other to obtain idle channels. Introducing directive antennas also increases the capacity, but the increase is at most only by a factor of $N_a/2$ (this is because the same carrier frequency can be used theoretically on every other antenna to avoid co-channel interference). However, when the users are in motion, frequent handoff operations (or carrier frequency change) are necessary.

Cellular Case Multiple base stations deployed to communicate with users are spatially distributed over a wide area. The area covered by a base station is called a cell. All the channels are grouped into F groups and each cell is assigned a different channel group as illustrated in Figure 13.14. F is called the cluster size. F cells that are assigned different channel groups form a cluster. The same channel groups are reused in different clusters. In this way, a wide area can be covered using a limited frequency bandwidth.

The capacity is defined as the number of available channels per cell. Unlike the single cell case, even in the FDMA and TDMA systems, the capacity is limited by the interference from co-channel cells. The capacities of FDMA and TDMA systems are given by

$$C_{\text{FDMA}} = C_{\text{TDMA}} = \frac{\mu G}{F} \tag{13.4}$$

which is one Fth of the capacity for the single cell case. For the hexagonal cell layout, F can take only limited integer numbers, that is, $1, 3, 4, 7, \ldots$ because of the constraint of $F = i^2 + j^2 + ij$, where i and j are integers; F is related to the cell radius, R, and distance, D, between the nearest co-channel cells by the following equation [9]:

$$F = \frac{1}{3}\left(\frac{D}{R}\right)^2 \tag{13.5}$$

where F is determined from the required signal-to-interference ratio (SIR), Λ, and the propagation parameters, that is, path-loss exponent β and shadowing standard deviation.

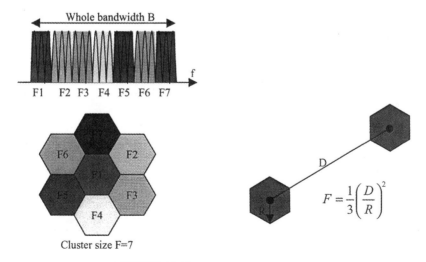

FIGURE 13.14 Frequency reuse concept.

For simplicity, let us assume that shadowing does not exist, that there are six users each in each of the neighboring six co-channel cells, and that they generate the same amount of interference to the desired cell. The worst case is each user is located at the cell edge. The SIR of the signal transmitted from the desired user is given by $(D/R - 1)^\beta$ and hence, D/R is obtained as

$$\frac{D}{R} = 1 + (6\Lambda)^{1/\beta} \tag{13.6}$$

where F is given by

$$F = \tfrac{1}{3}[1 + (6\Lambda)^{1/\beta}]^2 \tag{13.7}$$

When $\beta = 4$ and $\Lambda = 5\,\mathrm{dB}$, we obtain $F = 4$. As β decreases, the interference power decays more slowly as the distance increases, and hence the value of F becomes larger. When shadowing is taken into account, obtaining the value of F is complicated.

The CDMA system allows the use of the same carrier frequency at every cell, that is, $F = 1$. This is because the interference can be suppressed through the de-spreading process. A mobile station is connected to its nearest base station. However, as the user moves away from the base station and the received signal at the base station becomes weaker, the user is connected with multiple base stations to improve the transmission quality until the received signal at the new base station becomes sufficiently strong. This procedure is called soft handoff [4]. In the reverse link, the strongest signal received at multiple base stations is selected. On the other hand, in the forward link, the same signal is transmitted from multiple base stations and is received by the user terminal to be combined by the Rake combiner. This soft handoff is implemented in the DS-CDMA system; however, in FDMA and TDMA systems, this is quite difficult to realize because different base stations use different carrier frequencies.

In the DS-CDMA system, the interference from other cells can be viewed as that in a user's own cell. By introducing the ratio f of the interference from the other cells and that from the user's own cell [4], a similar capacity equation to the FDMA and TDMA systems can be formulated. Assuming an interference-limited condition, the capacity equation corresponding to Eq. (13.4) is given by

$$C_{\mathrm{CDMA}} \approx \frac{1.5G}{E_b/\eta_0} \frac{1}{F_{\mathrm{CDMA}}} \tag{13.8}$$
$$F_{\mathrm{CDMA}} = 1 + f$$

Thus F_{CDMA} can take any value, while F takes only limited integer numbers in the FDMA and TDMA systems. The equivalent frequency reuse factor, F_{CDMA}, of the CDMA system does not depend on the value of required E_b/η_0 but on the propagation parameters. As the value of path-loss exponent β decreases, the value of F_{CDMA} increases, thereby decreasing the capacity similarly to the FDMA and TDMA systems. If $\beta = 4$ and shadowing does not exist, $f = 0.44$ for a uniform spatial distribution of users [4] and thus $F_{\mathrm{CDMA}} = 1.44$. The capacity reduction from the single cell case is much smaller in the DS-CDMA system than in the FDMA and TDMA systems. This is a consequence of the single carrier frequency reuse.

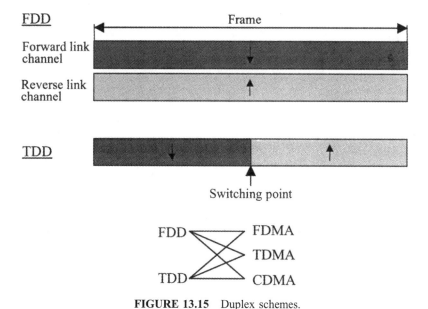

FIGURE 13.15 Duplex schemes.

13.3.5 Duplex Schemes

In two-way communication, a duplex scheme to allow simultaneous transmission and reception is necessary. This can be achieved based on either frequency division duplex (FDD) or time division duplex (TDD), as illustrated in Figure 13.15. In the FDD scheme, two separate frequency bands are prepared: One is for the forward link channel and the other is for the reverse link channel. On the other hand, in the TDD scheme, a single frequency band is used for both the forward and reverse link channels. Time is divided into two parts: One is for the reverse link channel and the other is for the forward link channel.

The FDD and TDD schemes can be combined with any multiple access technique. If the information data rate is the same for both the forward and reverse link channels, the TDD scheme requires a twofold wider channel bandwidth than the FDD scheme; however, the total bandwidth is the same for both the TDD and FDD schemes. The TDD scheme is more flexible in adapting to asymmetric traffic loads between the forward and reverse link channels. In Figure 13.15, the switching point is the middle of the frame; however, more of the time can be allocated to the forward link if its traffic load is heavier than the reverse link.

13.4 RANDOM MULTIPLE ACCESS

In the FDMA, TDMA, and CDMA systems, the channel is assigned to a user at the beginning of communication, and, once the channel is assigned, it is retained until communication is terminated. However, this demand assignment is inefficient for burst traffic such as in packet data communications. Random multiple access is used for this type of communications. Frequency efficiency is achieved through statistical multiplexing of many packets, which are randomly transmitted. In a random multiple access system, collision may occur among packets transmitted from different terminals. If the terminals involved in a collision retransmit the packets immediately after the collision, subsequent collisions will occur. To avoid this, retransmission scheduling is combined with a random multiple access technique, for example, the packet is transmitted after a random backoff time. It is generally adopted for channel acquisition in the demand-assign-based FDMA, TDMA, and CDMA systems.

13.4.1 ALOHA

ALOHA is the most popular random access technique. There are two types of ALOHA: pure ALOHA and slotted ALOHA. In pure ALOHA, a terminal transmits a packet immediately when the packet is generated. Therefore, collision occurs if two or more terminal packets overlap in time. A packet is successfully received only if other terminals do not transmit within one packet time interval from its start of transmission as illustrated in Figure 13.16. This time interval is called a vulnerable time interval. The vulnerable time interval of pure ALOHA is $2T$ seconds, where T is the packet length. Reducing the collision probability is possible by introducing some coordination. Time is divided into time slots, the length of which is equal to the packet length. All terminals are synchronized to time slots and are allowed to transmit their packets only at the beginning of each slot. This is called slotted ALOHA [10]. Collision happens only if more than one packet is generated in a vulnerable time interval of T, as illustrated in Figure 13.16.

The throughput, S, is an important performance measure of random multiple access. It is defined as the number of packets that are successfully transmitted per unit time. If the overall packet generation, including the retransmitted packets, is random, the probability of k packets generated within time interval τ is given by the Poisson distribution:

$$P_k(\tau) = \frac{(\lambda\tau)^k}{k!}\exp(-\lambda\tau) \tag{13.9}$$

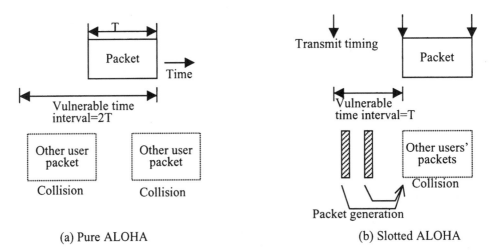

(a) Pure ALOHA   (b) Slotted ALOHA

FIGURE 13.16 ALOHA.

where λ is the overall packet generation rate per unit time. Letting $G = \lambda T$, which is called offered traffic, since collision does not occur if no packet is generated over a vulnerable time interval, the throughput is given by

$$S = \begin{cases} GP_{k=0}(2T) = G\exp(-2G) & \text{for pure ALOHA} \\ GP_{k=0}(T) = G\exp(-G) & \text{for slotted ALOHA} \end{cases} \tag{13.10}$$

The maximum throughput is $S_{\max} = 1/2e\ (=0.184)$ for the pure ALOHA system and $1/e\ (=0.368)$ for the slotted ALOHA system.

13.4.2 Carrier Sense Multiple Access

Packet collision is avoided if ongoing transmission is known to each terminal. Each terminal first listens to the channel before transmission and if the terminal senses that the channel is idle, the terminal transmits a packet, otherwise the terminal waits until the channel becomes idle. This is called carrier sense multiple access (CSMA). There are three types of CSMA: 1-persistent CSMA, nonpersistent CSMA, and p-persistent CSMA. In 1-persistent CSMA, if the terminal senses that the channel is busy, the terminal keeps probing the channel and transmits as soon as the channel state changes from "busy" to "idle." This leads to a collision if two or more packets are accumulated during a busy period. In nonpersistent CSMA, if the terminal senses that the channel is busy, the terminal probes the channel again after a random period of time. In p-persistent CSMA, time is slotted with the maximum propagation time delay. If the terminal senses that the channel is idle, a terminal transmits a packet with probability p or defers transmission until the next time slot with probability $(1-p)$ and probes the channel again. By choosing a small p value, the probability of more than one terminal initiating transmission when the channel becomes idle is minimized. The 1-persistent CSMA scheme is a special case of p-persistent CSMA.

Due to propagation time delay, carrier sensing is not ideal. The transmission from another terminal can be detected only after the propagation time delay from the initiation of a terminal's transmission. In other words, even if a terminal senses that the channel is idle, there is a possibility that another terminal has already initiated transmission of its packet and thus collision occurs. This is illustrated in Figure 13.17. The throughputs of nonpersistent and 1-persistent types of slotted CSMA are given by [11]:

$$S = \begin{cases} \dfrac{aG\exp(-aG)}{1-\exp(-aG)+a} & \text{for nonpersistent slotted CMSA} \\[4mm] \dfrac{G\exp[-G(1+a)][1+a-\exp(-aG)]}{(1+a)[1-\exp(-aG)]+a\exp(-(1+a)G)} & \text{for 1-persistent slotted CMSA} \end{cases} \tag{13.11}$$

where a is the propagation time delay normalized by the packet length. In Figure 13.18, the throughput-offered traffic performance is compared among the pure ALOHA system, slotted ALOHA system, nonpersistent CSMA system, and 1-persistent CSMA system. The maximum throughput of CSMA is larger than that of ALOHA but decreases as the propagation time delay increases.

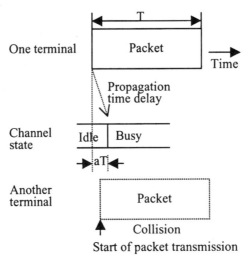

FIGURE 13.17   Collision in CSMA.

After a packet is successfully received, the receiving station transmits an acknowl-edgment packet. A terminal learns of a collision by the absence of an acknowledgment packet. In the CSMA system, collision can occur only if contending transmission starts within the propagation time delay after the start of transmission. Since the propagation time delay is generally much shorter than the packet length, that is, $a \ll 1$, aborting the collided transmission as soon as collision is detected can increase the throughput. In CSMA with collision detection (CSMA/CD), a transmitting terminal continues to probe the channel and if a collision is detected, it aborts the transmission. CSMA/CD is the most popular random-access technique for use as the media access protocol (MAC) of local area networks (LANs).

13.4.3   Idle Signal Casting Multiple Access

If the propagation time delay is short, CSMA provides significant improvement over ALOHA and is considered the ideal random multiple access technique for wireless packet networks. In the CSMA system, each terminal must be able to detect the transmissions of all other terminals. However, not all packets transmitted from different terminals can be sensed or terminals may be hidden from each other by buildings or some other obstacles. For example, both terminals i and j attempt to transmit to another terminal k; however, if terminals i and j cannot listen to each other, a collision may occur. This problem is known as the *hidden terminal problem*, which severely degrades the throughput of the CSMA system. Busy tone multiple access (BTMA) [12] is one solution to avoid this problem. When terminal k has detected a transmission, it transmits a busy tone in a separate busy-tone channel. All terminals within its communication range will receive the busy tone and refrain from transmission. Even if terminals i and j are hidden from each other, they can listen to each other in an indirect way, that is, by the way of busy-tone channel. This concept can be applied to a wireless communication system in which geographically distributed mobile terminals communicate with a base station.

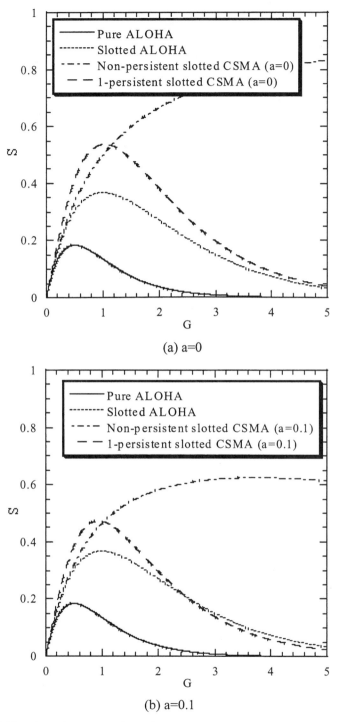

(a) a=0

(b) a=0.1

FIGURE 13.18 Throughput comparison of pure ALOHA, slotted ALOHA, nonpersistent CSMA, and 1-persistent CSMA.

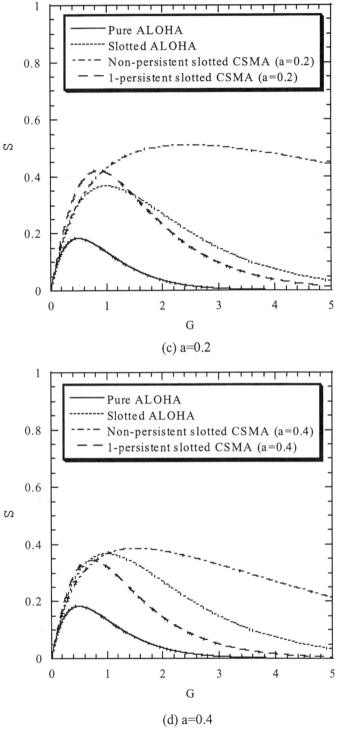

(c) a=0.2

(d) a=0.4

FIGURE 13.18 (*continued*)

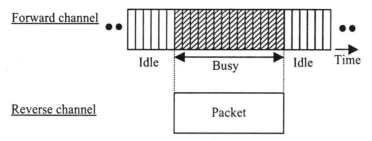

FIGURE 13.19 ICMA.

The idle signal casting multiple access (ICMA or ISMA) system [13] transmits an idle/busy signal from the base station to indicate the absence or the presence of another terminal's transmission. This is illustrated in Figure 13.19. When the base station starts to receive a packet from a terminal, it transmits a busy signal until the end of the transmission of that terminal on the forward link channel to stop the transmission from other terminals. When the base station receives no packet, it transmits an idle signal. A terminal having a packet to transmit waits until the idle signal is received. The ICMA and CSMA schemes are basically the same. In the CSMA system, each terminal must listen to all other terminals, while in the ICMA system, each terminal is informed from the base station of the other terminal's transmission. Similarly to the CSMA scheme, there are nonpersistent ICMA and 1-persistent ICMA schemes [14].

13.4.4 Random Multiple Access with Reservation

If the data traffic is continuous, the demand-assign-based multiple access techniques, FDMA, TDMA, and CDMA, can be used. At the beginning of communication, a terminal must acquire a communication channel. Channel acquisition is performed through a control channel based on random multiple access. On the other hand, if data traffic is bursty, random multiple access is used for communication. However, there is a case in which the data traffic is bursty, but a terminal has a sequence of packets. In this case, packet-by-packet pure random multiple access is not efficient. Time needed for successful transmission of all packets becomes long because all packets must be contended. To avoid this problem, only the first packet is transmitted based on pure random multiple access for channel reservation. If it is successful, the channel is reserved for that terminal and the rest of the packets can be transmitted without any contention. This is illustrated in Figure 13.20. After transmitting all packets, the terminal returns the reserved channel. The reservation can be based on the slotted ALOHA scheme. The difference between random multiple access with reservation and the demand-assign-based multiple access is that the same channel is used for both transmissions of the reservation request packet and communication packets.

13.4.5 Packet Reservation Multiple Access for Mixed Voice and Random Data Traffic

Wireless voice communication systems are generally based on demand-assign-based multiple access. However, with demand-assign-based multiple access, once communica-

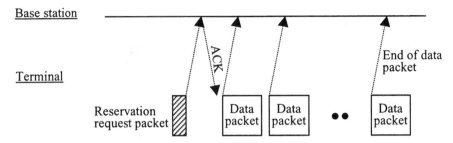

FIGURE 13.20 Random multiple access with reservation.

tion channel is assigned to a terminal, the channel is continuously occupied although there is no data. Packet reservation multiple access (PRMA) [15] allows a variety of information sources to share the same communication channel and obtain a statistical multiplexing effect. Speech terminals are classified as "periodic" and certain data terminals as "random." In the PRMA system, time is divided into frames, each of which consists of a fixed number of time slots. For the voice terminals, voice activity detection is adopted. The voice signal comprises a sequence of talk spurts. At the beginning of a talk spurt, the terminal transmits the first packet based on the slotted ALOHA. Once the packet is transmitted successfully, that terminal is allowed to use the same time slots in the succeeding frames (reservation is made). The reservation is kept until the end of the talk spurt. The status "reserved" or "unreserved" of each slot is broadcast from the base station. If the first packet is not successful within 32 ms, the terminal discards this packet and waits until the next slot for transmission of the next packet for reservation. The PRMA is different from other random multiple access with reservation, it tries until all the packets are successfully transmitted. Because of voice communication, the transmission delay is an important factor and must be within a certain delay to keep the voice communication quality at an acceptable level. On the other hand, a terminal of random data transmits packets in unreserved time slots.

13.4.6 Capture Effect

In general, random multiple access techniques are designed based on the assumption that if two or more contending packets are transmitted at the same time (overlapped in time), they collide and none of them are successfully received. However, in a wireless communication system, due to the random nature of the propagation environment and geographical distribution of terminals, the packets transmitted from different terminals are subjected to multipath fading, shadowing, distance-dependent path loss, and the like. Hence, packets are not necessarily received with equal powers by a base station. Owing to this power difference among the received packets, the receiver can capture the strongest of the overlapping packets. This is called the capture effect [16, 17]. The throughput improvement by capture depends on the statistical properties of the received powers of contending packets and hence, strongly depends on the propagation parameters.

13.4.7 Combination with Demand-Assign-Based Multiple Access

Any random multiple access technique can be combined with any demand-assign-based multiple access technique. As in Figure 13.1, when a terminal enters an active state, it requests a channel from a base station. The base station assigns one of multiple channels based on FDMA, TDMA, and CDMA. The assigned channel is shared by a group of active terminals. The terminal stays in the group until it leaves the active state. The channel assignment acts as admission control so that the channel does not become overloaded. During the active state, terminals in the same group access a base station based on a random access technique.

13.5 SUMMARY

The most important issues in wireless multiple access techniques are how flexible transmissions of various data rates can be made, and how efficiently the limited frequency resources can be utilized by as many users as possible. In this chapter, fundamentals of the multiple access techniques are introduced. The multiple access techniques are classified into two types: demand-assign-based multiple access and random multiple access techniques. The family of demand-assign-based multiple access includes FDMA, TDMA, and CDMA, and the family of random multiple access includes ALOHA, CSMA, ICMA, and PRMA. The type that is used depends on the type of traffic. If the data traffic is continuous and a very short transmission delay is required, for example, voice communication, demand-assign-based multiple access is applied. The demand-assign-based multiple access is inefficient for burst traffic such as in packet data communications. Random multiple access is used for this type of communications. However, any random multiple access technique can be combined with any demand-assign-based multiple access technique.

REFERENCES

1. W. T. Webb and L. Hanzo, *Modern Quadrature Amplitude Modulation*, IEEE Press, 1994.

2. J. G. Proakis, *Digital Communications*, McGraw-Hill, New York, 1989.

3. F. Adachi, K. Ohno, A. Higashi, T. Dohi, and Y. Okumura, "Coherent Multi-Code DS-CDMA Mobile Radio Access," *IEICE Trans. Commun.*, Vol. E79-B, pp. 1316–1325, Sept. 1996.

4. A. J. Viterbi, *CDMA: Principles of Spread Spectrum Communication*, Addison-Wesley, Reading, MA, 1995.

5. F. Adachi, M. Sawahashi, and K. Okawa, "Tree-Structured Generation of Orthogonal Spreading Codes with Different Lengths for Forward Link of DS-CDMA Mobile Radio," *IEE Electron. Lett.*, Vol. 33, pp. 27–28, Jan. 1997.

6. K. Okawa and F. Adachi, "Orthogonal Forward Link Using Orthogonal Multi-Spreading Factor Codes for Coherent DS-CDMA Mobile Radio," *IEICE Trans. Commun.*, Vol. E81-B, pp. 777–784, April 1998.

7. M. Schwarz, W. R. Bennett, and S. Stein, *Communication Systems and Techniques*, McGraw-Hill, New York, 1966.

8. R. L. Pickholtz, L. B. Milstein, and D. L. Schilling, "Spread Spectrum for Mobile Communications," *IEEE Trans. Veh. Tech.*, Vol. VT-40, pp. 313–322, May 1991.

9. W. C. Jakes, Jr., Ed., *Microwave Mobile Communications*, Wiley, New York, 1974.

10. L. G. Roberts, "ALOHA Packet System with and without Slots and Capture," *Comput. Commun. Rev.*, Vol. 5, pp. 28–42, Apr. 1975.

11. L. Kleinrock and F. A. Tobagi, "Packet Switching in Radio Channels: Part I—Carrier Sense Multiple-Access Modes and Their Throughput-Delay Characteristics," *IEEE Trans. Commun.*, Vol. COM-23, pp. 1400–1416, Dec. 1975.

12. F. A. Tobagi and L. Kleinrock, "Packet Switching in Radio Broadcast Channels: Part II—The Hidden Terminal Problem in Carrier Sense Multiple-Access and the Busy Tone Solution," *IEEE Trans. Commun.*, Vol. COM-23, pp. 1417–1433, Dec. 1975.

13. A. Murase and K. Imamura, "Idle-Signal Casting Multiple Access with Collision Detection (ICMA-CD) for Land Mobile Radio," *IEEE Trans. Veh. Tech.*, Vol. VT-36, pp. 45–50, Feb. 1987.

14. F. Adachi, K. Ohno, and M. Kitagawa, "Performance Analysis of ICMA/CD Multiple-Access for a Packet Mobile Radio," *Electron. Lett.*, Vol. 24, pp. 469–470, Apr. 1988.

15. D. J. Goodman, R. A. Valenzuela, K. T. Gayliard, and B. Ramamurthi, "Packet Reservation Multiple Access for Local Wireless Communications," *IEEE Trans. Commun.*, Vol. COM-37, pp. 885–890, Aug. 1989.

16. D. J. Goodman and A. A. M. Saleh, "Near/Far Effect in Local ALOHA Radio Communications," *IEEE Trans. Veh. Tech.*, Vol. VT-3, pp. 19–27, Feb. 1987.

17. J. C. Arnbak and W. van Blitterswijk, "Capacity of Slotted ALOHA in Rayleigh-Fading Channels," *IEEE J. Selected Areas Commun.*, Vol. SAC-5, pp. 261–269, Feb. 1987.

CHAPTER 14

Spatial-Temporal Signal Processing for Broadband Wireless Systems

DAVID FALCONER

14.1 INTRODUCTION: MOTIVATION AND CONFIGURATIONS FOR SPACE–TIME PROCESSING

Broadband digital multimedia communications is a fast-growing segment of the total traffic that will be carried by third-generation cellular and other wireless communications systems. Rapid growth and increasing demands for bandwidth and near-ubiquitous coverage, combined with the performance usually associated with wired or fibered systems pose difficult challenges for wireless system designers. Each broadband data signal (usually defined as being equal or greater than 2 Mb/s) will occupy a relatively large portion of the overall allocated system bandwidth; thus necessitating efficient frequency reuse among different users in the same small area. The use of "smart antennas"— generally consisting of arrays of antenna elements together with associated signal processing to control and combine the elements—can effectively meet many of these design challenges [1]. Adaptive antenna arrays [2–5], at either the transmitting end, the receiving end, or both, can reduce or eliminate the effects of fading and multipath delay spread. As well, by increasing the effective antenna aperture, they can increase the antenna gain. With sufficiently large separation among elements, they can counteract the effects of large-scale signal variations (shadowing). Through their interference and fading mitigation properties, adaptive arrays can also provide a useful complement to equalization and coding techniques, for example, reducing a code's minimum interleaving requirements in slowly fading channels [6].

A receiving antenna array comprises a set of individual antenna elements arranged in a two- or three-dimensional pattern, whose outputs are combined. Selection, equal gain, and maximal ratio combining are well-known approaches to providing diversity protection against fading and to minimize the effects of delay spread caused by multipath. In this chapter we are concerned with more general spatial processing, which can also reduce or eliminate interference from signals of other users of the same cellular communications

Wireless Communications in the 21st Century, Edited by Shafi, Ogose, and Hattori.
ISBN 0-471-155041-X © 2002 by the IEEE.

system. Since the capacity of cellular systems is mainly determined by their ability to withstand co-channel and adjacent channel interference, smart antenna arrays can have a direct and positive impact on system capacity. Spatial processing can also be combined with temporal processing such as time-domain filtering or equalization [7–10]. Advances in the technologies of digital signal processing, antenna fabrication, and radio front ends, which are paving the way toward "software radios," are also making relatively sophisticated spatial-temporal processing feasible and practical—at the transmitter, the receiver, or both. Ultimately *space division* multiple access (SDMA) is possible—in which different users' signals in the same vicinity, and using the same frequency band, can coexist without excessive interference.

Spatial processing (which can be thought of as spatial filtering) can shape an antenna pattern so as to emphasize desired signals and null out undesired signals (such as interferers or troublesome multipath components). It does this, for example, by appropriately weighting and combining the outputs of the individual antennas in the array. We shall see that spatial processing is complementary to temporal processing, which can eliminate or minimize intersymbol and/or co-channel interference from sampled channel responses. Spatial and temporal processing both separate and process signals based on differences among their "signatures" in space and time.

14.2 CHANNEL MODELS FOR MULTIELEMENT ARRAYS

A model, showing radio channels between a single transmitting antenna, designated i, and an array of N receiving antennas, is shown in Figure 14.1. The model assumes the use of a receiver and downconverter at the output of each receiving antenna element. The N-dimensional vector representing the complex baseband outputs of the receiving array is

$$\mathbf{r}_i(t) = \sum_l a_{il} \mathbf{h}_i(t - lT) + \mathbf{v}(t) \tag{14.1}$$

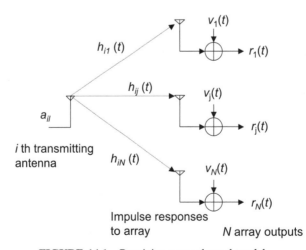

FIGURE 14.1   Receiving array channel model.

where $\{a_{il}\}$ are complex-valued data symbols transmitted at time lT from antenna i, $\mathbf{h}_i(t)$ is a complex vector impulse response from antenna i to the array, and $\mathbf{v}(t)$ is a vector representing additive noise.

The vector $\mathbf{h}_i(t)$ is also called the spatial-temporal signature [11] of the radio link between transmitter i and the array. For K transmitters ($i = 0, 1, 2, \ldots, K - 1$),

$$
\mathbf{r}(t) = \sum_{i=0}^{K-1} \sum_l a_{il} \mathbf{h}_i(t - lT) + \mathbf{v}(t)
$$

$$
= \sum_l \mathbf{h}(t - lT)\mathbf{a}_l + \mathbf{v}(t) \tag{14.2}
$$

where $\mathbf{h}(t) = [\mathbf{h}_0(t), \mathbf{h}_1(t), \ldots, \mathbf{h}_{K-1}(t)]$ is a N by K matrix channel impulse response, and

$$
\mathbf{a}_l = \begin{bmatrix} a_{0l} \\ a_{1l} \\ \vdots \\ a_{K-1,l} \end{bmatrix} \tag{14.3}
$$

The components of the vector impulse responses $\mathbf{h}_i(t)$ explicitly describe the properties of the radio channels between the ith transmitter and the array. Corresponding to the matrix impulse response $\mathbf{h}(t)$ is the matrix frequency response $\mathbf{H}(f)$. Note that the data symbols from different transmitters $\{a_{il}\}$ are assumed to be transmitted at the same symbol rate $1/T$, but are otherwise not necessarily synchronized. Small differences among symbol rates from different transmitters could be accounted for by corresponding slow time variations in the $\{\mathbf{h}_i(t)\}$, as long as the *average* symbol rates remain identical.

In the corresponding physical model, each radio channel in Figure 14.1 may represent the superposition of one or more physical radio paths with different delays and complex gains: a combination of direct (line-of-sight) paths, and/or reflections from scattering objects in the vicinity of the transmitter or receiver, and/or diffraction around obstacles. The signal from a given source may thus arrive at a given receiving antenna element from several different directions, with different delays and gains; the distribution of angles of arrival of a signal is called its angular spread. A large angular spread is generally found when the scattering/reflecting objects are close to or surrounding the receiving antenna. In this case, the outputs of receiving elements spaced by about $\lambda/2$ (where λ is the wavelength) will be nearly uncorrelated [12]. A small angular spread will require a larger spacing to achieve small correlation and is associated with environments where the scattering/reflecting objects are far from the receiver and subtend only a small angle from it. For a small angular spread, a correlation coefficient of less than 0.5 to 0.7 requires a minimum interelement spacing of roughly [12] $1.9\lambda/(\pi$ (angular spread in radians). The arrival of multiple versions of the same signal (multipath components) with different delays causes intersymbol interference if the delay differences are on the same order as a symbol interval or more.

Transmission over multiple uncorrelated channels provided by spatial or other forms of diversity is well known as an effective antidote to fading and multipath delay spread. Fading mitigation by spatial diversity is still effective even for correlation coefficients among diversity channels as high as 0.5 [12, 13].

Recently, there has been a renewed interest in spatial channel modeling as a result of the interest in applying spatial-temporal processing of the types discussed here. A number of analytical and measurement-based models have been developed [14] for correlation, spatial signature variation [11], and other statistical properties of signals received at antenna array elements. These models are useful for analyzing the performance and for simulating systems that use antenna arrays. In general, an array's ability to combat multipath and/or fading is maximized if the correlation among elements is as low as possible. For effective separation of interfering signals, desired and interfering signals should have very different spatial-temporal signatures, either because they have quite different directions of arrival or because their respective angular spreads, seen from the receiving array, are large and different.

14.3   RECEIVER SPACE–TIME PROCESSING

14.3.1   Linear Space–Time Processing

Optimal array-based multiuser detection receivers for minimizing error probability in the presence of Gaussian noise, channel dispersion, and interference, can be shown to include linear-processing front ends in the form of generalized spatial-temporal matched filter structures, followed by sampling, and nonlinear (generally exponentially complex) decision making [15]. Less general, but more practical, receiver structures are based on linear processing, together with sampling and decision making by simple quantization. Figure 14.2 shows a general linear space–time processing scheme for making a decision on the nth data symbol a_{0n}, from transmitter 0. The complex baseband output of each of N antenna array elements is fed to linear filters represented by a vector impulse response $\mathbf{w}_0(t)$, whose outputs are summed and sampled at the symbol rate (at times $\{0, T, 2T, \dots, nT\}$. The array elements are usually omnidirectional; but directional elements may be used as well, providing coverage over partially overlapping directional beam patterns fanning out from the receiver. In such a case, the array may be considered to provide a kind of direction-of-arrival sampling. Reference [16] shows that time and space

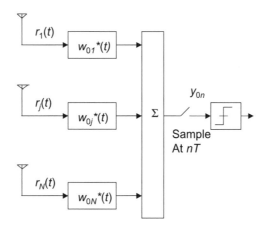

FIGURE 14.2   General linear space–time processor.

filtering and sampling operations provide a canonical space–time reception processing in terms of a fixed basis that is independent of the channel parameters.

In the general linear space–time receiver of Figure 14.2, the output on which the decision is based is

$$
y_{0n} = \int_{-\infty}^{\infty} \mathbf{w}_0(\tau)^H \mathbf{r}(nT - \tau)\, d\tau
$$

$$
= \sum_l \int_{-\infty}^{\infty} \mathbf{w}_0(\tau)^H \mathbf{h}(nT - lT - \tau)\mathbf{a}_l\, d\tau + \int_{-\infty}^{\infty} \mathbf{w}_0(\tau)^H \mathbf{v}(nT - \tau)\, d\tau \quad (14.4)
$$

Zero-Forcing Criterion [17] Under the zero-forcing criterion, y_{0n} should have no interference from other transmitters and no intersymbol interference; that is, $y_{0n} = a_{0,n-D} + v_{0n}$, where D is an integer delay chosen to make the filters $\{w_{ik}(t)\}$ causal, and

$$
v_{0n} = \int_{-\infty}^{\infty} \mathbf{w}_0(\tau)^H \mathbf{v}(nT - \tau)\, d\tau \quad (14.5)
$$

Henceforth we set $D = 0$ for notational simplicity (allowing noncausal filters). For additive white Gaussian noise with power spectral density $N_0/2$, the mean and variance of v_{0n} are, respectively, zero and

$$
N_0 \int_{-\infty}^{\infty} |\mathbf{w}_0(\tau)|^2\, d\tau = N_0 \int_{-\infty}^{\infty} |\mathbf{W}_0(f)|^2\, df. \quad (14.6)
$$

The condition for zero-forcing the co-channel and intersymbol interference is that the K-dimensional overall vector response $\mathbf{q}(t)$ satisfies

$$
\mathbf{q}(nT)^T \equiv \left[\int_{-\infty}^{\infty} \mathbf{w}_0(t)^H \mathbf{h}(nT - t)\, dt \right]^T = \begin{cases} [1, 0, 0, \ldots, 0] & \text{for } n = 0 \\ [0, 0, 0, \ldots, 0] & \text{for } n \neq 0 \end{cases} \quad (14.7)
$$

Taking Fourier transforms over the index n, we get the equivalent equation in the frequency domain

$$
\sum_m \mathbf{W}_0\left(f + \frac{m}{T}\right)^H \mathbf{H}\left(f + \frac{m}{T}\right) = [T, 0, 0, \ldots, 0] \quad (14.8)
$$

where $\mathbf{W}_0(f)$ and $\mathbf{H}(f)$ are the Fourier transforms of $\mathbf{w}_0(t)$ and $\mathbf{h}(t)$, respectively. If the channels and/or frequency responses are strictly bandlimited to say $|f| \leq B/T$, there are at least $\text{int}(2B)$ nonzero terms in the summation for any frequency f [and no more than $1 + \text{int}(2B)$], where $\text{int}(x)$ denotes the largest integer equal or less than x. Thus for any frequency f, the above matrix equation represents a system of K linear equations in $\text{int}(2B)$ unknown vectors $\mathbf{W}_0(f + m/T)$. Since each of these vectors has N components, there are a total of at least $N \, \text{int}(2B)$ unknowns in the K linear equations. A unique solution for any frequency f may be found, assuming the K columns of the matrix $\mathbf{H}(f)$ are linearly independent, if the number of interfering transmitters, K, satisfies

$$K = N \, \text{int}(2B) \tag{14.9a}$$

Figure 14.3 shows spectra with excess bandwidths of 0, 100, and 200%, and their interference suppression capabilities, corresponding to $B = \frac{1}{2}$, 1, and $\frac{3}{2}$, respectively. Conversely, there is no solution for a larger number of interfering transmitters, and there are in general an infinite number of possible solutions for a smaller number, that is, for

$$K < N \, \text{int}(2B) \tag{14.9b}$$

As a special case, we have the familiar result that for $B = \frac{1}{2}$ (the minimum Nyquist bandwidth), $N = 1$ (one antenna), and $K = 1$ (no interferers), a unique equalizer frequency response satisfying Eq. (14.8) is simply T times the inverse of the channel, $H_0(f)^{-1}$. In general, Eq. (14.9a) implies that the number of simultaneous users whose mutual interference can be mitigated by the system of Figure 14.2 is proportional to the number of antenna elements and also to the system's excess bandwidth. Interference-suppression capabilities for code division multiple access (CDMA) systems with finite numbers of filter taps have been obtained in [18].

The ability of $N + 1$ antenna elements to separate up to $N + 1$ users was pointed out by Winters and co-workers [19–21]. See also [22, 23]. The ability of a linear equalizer to suppress a number of interferers proportional to excess bandwidth was foretold in a study by Shnidman [24]. See also [25, 26]. That the number of separable (orthogonal) signals is

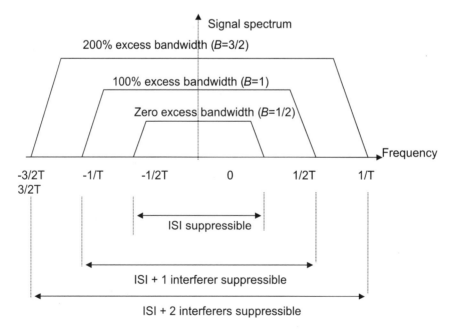

FIGURE 14.3 Illustration of the effect of excess bandwidth on interference suppression capability of a linear space–time processor.

proportional to the bandwidth/symbol rate ratio is a well-known result of signal theory and is the basis for spread spectrum CDMA multiple access schemes, as well as for frequency division multiple access (FDMA) and time division multiple access (TDMA) signal separation schemes for multiple access. In wireless systems, however, orthogonality among different users' transmitted signals is often lost as a result of channel multipath or frequency or time slot reuse.

What is the best way to maximize the system capacity, that is, the number of interfering users for which signal separation is possible: by increasing the number of antenna elements or by increasing the excess bandwidth? Increasing the number of antennas is clearly preferable since bandwidth is a finite resource. However, there are often physical and cost constraints that limit the number of antennas at a site. Thus the ability of linear equalization, coupled with excess bandwidth, to augment the interference suppression capability of an antenna array is very valuable. For TDMA and FDMA systems, the excess bandwidth is usually moderate (B is typically in the range of 0.5 to 1 for these systems), and thus temporal linear processing is often more useful for equalization than for interference suppression. However, temporal processing is very powerful against interference in direct sequence CDMA systems, where B is significantly greater than 1; for example, a binary CDMA system with a spreading gain of 32 would have $B = 16$, and up to 32 interferers per antenna element would theoretically be suppressible. However, these theoretical limits are only attainable if the set of Eq. (14.8) is nonsingular for any f. Furthermore, even if a zero-forcing solution exists, its filtering operation may cause significant enhancement of additive thermal noise. In practice, noise enhancement and digital signal processing and adaptation considerations limit the number of interferers that can be effectively suppressed to something on the order of half to three quarters of the maximum theoretical number. Nevertheless the effectiveness of linear filtering in removing interference from received CDMA signals leads to significant gains in system capacity and to reduced sensitivity to near-far interference effects. Furthermore, as will be seen later, supplementing temporal processing with spatial processing in CDMA systems furthers these benefits.

Before leaving the issue of theoretical capabilities of linear combining systems to combat multipath intersymbol interference (ISI), co-channel interference, and noise, we consider a common special case of linear combining in which the filters $\{w_{0k}(t)\}$ are replaced by memoryless complex weights $\{w_{0k}\}$; that is, the only "memory" available to the receiver results from any relative time delays in signals reaching the K antenna elements. It may appear that there is no capability to combat intersymbol interference in this case. However, that is not necessarily true, as shown by Clark and co-workers [27, 28]. The sampled array output is expressed, analogous to (14.7) as

$$\mathbf{q}(nT)^T = \mathbf{w}_0^H \mathbf{h}(nT) = \begin{cases} [1, 0, 0, \ldots, 0] & \text{for } n = 0 \\ [0, 0, 0, \ldots, 0] & \text{for } n \neq 0 \end{cases} \qquad (14.10)$$

If the $\mathbf{h}_i(t)$ are all time limited to say $|t| \leq ST$, then there are $2S + 1$ nonzero samples of each $\mathbf{h}_i(t)$, and (14.10) represents a set of $(2S + 1)K$ equations in the N unknowns $w_{01}, \ldots, w_{0N}$. Thus the ISI and interference can be eliminated by this "memoryless" array, provided that

$$N \geq (2S + 1)K \qquad (14.11)$$

Thus the required number of antenna elements is proportional to the time duration of the multipath impulse responses and to the total number of users. Clark et al. [28] showed numerical results that demonstrated the ability of memoryless arrays to effectively mitigate intersymbol interference.

Minimum Mean-Squared Error Criterion   The above zero-forcing results illustrate the ability of linear equalizers and/or linear spatial combiners to completely suppress ISI and co-channel interference up to limits imposed by bandwidth, number of antenna elements, and number of interferers. However, noise enhancement and adaptation can be problematic for receivers based on zero forcing. Minimization of the total mean-squared error (MSE), consisting of thermal noise and residual ISI and co-channel interference at the equalizer or combiner output, is usually a more useful criterion for evaluation and adaptation purposes. The analytical problem can be formulated as minimization of the MSE of one user, say user "0":

$$\text{MSE} = E|y_{0n} - a_{0n}|^2$$

The optimum set of linear combining filters turns out to be representable as a bank of filters matched to the components of the $\{\mathbf{h}_k(t)\}$ channel responses, whose outputs are sampled at T second intervals, and routed to and subsequently combined by sets of transversal filters [25]. A general expression for the minimum total MSE for an N-input, N-output linear system was derived by Salz [23]. A related zero-forcing problem, minimizing the output noise variance under the zero-forcing constraint (14.8) for a K-input, K-output linear system was solved by Van Etten [22]. In all of these results, no constraint was placed on the memory or complexity of the filter or filters $w_k^{(l)}(t)$. Practical equalizers

$$\mathbf{w} = \begin{bmatrix} \mathbf{w}_{-M} \\ \vdots \\ \mathbf{w}_M \end{bmatrix}$$

based on transversal filters, have finite numbers of tap coefficients, and, following fixed antialiasing filters, operate at sampling rates that equal the symbol rate or a multiple of it. For a linear equalizing combiner, with say $2M + 1$ taps per antenna element, and with a sampling rate of $1/\Delta$, the output corresponding to Eq. (14.4) is

$$y(iT) = \mathbf{w}^H \mathbf{r}_i \qquad (14.12)$$

where

$$\mathbf{w} = \begin{bmatrix} \mathbf{w}_{-M} \\ \vdots \\ \mathbf{w}_M \end{bmatrix} \qquad (14.13)$$

and

$$\mathbf{r}_i = \begin{bmatrix} \mathbf{r}(iT + M\Delta) \\ \vdots \\ \mathbf{r}(iT - M\Delta) \end{bmatrix} \qquad (14.14)$$

are $(2M + 1)N$-dimensional vectors representing the tap coefficients and the channel output samples, respectively. Minimization of the MSE expression for user 0 results in

$$\mathbf{w}_{\mathrm{opt}} = \mathbf{R}^{-1}\mathbf{v} \qquad (14.15)$$

where $\mathbf{R}$ is the channel output autocorrelation matrix,

$$\mathbf{R} = E(\mathbf{r}_i \mathbf{r}_i^H) = \begin{bmatrix} \mathbf{R}_{-M,-M} & \cdots & \mathbf{R}_{-M,M} \\ \mathbf{R}_{M,-M} & \cdots & \mathbf{R}_{M,M} \end{bmatrix} \qquad (14.16a)$$

where $\mathbf{R}_{ij}$ are N-by-N square matrices:

$$\mathbf{R}_{ij} = \sum_n \mathbf{h}(i\Delta - nT)\mathbf{h}(j\Delta - nT)^H + \frac{2N_0 B_r}{T}\mathbf{I}\delta_{ij}, \qquad (14.16b)$$

where $\mathbf{I}$ is an identity matrix, δ_{ij} is the Kronecker delta function, and $\mathbf{v}$ is the desired channel propagation vector:

$$\mathbf{v} = E(\mathbf{r}_i a_{0i}^*) = \begin{bmatrix} \mathbf{h}_0(M\Delta) \\ \vdots \\ \mathbf{h}_0(-M\Delta) \end{bmatrix} \qquad (14.17)$$

assuming uncorrelated unit variance data symbols and white noise. The minimum MSE (MMSE) is then

$$\mathrm{MMSE} = 1 - \mathbf{v}^H \mathbf{R}^{-1} \mathbf{v}. \qquad (14.18)$$

With the addition of decision feedback, $\mathbf{w}$ is augmented by F complex-valued feedback coefficients, and $\mathbf{r}_i$ is augmented by the previous F decisions $\{a_{0n}\}_{n=1}^{F}$.

The optimum set of tap coefficients and minimum MSE is still given by (14.15) and (14.18), respectively, but

$$\mathbf{v} = \begin{bmatrix} \mathbf{h}_0(M\Delta) \\ \vdots \\ \mathbf{h}_0(-M\Delta) \\ \cdots\cdots\cdots \\ 0 \\ \vdots \\ 0 \end{bmatrix} \qquad (14.19)$$

and

$$
\mathbf{R} = \begin{bmatrix} & \vdots & \\ \mathbf{R}_{11} & \vdots & \mathbf{R}_{12} \\ & \vdots & \\ \cdots & \cdots & \cdots & \vdots & \cdots \\ & \vdots & \\ \mathbf{R}_{21} & \vdots & \mathbf{R}_{22} \end{bmatrix} \tag{14.20}
$$

where $\mathbf{R}_{11}$ has the same form as (14.16a), $\mathbf{R}_{22}$ is an F-by-F identity matrix, and

$$
\mathbf{R}_{12} = \mathbf{R}_{21}^{H} = \begin{bmatrix} \mathbf{h}_0(M\Delta + T) & \cdots & \mathbf{h}_0(M\Delta + FT) \\ \vdots & & \vdots \\ \mathbf{h}_0(-M\Delta + T) & \cdots & \mathbf{h}_0(-M\Delta + FT) \end{bmatrix}. \tag{14.21}
$$

Max. SINR and Error Probability   The output signal-to-interference-plus-noise ratio (SINR) is another measure of the performance of a spatial and/or temporal processing receiver that suppresses interference and noise. It is defined as the ratio of the squared magnitude of the desired signal to the sum of the mean-squared values of the total interference and noise, both measured at the sampling instant at the output of the linear combiner. It can be shown, by application of the matrix inversion lemma, that the set of tap coefficients that minimizes MSE, given by (14.15), also maximizes SINR [21]. The maximum SINR is given by

$$
\text{Max.SINR} = \frac{1 - \text{MMSE}}{\text{MMSE}} \tag{14.22a}
$$

where MMSE is given by (14.18). It can be shown by the matrix inversion lemma that another formula equivalent to (14.22a) is [21]

$$
\max \text{SINR} = \mathbf{v}^{H} \mathbf{R}_{1+N}^{-1} \mathbf{v} \tag{14.22b}
$$

where $\mathbf{R}_{1+N}$ is the covariance matrix of the interference plus noise.

A formulation for the probability distribution for SINR was obtained by Shah and Haimovich [29] for the case of $K > N$ equal power, Rayleigh-faded interferers, with no additive noise. This result was generalized to include the effects of noise and unequal power interferers by Gao et al. [30].

Symbol error probability is also an important performance measure. A simple and useful Chernoff-type upper bound on the symbol error probability is [20]

$$
P_e \leq \exp\left(-\frac{1}{2\text{MMSE}} \right) \tag{14.23}
$$

where MMSE, given by (14.18), applies for unit data symbol variance. A relatively simple upper bound on bit error probability for spatial processing, accurate for low bit error rates, was given by Winters and Salz [31]. They bound (14.23) in terms of the determinant of $\mathbf{R}_{I+N}$, and further simplify the bound for up to seven antennas in terms of interferer powers.

14.3.2 Nonlinear Space–Time Processing

Maximum likelihood is an even more powerful detection approach for temporal and spatial anti-interference processing [32, 33]. However, for receivers with multielement antenna arrays, the differences in performance among linear, decision feedback, and maximum-likelihood detectors are typically less than they would be for non-spatial-processing receivers operating in the presence of severe multipath [34]. This is because the spatial processing adds extra degrees of freedom. For example, suppose the jth frequency response component $H_{0j}(f)$ of $\mathbf{H}_0(f)$ has spectral nulls, which would be best handled by a decision feedback equalizer or maximum-likelihood sequence detector if there is only one antenna element j; but it is unlikely that the N responses to N antenna elements will have the same nulls, and thus linear spatial-temporal processing might yield almost equivalent performance to that of decision feedback processing, if previous decisions of the desired signal are fed back. An asymptotic result shows that an optimum linear N-branch space–time receiver requires only one additional diversity branch to achieve maximum-likelihood receiver performance [35] against multipath.

However, if previous decisions of the *interfering* signals are also fed back (some or all of these would generally be available at a cellular radio base station that is simultaneously receiving signals from many terminals), substantial gains in interference suppression performance can result. This is called *centralized decision feedback* [36–38] and is a form of *interference cancellation* [39]. Figure 14.4 shows a centralized decision feedback receiver for signals from interfering sources 0 and 1. If say, source 0's signal is substantially larger, source 1's decision could be delayed so that essentially all of source 0's interference is canceled from it before a decision is made, as suggested in Figure 14.4.

An important design question for linear or nonlinear space–time processing is how many tap coefficients to provide for each of the N forward filters. A very useful rule-of-thumb result was recently obtained in [40]: the number of tap coefficients for which the MMSE performance approaches that of infinite-tap processors. Analysis in [40], supported by simulations, shows that for a decision feedback equalizer structure, in which the number of antennas exceeds the number of interferers, the forward filters should each span a number of symbol intervals roughly equal to $D[1 + (2K - 1)\varphi]$, where D is the span of the delay spread in symbol intervals, K is the total number of interfering signals (including the desired signal), and φ is the input signal-to-noise ratio (SNR) (in decibels) divided by 10. Of this total, the causal and anticausal portions are, respectively, $C = D(K - 1)\varphi$ and $A = D(1 + K\varphi)$. The number of feedback coefficients should be $D + C$. These numbers were shown to also be valid for practical maximum-likelihood sequence estimation reception and (approximately) for linear reception. Note that they are independent of the number of antenna elements. In the next section, we consider some applications of these results.

14.3.3 Applications of Receiver Space–Time Processing

The capacity of cellular systems increases with the number of interfering signals that the receivers in the system can tolerate with adequate performance. Winters [19, 21] showed

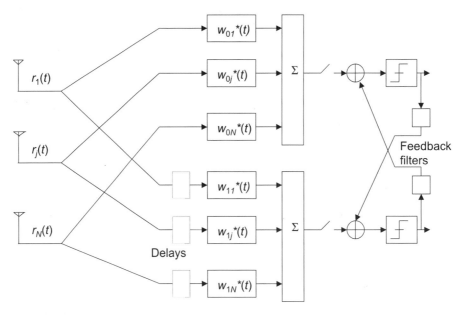

FIGURE 14.4   Illustration of centralized decision feedback for $K = 2$ signals, with full cancellation of signal "0" from delayed signal "1."

that a receiver with an array of N antennas can theoretically tolerate (suppress) up to $N - 1$ interferers. The capacity is thus proportional to N [20]. Furthermore, for independent flat Rayleigh fading on each of the N paths, $K + N$ antennas will suppress up to $K - 1$ interferers, and also provide up to N-fold path diversity for each user [20]. Correlations of up to 0.5 among the N paths was shown [41] to have only a minor effect. The ability of adaptive arrays to increase capacity in TDMA cellular systems has been confirmed in laboratory and field experiments [20, 42, 43].

Direct sequence CDMA cellular systems are robust to interference, but their capacity can be significantly improved by appropriate time-domain processing. Verdú [15] and others have derived optimum and suboptimum receivers for multiple user CDMA systems. Linear or decision feedback equalization is one form of time-domain processing that has been found effective in suppressing interference (and thereby increasing system capacity) in CDMA systems with short spreading codes [44–46]. Short spreading codes are those whose length equals a bit interval or a small multiple of it, and which do not change from bit to bit. The interference suppression capability of linear filtering used with direct sequence CDMA follows as a consequence of Eq. (14.9) from its large excess bandwidth (approximately equal to the spreading gain). Short-code CDMA systems using adaptive equalization have been shown to approximately double or triple the capacity of systems using conventional (matched filter) reception, and also include the function of RAKE-type reception in the presence of multipath [44–47]. Furthermore, such systems generally have much less sensitivity to variations in interferers' power due to imperfect power control than do conventional direct-sequence CDMA systems.

Figure 14.5 shows a realization of linear filtering applied to a direct sequence CDMA receiver, for the case where multipath delay spread and interference span up to two bit intervals, and therefore the linear filtering extends over two bit intervals. Close examina-

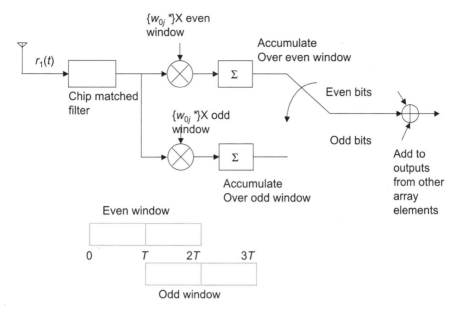

FIGURE 14.5   Space–time processing of CDMA signal where filter memory $=2$ bit intervals (2T).

tion of the processing depicted in this figure reveals that it is equivalent to transversal filtering with complex tap coefficients $\{w_{0j}^*\}$, spanning two bit intervals (2T). It is also equivalent to conventional RAKE receiver processing [13], with one principal difference: In the RAKE receiver, the $\{w_{0j}^*\}$ would be replaced by linear combinations of ± 1 spreading code chip values, with weights determined by estimated multipath component gains. This equivalence indicates that temporal linear processing to deal jointly with multipath and interference need not be significantly more complex than conventional RAKE receiver techniques based on correlation.

Combining time-domain processing of CDMA with antenna array processing as in Figure 14.2 yields further capacity benefits [33, 48], primarily because of the exploitation of the extra degrees of freedom mentioned earlier. In effect, spatial and temporal processing are complementary to each other. For example, interference or multipath components from similar directions may have similar spatial signatures, but can be eliminated by temporal processing, since their temporal signatures will likely be different. Interfering components with similar temporal signatures (e.g., interferers with highly correlated spreading code signatures) but different directions of arrival may be eliminated by the spatial processing. It is worth noting that the benefits of adaptive array processing by itself, without adaptive temporal processing to combat interference, may be minimal in CDMA systems with large spreading gains. Instead, switched-beam antenna systems are used with conventional CDMA systems. The reason is that the number of interferers is usually much larger than the practical number of adaptive antenna elements that can be implemented [1]. Figure 14.5 illustrates spatial-temporal processing through the combination of other array outputs.

The dependence of MMSE on the number of antenna elements and on the number of adaptive tap coefficients at the output of each element is illustrated [48] in Figures 14.6

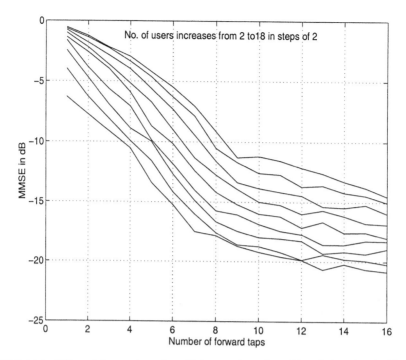

FIGURE 14.6 Effect of the number of coefficients and number of interfering signals. Number of antenna elements = 4 (from [48]). Curves from bottom to top represent numbers of interfering users from 2 to 18 in steps of 2.

and 14.7. In these examples, the spreading gain is 8; short spreading codes are randomly chosen for each user; each interfering user's signal is received at each antenna element over 3 independent Rayleigh-distributed multipath components with equal average power, with overall delay spread ≤ 6 chips; interferers' directions of arrival are random; and E_b/N_0 of each received signal at each element is 15 dB. The sampling rate into each forward filter is the chip rate. Figure 14.6 shows MMSE versus number of forward tap coefficients per element, as the number of interfering user signals increases from 2 to 18, when there are 4 antenna elements. It indicates that an appropriate choice for the number of forward tap coefficients is approximately the number of chips per bit plus the maximum expected delay spread, in chip intervals. For a typical CDMA operating point, with an MMSE of about -10 dB, Figure 14.7 shows the rapid increase in the number of tolerable interfering signals as the number of antenna elements is increased; this number approaches the product of the spreading gain and the number of elements.

Figure 14.8 shows the significant reduction in MMSE that is possible from centralized decision feedback processing [38], as compared to linear and conventional decision feedback processing. The conditions are the same as in the above example, with the exception of E_b/N_0, which is 18 dB. Similar results for non-CDMA systems are shown in [37], which also derives a closed-form asymptotic expression for MMSE for infinite-length temporal processing. An extension to partial centralized decision feedback connectivity, where, for example, out-of-cell interferers are dealt with by forward temporal and spatial processing but not by feedback processing, is found in [49].

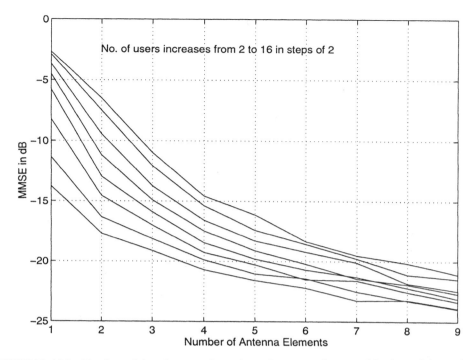

FIGURE 14.7   Number of interferers and number of antenna elements. Number of forward taps = 14 (from [48]).

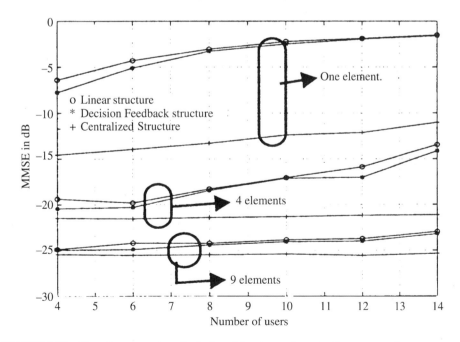

FIGURE 14.8   Capacity comparison for 1, 4, and 9 antenna elements: linear space–time processor, decision feedback space–time processor, and centralized decision feedback space–time processor; 8 forward taps and 5 feedback taps (from [38]).

14.4 RECENT SPACE–TIME WIRELESS COMMUNICATION ARCHITECTURES

Recently, much interest has centered on the simultaneous use of multiple transmitting and receiving antennas as a means of increasing the capacity of restricted bandwidth links, almost without limit [50–52]. The application generally involves a single wireless link between a transmitting end and a receiving end. With K distinct parallel data streams transmitted from K antennas (all at the same physical location), and with $N \geq K$ receiving antenna elements, whose outputs are linearly combined into K receivers, it is possible to create essentially K "parallel data pipes," each carrying independent data [52]. The total end-to-end bit rate, and therefore the system capacity, is increased by a factor of K without any bandwidth expansion. However, for this capacity expansion to be realized, the transmission channels between each pair of transmitting and receiving antennas should be independent. This will be possible if the antenna elements are spaced at least half a wavelength apart, and if dense scattering objects cause independent fading on all the paths. The system can use linear spatial, and if necessary temporal processing, at the receiving array similar to what has been previously described, with a separate coefficient vector $\mathbf{w}_k$, $k = 1, 2, \ldots, K$, optimized according to either a zero-forcing or an MMSE criterion, for each of the K receivers. The receiver performance is further enhanced by the use of nonlinear cancellation: The K data symbols are detected sequentially in order of decreasing output SINR, and as each one is detected, its response is subtracted from the inputs, thus reducing the interference to later detected symbols. The idea has been successfully demonstrated in hardware [52] in an indoor office environment. The "bandwidth efficiency" (bits/seconds/hertz) possible with this approach is theoretically limited only by the number of parallel transmitters and antennas that can be deployed at each end of a link.

The approach has also been generalized to "space–time" coding, in which block or convolutional codes, and decoding is applied across the parallel streams of data, again using a combination of linear receiver spatial processing and an interference cancellation approach [50, 51, 53–55].

14.5 ADAPTATION ISSUES

Adaptation of equalizer and antenna array tap coefficients can be based on using receiver decisions in a "decision-directed" mode as a reference desired output. Initially, with all tap coefficients set to arbitrary values, say zero, a known "training" sequence of data may be transmitted and supplied as a reference, until the receiver decisions are sufficiently reliable to use as a reference. Alternatively, directions of arrival, and hence spatial signatures, for individual received signal components can be estimated using a variety of methods such as MUSIC or ESPRIT [5], but these approaches lose effectiveness if the number of multipath components associated with each received signal is large.

A likely candidate for a training-based adaptation algorithm is the simple LMS (least mean squares) algorithm, which continually adjusts the adaptive coefficients in the direction of reduced mean-squared error [56]. At time nT, the coefficient vector is updated as

$$\mathbf{w}_0(n + 1) = \mathbf{w}_0(n) - \mu_n e(n)\mathbf{r}_n \tag{14.24}$$

where $e(n) = y(nT) - a_{0n}$ is the error at time nT, and μ_n is an appropriately normalized step size scalar parameter (proportional to the inverse of the current estimated input signal variance).

Convergence of the LMS algorithm to the vicinity of the minimum mean-squared error typically requires a number of training symbols (iterations) equal to at least 10 times the number of coefficients being adapted; for arrays with significant correlation among element outputs, much longer convergence times are required [56]. The LMS algorithm is best suited to applications that permit long training times and to slowly varying channels.

An alternative to this LMS adaptation approach is to take a least squares approach; that is, find coefficients that directly minimize the sum of squares of errors between desired output data symbols and spatial-temporal array outputs. In this approach we *estimate* the channel autocorrelation matrix $\mathbf{R}$ and the desired channel propagation vector $\mathbf{v}$, from processing and time averaging the appropriate products of channel outputs and receiver decisions [56]:

$$\hat{\mathbf{R}} = \frac{1}{TB}\sum_{i=1}^{TB}\mathbf{r}_i\mathbf{r}_i^H \quad \text{and} \quad \hat{\mathbf{v}} = \frac{1}{TB}\sum_{i=1}^{TB}\mathbf{r}_i a_{0i}^* \quad (14.25)$$

where TB is the length of the training period, in symbols, and computing the tap coefficients from (14.15); that is,

$$\hat{\mathbf{w}} = \hat{\mathbf{R}}^{-1}\hat{\mathbf{v}} \quad (14.26)$$

This is called DMI (direct matrix inversion) and is performed once (at the end of each training period) to compute the coefficients relevant to that ing period. The length of the training period, measured in data symbols, needed for convergence to within 1 to 3 dB of the minimum mean-squared error, is roughly twice the number of coefficients to be adapted [56]. At the end of the training period, the adaptation could be switched to decision-directed LMS, in order to track slow channel variations. The convergence of the DMI algorithm is much faster than that of LMS. However, it is more complex. If the solution of (14.25) is implemented with Cholesky factorization, the total number of complex multiplications, including those for estimating $\mathbf{R}$ and $\mathbf{v}$ can be shown to be on the order of $M^3/6 + 6M^2$, where M is the total number of coefficients, assuming that the training period is $2M$ symbols. For the LMS algorithm, the total number of multiplies would be about $2M$ times the number of symbols in the training period. If the required training period for the LMS algorithm to converge is very long, the total complexity (as measured by number of multiplications) could be similar for the two algorithms, or even greater for LMS.

The recursive least squares (RLS) algorithm is a recursive version of DMI in which the coefficients are updated once per training symbol so as to minimize the sum of squares of errors up to that time [56]. If the coefficients are required to be estimated only once during the training period, the DMI approach uses fewer computations than the RLS approach. For a version of RLS that uses a rectangular sliding window. Reference [57] demonstrates that estimation of $\mathbf{R}_{I+N}$ and use of (14.22b) instead of $\mathbf{R}$ and (14.22a) results in more accurate tracking of time variations.

For high-bit-rate applications, where the multipath delay spread extends over tens or hundreds of symbol intervals, the total number of required coefficients M, and hence the complexity of temporal processing is high. In this case, frequency domain receiver space–time processing, including adaptation, can yield significant simplification [35]. The simplification is achieved by using FFT operations to replace time-domain filtering, and through independent adaptation of each frequency component.

Note that the actual autocorrelation matrix $\mathbf{R}$, given by (14.16b), depends only on desired and interfering channel responses and noise, and its estimate is obtained by time-averaging products of delayed array element outputs. However, the actual propagation vector $\mathbf{v}$, given by (14.17), depends only on the desired channel response, and its estimate is obtained by time-averaging products of array outputs with training data symbols; that is, estimation of $\mathbf{v}$ requires training symbols, while $\mathbf{R}$ does not. In principle, then, it seems that $\mathbf{v}$ could be estimated during the training period, while $\mathbf{R}$ could be estimated over longer periods to achieve a more accurate estimate, such as when new interferers appear or old ones disappear. Unfortunately, using different data to estimate $\mathbf{R}$ and $\mathbf{v}$ turns out to produce poor estimates of the coefficients, unless the estimation periods are impractically long, and conditions do not vary with time; the reason is that this potential approach does not give a least squares solution for the given data. A partial solution to this problem is to take a subspace processing approach [58] in which $\mathbf{v}$ is estimated from the training symbols in combination with a subspace decomposition of the estimate of $\mathbf{R}$, which has been obtained by time averaging from a longer block of received data. The result is a robust array adaptation algorithm that allows estimation of the optimal coefficients from an arbitrarily long sequence of array channel outputs and a relatively short training sequence.

The robustness to suddenly appearing and disappearing interferers, of adaptive space–time processing in CDMA systems is shown in [59]. In comparison to their effects on a CDMA receiver with adaptive temporal-only processing, newly appearing interferers cause a much lower mean-squared error, and their cumulative effect is minimal as shown in Figure 14.9.

While it is likely that most applications of spatial-temporal processing of digital communications signals will employ known training sequences for adaptation and synchronization, *blind* adaptation (without the use of a training sequence) may also be useful, for example, in broadcast situations, where receivers may have to recover from disruptions. Constant modulus blind algorithms are relatively simple and robust, but converge slowly. Faster convergence, at the expense of higher signal processing complexity, is achieved with oversampling, cyclostationarity-exploiting algorithms, and subspace approaches. A good survey and comparison of these approaches is found in [10].

An example of a blind array adaptation approach that exploits cyclostationarity is found in [60]. In this approach, users' signals are separated by small frequency offsets (much less than signal bandwidth), and a reference signal used for DMI adaptation is just the complex output of one of the array elements. The adaptation uses a weighted sum of the results of several such reference signals. Adaptation is relatively fast, and can be accelerated to decision-directed adaptation once the array coefficients start to converge.

14.6 TRANSMITTER SPACE–TIME PROCESSING

Up to this point our discussion has mainly been relevant to space–time processing at a *receiver*. However, transmitter space and/or time processing can also be done at

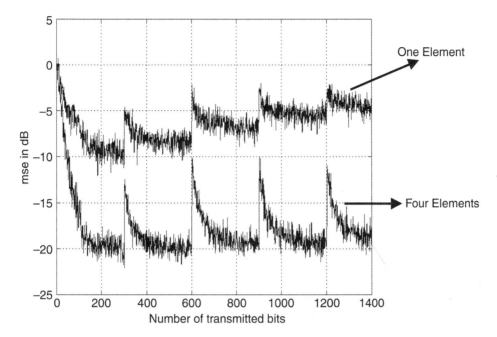

FIGURE 14.9 Effect of sudden birth of interference. The system starts with six users, after which a new user is added every 300 symbols (from [59]).

transmitters; an important application is to cellular wireless systems, where size and cost constraints limit or preclude array processing at mobile terminals, thus leaving the adaptive processing of downlink (base to mobile) signals to be done at the base transmitters. Assuming that transmission paths are linear, and that their vector impulse responses are known, or can be estimated, one can show that linear space and/or time processing can be done at transmitters, using zero-forcing or MMSE optimization criteria. If the channels are *reciprocal*, which means that their uplink (mobile to base) and downlink responses are identical, then the optimum transmitter space–time coefficients can be derived from (in fact they are identical to) the corresponding uplink coefficients. The assumption of reciprocity is valid if the same antenna elements are used for transmitting and receiving, and if uplink and downlink transmissions take place in the same frequency band, close enough in time that channels do not change significantly between uplink and downlink transmissions. Time division duplex (TDD) systems with burst durations smaller than the channel's coherence time satisfy reciprocity. FDD (frequency division duplex) systems generally do not, since widely separated frequency bands are used for uplink and downlink transmission.

A simple example can illustrate relationships between spatial processing at transmitters and receivers, under the assumption of reciprocity. Consider first two base stations, designated 1 and 2, which are receiving signals from two mobiles, also designated as 1 and 2; mobile 1 is transmitting to base station 1 and provides interference at base station 2; mobile 2 is transmitting to base 2 and provides interference at base 1. The two base stations each have antenna N-dimensional array coefficient vectors $\mathbf{w}_1$ and $\mathbf{w}_2$, respectively. As shown in Figure 14.10, the complex vector responses among the two mobiles and two base stations are designated $\mathbf{h}_{11}$, $\mathbf{h}_{12}$, $\mathbf{h}_{21}$, and $\mathbf{h}_{22}$. If mobile 1 is transmitting data

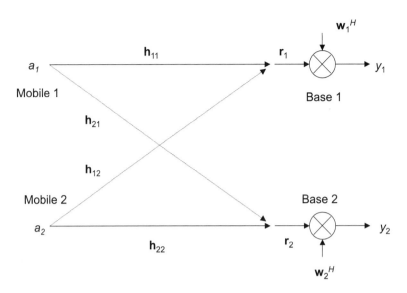

FIGURE 14.10 Mobile-to-base transmission. Processing at base receivers.

symbol a_1, and mobile 2 is transmitting a_2, the two array input vectors $\mathbf{r}_1$ and $\mathbf{r}_2$ at the two base stations are given in vector form as

$$\begin{pmatrix} \mathbf{r}_1 \\ \mathbf{r}_2 \end{pmatrix} = \begin{pmatrix} \mathbf{h}_{11} & \mathbf{h}_{12} \\ \mathbf{h}_{21} & \mathbf{h}_{22} \end{pmatrix} \begin{pmatrix} a_1 \\ a_2 \end{pmatrix} \tag{14.27}$$

The corresponding two array outputs, y_1 and y_2, are

$$\begin{pmatrix} y_1 \\ y_2 \end{pmatrix} = \begin{pmatrix} \mathbf{w}_1^H & \mathbf{0}^H \\ \mathbf{0}^H & \mathbf{w}_{12}^H \end{pmatrix} \begin{pmatrix} \mathbf{r}_1 \\ \mathbf{r}_2 \end{pmatrix} = \begin{pmatrix} \mathbf{w}_1^H & \mathbf{0}^H \\ \mathbf{0}^H & \mathbf{w}_{12}^H \end{pmatrix} \begin{pmatrix} \mathbf{h}_{11} & \mathbf{h}_{12} \\ \mathbf{h}_{21} & \mathbf{h}_{22} \end{pmatrix} \begin{pmatrix} a_1 \\ a_2 \end{pmatrix} \tag{14.28}$$

The following zero-forcing solution will eliminate interference and yield the desired outputs $y_1 = a_1$ and $y_2 = a_2$ if

$$\mathbf{w}_1^H \mathbf{h}_{11} = 1; \qquad \mathbf{w}_1^H \mathbf{h}_{12} = 0; \qquad \mathbf{w}_2^H \mathbf{h}_{21} = 0 \qquad \mathbf{w}_2^H \mathbf{h}_{22} = 1 \tag{14.29}$$

Figure 14.11 shows the corresponding situation in which base stations 1 and 2 are transmitting to mobiles 1 and 2, respectively. The complex vector responses are the same, assuming reciprocity. If the same coefficient vectors are used, as shown in the figure, the two scalar outputs, z_1 and z_2, at the two mobiles are

$$(z_1 \quad z_2) = (a_1 \quad a_2) \begin{pmatrix} \mathbf{w}_1^H & \mathbf{0}^H \\ \mathbf{0}^H & \mathbf{w}_{12}^H \end{pmatrix} \begin{pmatrix} \mathbf{h}_{11} & \mathbf{h}_{12} \\ \mathbf{h}_{21} & \mathbf{h}_{22} \end{pmatrix} = (a_1 \quad a_2) \tag{14.30}$$

if the two transmitting coefficient vectors are the same as the optimal receiving coefficient vectors given by Eq. (14.28). While this example applies to zero-forcing optimization, MMSE optimization will behave in essentially the same fashion.

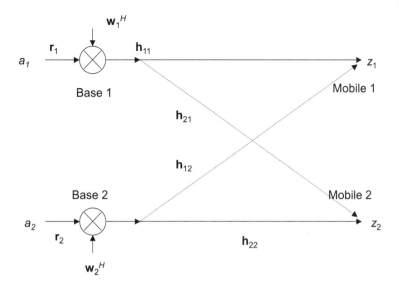

FIGURE 14.11 Base-to-mobile transmission. Processing at base transmitters.

Thus when there is true reciprocity, as in a TDD system, base stations can adapt their arrays to minimize co-channel and intersymbol interference in the *received* signals, as in Figure 14.10, and then use the same array coefficient vectors when transmitting, as in Figure 14.11. As illustrated in the above example, this will shape the transmitting antenna arrays' beam patterns in such a way that downlink interference from base stations to mobiles is minimized or eliminated. Note that adaptation can be done during the receive (uplink) mode at each separate base station, without requiring coordination among base stations.

For nonreciprocal FDD systems, base station transmitter array processing, as in Figure 14.11 can also eliminate downlink interference if the coefficient vectors satisfy (14.28), but in this case, the complex vector responses $\{\mathbf{h}_{ij}\}$ are different in the uplink and downlink directions. Application of (14,29) then requires separate estimation, using training sequences, of the downlink $\{\mathbf{h}_{ij}\}$, which can be complex and require significant feedback signaling among base stations and mobiles [10, 61]. In practice, transmitter-only spatial processing in such multibase station systems can make use of switching among directional transmit antenna beams at each base station. Alternatively, downlink adaptive approaches can be used averaging of data collected on the uplink. For example in [62], time averaging is used to remove uplink frequency-specific characteristics from estimates of the desired signal and interference covariance matrices. The results are used to determine downlink array coefficients that maximize average desired signal power while constraining average interference plus noise power to be below a fixed level. A related approach, involving a transformation of the uplink covariance matrix is described in [63].

14.7 CONCLUSIONS AND FUTURE APPLICATIONS

Adaptive spatial and temporal linear combining both have the ability to minimize intersymbol interference arising from multipath, and co- and adjacent-channel interference in multiuser wireless digital communications. They can be used singly or together to allow a number of users to share the same time, bandwidth, and space. We have seen that the number of interfering signals that can be separated at a receiver is proportional to the number of antenna elements and also to the received signals' excess bandwidth. We have mainly considered *linear* space–time processing methods here since their implementation and adaptation is relatively simple. However, nonlinear interference cancellation techniques will also be useful, especially at cellular base stations for removing the strongest interferers.

Smart antennas have a capacity-multiplying effect in cellular systems that employ them. They make possible space division multiple access (SDMA) in which individual channels may be reused by different users within a cell, without performance loss due to interference. Smart antennas have been proposed for capacity and reliability enhancement in third-generation wireless systems [64]. Indoor and outdoor broadband cellular systems will also benefit from the use of smart antennas, starting first with directive, switched-beam [65, 66] and eventually incorporating full array adaptation and temporal processing [67] for interference and multipath mitigation. The interference-elimination properties of spatial-temporal processing will also likely find application in *unlicensed* wireless broadband systems, such as those expected to operate in three recently allocated 100-MHz-wide U-NII (unlicensed national information infrastructure) bands in the United States [68]. Users of these bands will operate in an interference environment that is essentially uncontrolled, except for some basic rules on transmitted power, power spectral density, and antenna gain. Array processing, coupled with other techniques such as coding and smart multiple access protocols, will be essential, once the U-NII bands become heavily used.

A significant remaining problem area is implementation complexity. For example, the realization of spatial-temporal processing at a receiver generally requires a separate radio frequency (RF) front end, together with a downconverter and analog-to-digital (A/D) converter, to allow for digital processing of the baseband or IF output of each antenna element [69]. An alternative approach—direct weighting and combining at RF—is not supported by mature, cost-effective technology at this point. Thus the development of low-cost, low-power single-chip radio realizations are of prime importance.

Acknowledgments It is a pleasure to acknowledge very helpful discussions and suggestions from Lek Ariyavisitakul of Home Wireless Networks, Abdelgader Legnain and Sébastien Roy of Carleton University, Mansoor Shafi of Telecom New Zealand, and John Thompson of the University of Edinburgh.

REFERENCES

1. J. H. Winters, "Smart Antennas for Wireless Systems," *IEEE Personal Commun.*, Vol. 5, No. 1, pp. 23–27, Feb. 1998.

2. W. F. Gabriel, "Adaptive Arrays—An Introduction," *Proc. IEEE*, Vol. 64, No. 2, pp. 239–272, Feb. 1976.

3. D. H. Johnson and D. E. Dudgeon, *Array Signal Processing*, Prentice-Hall, Englewood Cliffs, NJ, 1993.

4. L. C. Godara, "Applications of Antenna Arrays to Mobile Communications, Part I: Performance Improvement, Feasibility and System Considerations," *Proc. IEEE*, Vol. 85, No. 7, pp. 1029–1060, July 1997.

5. L. C. Godara, "Applications of Antenna Arrays to Mobile Communications, Part II: Beam-Forming and Direction-of-Arrival Considerations," *Proc. IEEE*, Vol. 85, No. 8, pp. 1193–1245, Aug. 1997.

6. C. L. Despins, D. D. Falconer, and S. A. Mahmoud, "Compound Strategies of Coding Equalization and Space Diversity for Wideband TDMA Indoor Wireless Channels," *IEEE Trans. Veh. Tech.*, Vol. 41, No. 4, pp. 369–379, Nov. 1992.

7. P. Monsen, "Feedback Equalization for Fading Dispersive Channels," *IEEE Trans. Infor. Theory*, Vol. 17, No. 1, pp. 56–64, Jan. 1971.

8. P. Monsen, "MMSE Equalization of Interference on Fading Diversity Channels," *IEEE Trans. Commun.*, Vol. COM-32, No. 1, pp. 5–12, Jan. 1984.

9. A. J. Paulraj and B. C. Ng, "Space-Time Modems for Wireless Personal Communications," *IEEE Personal Commun.*, Vol. 5, No. 1, pp. 36–48, Feb. 1998.

10. A. J. Paulraj and C. B. Papadias, "Space-Time Processing for Wireless Communications," *IEEE Signal Proc. Mag.*, pp. 49–83, Nov. 1997.

11. S.-S. Jeng, G. Xu, H.-P. Lin, and W. J. Vogel, "Experimental Studies of Spatial Signature Variation at 900 MHz for Smart Antenna Systems," *IEEE Trans. Ant. Propagation*, Vol. 46, No. 7, pp. 953–962, July 1998.

12. W. C. Jakes, Jr., *Microwave Mobile Communications*, Wiley, New York, 1974.

13. T. S. Rappaport, *Wireless Communications*, Prentice-Hall, Englewood Cliffs, NJ, 1996.

14. R. B. Ertel, P. Cardieri, K. W. Sowerby, T. S. Rappaport, and J. H. Reed, "Overview of Spatial Channel Models for Antenna Array Communication Systems," *IEEE Personal Commun.*, Vol. 5, No. 1, pp. 10–22, Feb. 1998.

15. S. Verdú, "Multiuser Detection," *Advances in Statistical Signal Processing*, Vol. 2, JAI Press, 1993, pp. 369–409.

16. A. M. Sayeed, E. N. Onggosanusi, and B. D. Van Veen, "A Canonical Space-Time Characterization of Mobile Wireless Channels," *IEEE Commun. Lett.*, Vol. 3, No. 4, pp. 9–96, April 1999.

17. D. D. Falconer, M. Abdulrahman, N. W. K. Lo, B. R. Petersen, and A. U. H. Sheikh, "Advances in Equalization and Diversity for Portable Wireless Systems," *Digital Signal Processing*, Vol. 3, Academic, 1993, pp. 148–162.

18. H. Liu and M. D. Zoltowski, "Blind Equalization in Antenna Array CDMA Systems," *IEEE Trans. Sig. Proc.*, Vol. 45, No. 1, pp. 161–172, Jan. 1997.

19. J. H. Winters, "Optimum Combining for Indoor Radio Systems with Multiple Users," *IEEE Trans. Commun.*, Vol. COM-35, No. 11, pp. 1222–1230, Nov. 1987.

20. J. H. Winters, J. Salz, and R. D. Gitlin, "The Impact of Antenna Diversity on the Capacity of Wireless Communication Systems," *IEEE Trans. Commun.*, Vol. 42, No. 2/3/4, pp. 1740–1751, Feb./Mar./April 1994.

21. J. H. Winters, "Optimum Combining in Digital Mobile Radio with Cochannel Interference," *IEEE J. Selected Areas in Commun.*, Vol. SAC-2, pp. 528–539, July 1984.

22. W. Van Etten, "An Optimum Linear Receiver for Multiple Channel Digital Transmission Systems," *IEEE Trans. Commun.*, Vol. COM-23, No. 8, pp. 828–834, Aug. 1975.

23. J. Salz, "Digital Transmission over Cross-Coupled Linear Channels," *AT&T Tech. J.*, Vol. 64, No. 6, pp. 1147–1159, Aug. 1985.

24. D. A. Shnidman, "A Generalized Nyquist Criterion and an Optimum Linear Receiver for a Pulse Modulation System," *Bell Sys. Tech. J.*, Vol. 46, No. 9, pp. 2163–2177, Nov. 1967.

25. A. R. Kaye and D. A. George, "Transmission of Multiplexed PAM Signals over Multiple Channel and Diversity Systems," *IEEE Trans. Commun.*, Vol. COM-18, No. 5, pp. 520–526, Oct. 1970.

26. B. R. Petersen and D. D. Falconer, "Suppression of Intersymbol, Adjacent Channel and Co-Channel Interference by Equalizers and Linear Combiners," *IEEE Trans. Commun.*, Vol. 42, No. 12, pp. 3109–3118, Dec. 1994.

27. M. V. Clark, Diversity and Equalization in Digital Cellular Radio, P.h.D Thesis, Dept. of Electrical and Electronic Engineering, University of Canterbury, Christchurch, New Zealand, May 1992.

28. H. V. Clark, L. J. Greenstein, W. K. Kennedy, and M. Shafi, "MMSE Diversity Combining for Wide-Band Digital Cellular Radio," *IEEE Trans. Commun.*, Vol. 40, No. 6, pp. 1128–1135, June 1992.

29. A. Shah and A. M. Haimovich, "Performance Analysis of Optimum Combining in Wireless Communications with Rayleigh Fading and Cochannel Interference," *IEEE Trans. Commun.*, Vol. 46, No. 4, pp. 473–479, April 1998.

30. H. Gao, P. J. Smith, and M. V. Clark, "Theoretical Reliability of MMSE Linear Diversity Combining in Rayleigh-Fading Additive Interference Channels," *IEEE Trans. Commun.*, Vol. 46, No. 5, pp. 666–672, May 1998.

31. J. H. Winters and J. Salz, "Upper Bounds on the Bit Error Rate of Optimum Combining in Wireless Systems," *IEEE Trans. Commun.*, Vol. 46, No. 12, pp. 1619–1624, Dec. 1998.

32. S. Verdú, "Minimum Probability of Error for Asynchronous Gaussian Multiple-Access Channels," *IEEE Trans. Inform. Theory*, Vol. IT-32, No. 1, pp. 85–96, Jan. 1986.

33. S. Miller and S. Schwartz, "Integrated Spatial-Temporal Detectors for Asynchronous Gaussian Multiple Access Channels," *IEEE Trans. Commun.*, Vol. 43, No. 2/3/4, pp. 396–411, Feb./Mar./Apr. 1995.

34. M. V. Clark, L. J. Greenstein, W. K. Kennedy, and M. S. Shafi, "Optimum Linear Diversity Receivers for Mobile Communications," *IEEE Trans. Veh. Tech.*, Vol. 43, No. 1, pp. 47–56, Feb. 1994.

35. M. V. Clark, "Adaptive Frequency-Domain Equalization and Diversity Combining for Broadband Wireless Communications," *IEEE J. Selected Areas in Commun.*, Vol. 16, No. 8, pp. 1385–1395, Oct. 1998.

36. A. Duel-Hallen, "Decorrelating Decision-Feedback Multiuser Detector for Synchronous Code-Division Multiple-Access Channel," *IEEE Trans. Commun.*, Vol. 41, No. 2, pp. 285–290, Feb. 1993.

37. S. Roy and D. D. Falconer, "Multi-User Decision-Feedback Space-Time Equalization and Diversity Reception," *Proc. IEEE Veh. Tech. Conf.*, Houston, May, 1999.

38. A. M. Legnain, D. D. Falconer, and A. U. H. Sheikh, "Centralized Decision Feedback Adaptive Combined Space-Time Detector for CDMA Systems," *Proc. Int. Conf. on Commun.*, Vancouver, June 1999.

39. R. Kohno, H. Imai, M. Hatori, and S. Pasupathy, "Combination of an Adaptive Array Antenna and a Canceller of Interference for Direct Sequence Spread Spectrum Multiple Access System," *IEEE J. Selected Areas in Commun.*, Vol. 8, No. 4, pp. 675–682, May 1990.

40. S. I. Ariyavisitakul, J. H. Winters, and I. Lee, "Optimum Space-Time Processors with Dispersive Interference: Unified Analysis and Required Filter Span," *IEEE Trans. Commun.*, Vol. 47, No. 7, pp. 1073–1083, July 1999.

41. J. Salz and J. H. Winters, "Effect of Fading Correlation on Adaptive Arrays in Digital Wireless Communications," *Proc. IEEE Int. Conf. on Commun.*, Geneva, May, 1993.

42. S. Anderson, B. Hagerman, H. Dam, U. Forssén, J. Karlsson, F. Kronestedt, S. Mazur, and K. J. Molnar, "Adaptive Antennas for GSM and TDMA Systems," *IEEE Personal Commun.*, Vol. 6, No. 3, pp. 74–86, June 1999.

43. R. L. Cupo, G. D. Golden, C. C. Martin, K. L. Sherman, N. R. Sollenberger, J. H. Winters, and P. W. Wolniansky, "A Four-Element Adaptive Antenna for IS-136 PCS Base Stations," *Proc. IEEE Veh. Tech. Conf.*, Phoenix, May 1997, pp. 1577–1581.

44. M. Abdulrahman, A. U. H. Sheikh, and D. D. Falconer, "Decision Feedback Equalization in CDMA in Indoor Wireless Communication," *IEEE J. Selected Areas in Commun.*, Vol. 12, No. 4, pp. 698–706, May 1994.

45. P. B. Rapajic and B. Vucetic, "Adaptive Structures for Asynchronous CDMA Systems," *IEEE J. Selected Areas in Commun.*, Vol. 12, No. 4, pp. 685–697, May 1994.

46. U. Madhow and M. L. Honig, "MMSE Interference Suppression for Direct Sequence CDMA," *IEEE Trans. Commun.*, Vol. 42, No. 12, pp. 3178–3188, Dec. 1994.

47. S. L. Miller, "An Adaptive Direct-Sequence Code Division MultipleAccess Receiver for Multiuser Interference Rejection," *IEEE Trans. Commun.*, Vol. 43, No. 2/3/4, 1995, pp. 1746–1755, Feb./Mar./Apr.

48. A. M. Legnain, D. D. Falconer, and A. U. H. Sheikh, "Performance of New Adaptive Combined Space-Time Receiver for Multiuser Interference Rejection in Multipath Slow Fading CDMA Channels," *Proc. IEEE Globecom 98*, Sydney, Nov. 1998.

49. S. Roy and D. D. Falconer, "Multi-User Decision-Feedback Space-Time Processing with Partial Cross-Feedback Connectivity," *IEEE Veh. Techl. Conf.*, Tokyo, May 2000.

50. G. J. Foschini, "Layered Space-Time Architecture for Wireless Communication in a Fading Environment When Using Multi-Element Antennas," *Bell Labs Tech. J.*, Vol. 1, No. 2, Autumn, 1996.

51. G. J. Foschini and M. J. Gans, "On Limits of Wireless Communications in a Fading Environment When Using Multiple Antennas," *Wireless Personal Commun.*, Vol. 6, No. 3, pp. 311–335, 1998.

52. P. W. Wolniansky, G. J. Foschini, G. D. Golden, and R. A. Valenzuela, "V-BLAST: An Architecture for Realizing Very High Data Rates Over the Rich-Scattering Wireless Channel," *Proc. ISSSE '98*, Pisa, Italy, Oct. 1998.

53. V. Tarokh, A. F. Naguib, N. Seshadri, and A. R. Calderbank, "Combined Array Processing and Space-Time Coding," *IEEE Trans. Inform. Theory*, Vol. 45, No. 4, pp. 1121–1128, May 1999.

54. A. F. Nagub, V. Tarokh, N. Seshadri, and A. R. Calderbank, "A Space-Time Coding Modem for High Data Rate Wireless Communications," *IEEE J. Selected Areas in Commun.*, Vol. 16, No. 2, pp. 744–765, March 1998.

55. S. K. Ariyavisitakul, "Turbo Space-Time Processing to Improve Wireless Channel Capacity," *IEEE Trans. Commun.*, to appear.

56. S. Haykin, *Adaptive Filter Theory*, 3rd ed., Prentice-Hall, Englewood Cliffs, NJ, 1996.

57. T. Boros, G. G. Raleigh, and M. A. Pollack, "Adaptive Space-Time Equalization for Rapidly Fading Communication Channels," *Proc. Globecom '96*, London, pp. 984–989, Nov. 1996.

58. X. Wang and H. V. Poor, "Robust Adaptive Array for Wireless Communications," *IEEE J. Selected Areas in Commun.*, Vol. 16, No. 8, pp. 1352–1366, Oct. 1998.

59. A. M. Legnain, D. D. Falconer, and A. U. H. Sheikh, "Effect of Time-Varying User Population on the Performance of the Adaptive Combined Space-Time Receiver for DS-CDMA Systems," *Proc. IEEE Wireless Communication and Networking Conference*, New Orleans, Sept., 1999.

60. J. Cui, D. D. Falconer, and A. U. H. Sheikh, "Blind Adaptation of Antenna Arrays Using a Simple Algorithm Based on Small Frequency Offsets," *IEEE Trans. Commun.*, Vol. 46, No. 1, pp. 61–70, Jan. 1998.

61. D. Gerlach and A. Paulraj, "Adaptive Transmitting Antenna Arrays with Feedback," *IEEE Signal Proc. Lett.*, 1, pp. 150–152, Oct. 1994.

62. H. Asakura and T. Matsumoto, "Cooperative Signal Reception and Down-Link Beamforming in Cellular Mobile Communications," *IEEE Trans. Veh. Tech.*, Vol. 48, No. 2, pp. 333–341, Mar. 1999.

63. K. Hugl, J. Laurila, and E. Bonek, "Downlink Beamforming for Frequency Division Duplex Systems," *Proc. Globecom '99*, Rio de Janeiro, Dec. 1999, pp. 2097–2101.

64. R. D. Carsello et al., "IMT-2000 Standards: Radio Aspects," *IEEE Personal Commun.*, Vol. 4, No. 4, pp. 30–40, Aug. 1997.

65. D. Buchholtz et al., "Wireless In-Building Network Architecture and Protocols," *IEEE Network*, Vol. 5, No. 6, pp. 31–38, Nov. 1991.

66. S. Q. Gong and D. D. Falconer, "Cochannel Interference in Cellular Fixed Broadband Access with Directional Antennas," *Wireless Personal Commun.* (Kluwer) Vol. 10, No. 1, pp. 103–117, June 1999.

67. M. Chiani and A. Zanella, "Spatial and Temporal Equalization for Broadband Wireless Indoor Networks at Millimeter Waves," *IEEE J. Selected Areas in Commun.*, Vol. 17, No. 10, pp. 1725–1734, Oct. 1999.

68. Federal Communications Commission, "Amendment of the Commission's Rules to Provide for Operation of Unlicensed NII Devices in the 5 GHz Frequency Range," Report and Order, FCC-97-005, Gen. Docket No. 96–102, Jan. 9, 1997.

69. J.-Y. Lee, H.-C. Liu, and H. Samueli, "A Digital Adaptive Beamforming QAM Demodulator IC for High Bit Rate Wireless Communications," *IEEE J. Solid State Circuits*, Vol. 33, No. 3, pp. 367–377, March 1998.

CHAPTER 15

Interference Cancellation and Multiuser Detection

RYUJI KOHNO

15.1 INTRODUCTION

The standard for wireless local area networks (LANs) such as IEEE802.11 [1] and Bluetooth has been accelerating research and development of spread spectrum (SS) systems for consumer and personal communications, since its main regulation is to use SS modulation such as direct-sequence (DS), frequency hopping (FH), or their hybrid.

Many advantages of SS techniques such as robustness against interference and noise, less interference to other systems, and efficient use based on code division multiple access (CDMA) of frequency spectrum reveal wide wireless applications for commercial use [2–5]. The U.S. Federal Communications Commission (FCC) opened the industrial, scientific, and medical (ISM) frequency band for unlicensed operation of SS systems in 1985 and modified its rules to permit wider bandwidth in 1990 [6].

Moreover, SS techniques have a shot at becoming a mass-market technology in the digital standard for cellular telephones. Both CDMA and time division multiple access (TDMA) were proposed as the transmission standard for digital cellular telephone systems and products. Finally, CDMA has been selected to be the third generation of international mobile telecommunications, that is, IMT-2000. CDMA offers more capacity than first-generation TDMA. That improvement of capacity depends on some additional techniques such as power control, voice activation, soft handoff, and sectorization [7].

Efficient utilization of available frequency spectra is of major importance in not only wireless LANs and digital cellular telephones but also other consumer and personal communications. Since mature techniques to improve efficiency of frequency spectra and capacity of CDMA more are established, CDMA based on SS techniques won the battle.

This chapter gives an overview of several recently developed techniques to improve capacity of CDMA. In particular, we will focus on cancellation of co-channel interference (CCI) or multiuser interference (MUI) in CDMA, which restricts the number of available users who can simultaneously access a channel or the capacity of CDMA [8, 9]. Before

Wireless Communications in the 21st Century, Edited by Shafi, Ogose, and Hattori.
ISBN 0-471-155041-X © 2002 by the IEEE.

discussing CCI cancellation, we will briefly review optimum receiver for CDMA from an information-theoretic viewpoint. Since optimum detection for synchronous CDMA systems or M-ary SS ones can be reduced to an optimum coding and decoding problem for a multiple channel in coding theory, we will discuss optimum and suboptimum multiuser detection for asynchronous CDMA systems in which transmission of every user is not synchronized one another.

In wireless personal, indoor and mobile communication channels, it is difficult to carry out optimum detection for asynchronous CDMA due to unknown and time-varying characteristics of the channels. Therefore, suboptimum multiuser detection based on CCI cancellation is more attractive. Since CCI of DS/CDMA and that of FH/CDMA have different origins, we will discuss several schemes of CCI cancellation based on temporal filtering, spatial filtering, and coding techniques for DS/CDMA and FH/CDMA individually.

15.2   CDMA SYSTEM MODEL

In this section, a model of CDMA is described to clarify a problem in CDMA. For the sake of simplicity, we formalize a DS/CDMA system model.

Figure 15.1 shows a model of DS/CDMA system, where M users are transmitting individually spread spectrum signals $S_m(t)$, $m = 1, 2, \ldots, M$ at the same time and in the same frequency band. This system can also be considered as the reverse link or uplink of a single cell in a cellular mobile communication system [10].

To spread a signal spectrum in a transmitter of the mth user, the message sequence signal $d_m(t)$ is directly multiplied by signature, spreading, or pseudo-noise (PN) sequence signal $C_m(t)$, which has a much higher symbol rate than $d_m(t)$. Signal $S_m(t)$ can be written as

$$S_m(t) = \text{Re}\{d_m(t)C_m(t)\exp[\,j\omega_c t + \theta_m]\} \tag{15.1}$$

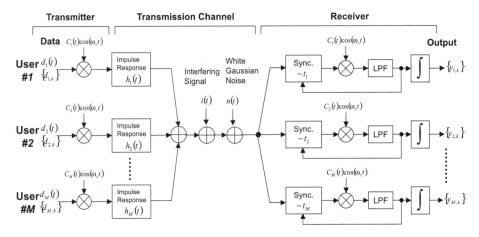

FIGURE 15.1   Model of a DS/CDMA system.

where $j = \sqrt{-1}$, ω_c and θ_m are the carrier frequency and the phase offset, respectively; $d_m(t)$ and $C_m(t)$ are expressed by

$$d_m(t) = \sum_{k=-\infty}^{\infty} d_{m,k} U_{T_d}(t - kT_d) \tag{15.2}$$

$$C_m(t) = \sum_{j=-\infty}^{\infty} C_{m,j} U_T(t - jT) \tag{15.3}$$

where $U_T(t)$ is the unit rectangular pulse defined as $U_T(t) = 1$ for $(0 \leq t \leq T)$ and $U_T(t) = 0$ otherwise. The term $d_{m,k}$ is a symbol of the message sequence $\{d_{m,k}\}$ of the mth user at instant kT_d; $C_{m,j}$ is a symbol of the spreading sequence $\{C_{m,j}\}$ assigned to the mth user at instant jT. In the case of binary phase shift keying (BPSK) $d_{m,k}$, $C_{m,j} \in \{+1, -1\}$. Here, T is the duration of a "chip" in the spreading sequences, while T_d is that of a bit of message sequences. We assume that T_d is a multiple of T and define $N = T_d/T$.

In a receiver, multiplexed SS signals from all users are received and multiplied again by the same spreading sequence signal used in transmitter, so as to transform the spread spectrum of the desired SS signal into its original narrow band. The received signal $r(t)$ can be represented as

$$r(t) = \sum_{m=1}^{M} \int_{-\infty}^{\infty} h_m(\tau) S_m(t - \tau)\, d\tau + i(t) + n(t) \tag{15.4}$$

where $h_m(t)$ is an impulse response of the transmission channel from the mth user to the receiver, $i(t)$ is an interfering or jamming signal, and $n(t)$ is channel noise. In multipath fading channels, the impulse response of the mth user's link takes on the form

$$h_m(t) = \sum_{\lambda=1}^{L_m} g_{m,\lambda} \delta(t - t_{m,\lambda}) e^{j\theta_{m,\lambda}} \tag{15.5}$$

Each path's excess path delay, $t_{m,\lambda}$, is organized in order of increasing magnitude with λ such that $t_{m,1} = 0$ and $T_C \leq t_{m,2} \leq \cdots \leq t_{m,L_m} < \Delta$. Moreover, each is uniformly distributed over the interval $[T_C, \Delta]$ where Δ represents the maximum excess path delay possible; $g_{m,\lambda}$ is the path amplitude from the λth path and $\theta_{m,\lambda}$ is the path phase.

All the multiplexed SS signals except the desired SS signal interfere with the desired SS signal, due to the cross correlation of the different spreading sequences assigned to individual users, and is described as co-channel interference (CCI).

The output signal of the correlator for the nth user can be derived by integrating $r(t)C_n(t)$ over $T_d = NT$. When spreading sequence acquisition is performed, the sample value $y_{n,k}$ of the output at instant kT_d can be represented in the form

$$
\begin{aligned}
y_{n,k} = {}& h_{n,0} d_{n,k} \\
& + \sum_{i \neq 0} \sum_{j=0}^{N-1} h_{n,j+iN} \Theta_{n,n}(j) d_{n,k-i} \\
& + \sum_i \sum_{j=0}^{N-1} \left[\begin{aligned} &\sum_{m=0}^{n-1} h_{m,j+iN}\{R_{n,m}(j+j_m)d_{m,k-i-1} + \hat{R}_{n,m}(j+j_m)d_{m,k-i}\} \\ &+ \sum_{m=n+1}^{M} h_{m,j+iN}\{R_{n,m}(j+j_m)d_{m,k-i} + \hat{R}_{n,m}(j+j_m)d_{m,k-i+1}\} \end{aligned} \right] \\
& + i_{n,k} + n_{n,k}
\end{aligned}
\tag{15.6}
$$

where $h_{m,j} = h_m(jT)$. It is assumed that $j_m = (t_{n,0} - t_{m,0})/T \bmod N$. The functions $R_{n,m}(j)$ and $\hat{R}_{n,m}(j)$ are the partial cross-correlation functions between the spreading sequences of the nth and the mth user [9].

In Eq. (15.6), the first term of the right-hand side indicates the desired baseband data, the second term, the intersymbol interference (ISI) in the desired user's signal, the third, the CCI due to the undesired user, the fourth, the interfering signal component, and the fifth, the noise component. The magnitude of the co-channel interference is determined by either the periodic or even cross-correlation function:

$$\Theta_{n,m}(j) = R_{n,m}(j) + \hat{R}_{n,m}(j) \quad \text{and} \quad \hat{\Theta}_{n,m}(j) = R_{n,m}(j) - \hat{R}_{n,m}(j)$$

depending on $d_{m,k} = d_{m,k-1}$ and $d_{m,k} = -d_{m,k-1}$, respectively.

In the presence of CCI, the cross correlation between the spreading sequences of desired and undesired users prevents acquisition and tracking of the spreading sequence and restricts the number of users simultaneously accessing the channel. This is because side lobes of the output signal of the correlator hamper the detection of the main lobe of autocorrelation and the side lobes increase with the number of available users.

15.3   MULTIUSER DETECTION FOR CDMA

Note from the preceding discussion that CCI due to cross correlation of spreading sequences hampers establishment of acquisition and tracking and limits the capacity of CDMA.

CCI results in the "near–far" problem that relates to the problem of very strong signals from undesired users at a receiver swamping out the effects of weaker signals from desired user [2]. Therefore, CDMA receivers should be designed so as to improve the synchronization and the capacity against the near–far problem.

15.3.1   Single-User Receivers and Multiuser Receivers

In general, CDMA receiver structures can be classified as either multiuser or single-user structures. The conventional single-user detector assigned to each user is described in the previous section. Another example is the multipath combining receiver by Lehnert and Pursley [11]. Optimum single-user detector was proposed by Poor and Verdu [12]. The conventional single-user receiver decides the correlator output for each user only in the interval corresponding to the data symbol individually among multiple access users. In the decision, CCI is considered as an undesired component such as noise.

From an information-theoretic viewpoint, however, we can utilize CCI as redundant information that multiple access users are sharing in a common channel or a multiuser channel, so as to afford performance gains over the conventional single-user receiver. The receiver by which all multiple access users' signals can be correctly detected using CCI is called the multiuser receiver. The multiuser receiver can be used at a cell site of a base station to detect all multiple access users' signals.

In contrast with the single-user receiver, for multiuser receivers, the decision of all the users' data is an interdependent process. Although their structures are much more complicated, their bit error rate (BER) performance can greatly exceed that of the conventional correlator. These receivers vary in complexity from the optimum multiuser detector presented independently by Kohno et al. [13, 14] and Verdu [15] to the suboptimum detectors developed independently by Kohno et al. [16, 17], Masamura [18], Xie et al. [19], and Varanasi and Aazhang [20].

Apart from Kohno et al. [17], these detectors were all designed for the additive white Gaussian noise (AWGN) channel. On the other hand, the receiver of Kohno et al. [17], which adaptively cancels the CCI, was designed for time-varying baseband channels. Yoon et al. [21] proposed multiuser receiver that can be considered as a continuation of the research of Kohno et al. [17] and introduced a cascade of CCI cancellers and its application to multipath fading channels [21–23] independent of Grant et al. [24].

The above-mentioned receivers have been discussed mainly for usual DS/CDMA in which all simultaneous transmitted signals are received as the sum of their envelopes. It is called an *ADDER channel*. On the other hand, FH/CDMA based on multilevel frequency shift keying (FSK) developed by Goodman et al. [25] is considered an *OR channel*, whose receiver has no information erasure due to frequency collision or hit because an energy detector outputs total energy of multiple access users' transmitted signals in individual frequency bands. For such an OR channel, we can improve the capacity of CDMA by applying appropriate error-correcting coding and decoding [26, 27] and canceling CCI [28] different from them for the ADDER channel. These techniques will be introduced in Section 15.5. Kawahara and Matsumoto [29] introduced similar *OR logic operation* into DS/CDMA so as to utilize efficient decoding for the multiuser channel.

15.3.2   Optimal Multiuser Receiver

Without loss of generality and for the sake of notational simplicity, (15.6) can be rewritten if there is no intersymbol interference or multipath distortion in a channel, no narrowband interference, and only two users accessing the channel:

$$y_{1,k} = h_{2,0}R_{1,2}\mathbf{d_{2,k-1}} + h_{1,0}\mathbf{d_{1,k}} + h_{2,0}R_{2,1}\mathbf{d_{2,k}} + n_{1,k} \tag{15.7}$$

$$y_{2,k} = h_{1,0}R_{2,1}\mathbf{d_{1,k}} + h_{2,0}\mathbf{d_{2,k}} + h_{1,0}R_{1,2}\mathbf{d_{1,k+1}} + n_{2,k} \tag{15.8}$$

where $R_{1,2}$ and $R_{2,1}$ are partial cross-correlation factors. It is noted from these equations that the correlator outputs $y_{1,k}$ and $y_{2,k}$ can be considered as multidimensional finite-state machine outputs corrupted with noise or multidimensional analog convolutional encoder output with noise. Figure 15.2 illustrates its equivalent multidimensional tapped delay lines.

Therefore, we can achieve optimum multiuser detection for CDMA by using a Viterbi algorithm that is well known as an efficient maximum-likelihood sequence detection for a convolutional code, a trellis-coded modulation and so on. In general, the optimum K-user receiver consists of a bank of single-user correlators in Figure 15.1 followed by a Viterbi algorithm whose complexity per binary decision is $O(2^K)$ [13–15]. Figure 15.3 shows worst-case and average error probabilities achieved by conventional single-user and optimum multiuser receivers with three active users using m sequences of period 31.

15.4   CO-CHANNEL INTERFERENCE CANCELLATION FOR DS/CDMA

The optimum multiuser receiver affords much performance gain over conventional single-user one but expends some possible disadvantages such as higher complexity and the need

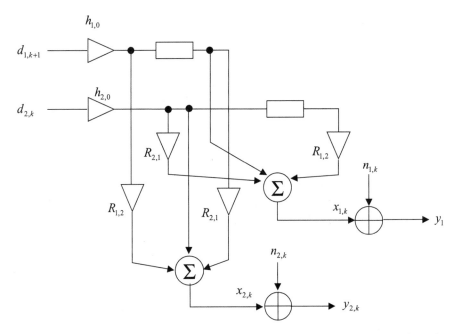

FIGURE 15.2 Equivalent tapped delay line model of asynchronous DS/CDMA channel (two users).

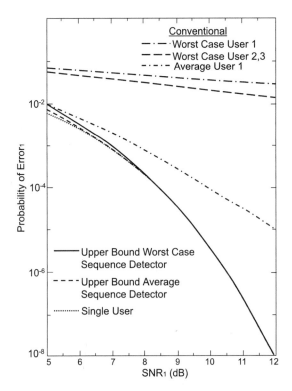

FIGURE 15.3 Error probabilities of conventional and optimum multiuser receivers (three users, m-sequence of period 31).

to know the relative characteristics of the multiuser channel, for example, amplitudes, path delays, and path phases in (15.5). For indoor and mobile radio communications, implementation complexity should be reduced to a feasible level even if the performance degrades slightly from an optimum one. Estimation of the channel characteristics used to be more difficult due to its time variance. In fact, presently developed CDMA cellular radio networks do not include the optimum mutiuser receiver.

In this section, CCI cancellation techniques, which carry out suboptimum but more practical multiuser receivers, will be explained. There are several methods that strive to reduce or remove CCI. The design of spreading sequences attempts to produce sets of sequences with optimum correlation properties [9, 30, 31]. To further aid the receiver in canceling CCI, temporal, spatial signal processing and their combination can be employed.

15.4.1 CCI Cancellation by Temporal Adaptive Filtering

Single-Stage CCI Canceller   The receiver must obtain initial data estimates for use in CCI cancellation. We simply perform a hard decision upon the kth user's correlator output $y_{k,n}$ in (15.6) as shown in Figure 15.1; $\hat{b}_n^{(k)} = \mathrm{sgn}(y_{k,n})$. Each initial data estimate, $\hat{b}_n^{(k)}$, of the kth user's originally transmitted data $d_{k,n}$ at instant $nT_b(T_b = T)$, is first multiplied by its respective synchronized spreading sequence and carrier signal to create a replica of the originally transmitted signal, as shown in Figure 15.4. Then the signal is passed through a transversal filter (TF) that emulates channel $h_k(t)$. The structure of the TF is shown in Figure 15.5. Each tap coefficient, at the appropriate excess path delay, $t_{k,\lambda}$, is $G_{k,\lambda} = g_{k,\lambda} \cos \phi_{k,\lambda}$. The estimation of the channel parameters may be feasible by an adaptive updating algorithm [17, 22, 23].

The output of the filter consists of two signals: the sum, $F^{(k)}$, of the kth user's faded terms ($\lambda \geq 2$) and the full replica of his received signal component, which is the sum of $F^{(k)}$ and his first path's signal ($\lambda = 1$), $D^{(k)}$. By adding the mth user's own faded terms, $F^{(m)}$, with all the other users' $F^{(k)} + D^{(k)}$, the canceller can create a replica of the CCI of

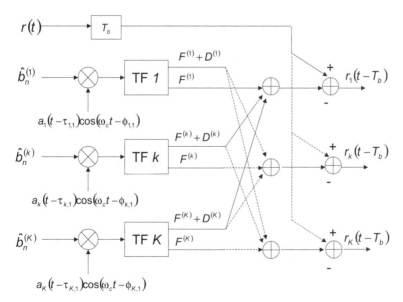

FIGURE 15.4   CCI canceller.

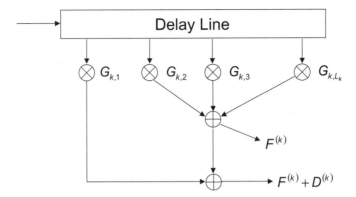

FIGURE 15.5 kth transversal filter (TF).

the mth user (including his intersymbol interference). How closely it resembles the actual CCI depends on two things: the number of correct data estimates and the accuracy of the channel estimates. The CCI replica of each user is subtracted from the original received signal, $r(t)$, to produce a "cleaned" received signal, $r_m(t)$, for the mth user. This signal is passed to a second set of correlators, as shown in Figure 15.6, to obtain a second set of data estimates. The accuracy of the canceller depends on the ratio of the number of correct data estimates to the total number data estimates. As long as the ratio is high, more of the CCI can be removed instead of being added.

Cascade of CCI Canceller Next, a cascade of such CCI canceller and data decision stages (Figures 15.4 and 15.6) can be constructed as shown in Figure 15.7. The decisions of the final stage give the final data decisions for the receiver. Intuitively, if the number of incorrect data decisions at the first stage in comparison to that at the initial data estimation has been reduced, more accurate replicas can be reconstructed at the CCI canceller of the second stage. In repeating this process, at a certain stage, the data decisions should converge to the actually transmitted data values.

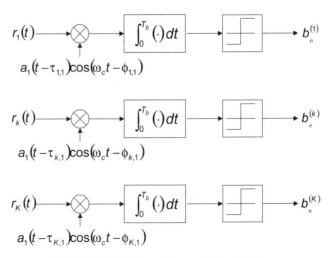

FIGURE 15.6 Correlator and data decision.

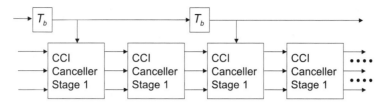

FIGURE 15.7 Cascade of CCI cancellers.

BER Performance The probabilities of bit error in the cascade of CCI cancellers, which was calculated from the theoretical BER analysis and from computer simulations, are shown in Figures 15.8 to 15.10. The system, channel, and simulation parameters are listed in [32].

15.4.2 CCI Cancellation by Spatial Adaptive Filtering

An adaptive array antenna is useful in suppressing interfering signals because it can adaptively control directivity of the antenna by updating weights of element antennas even if the desired signal's arrival angle is unknown [33]. It is usually assumed that the reference signal is correct, and the array can suppress interference effectively enough to obtain correct reference signal. However, since the weights are updated by the error signal or the difference between the reference signal and the antenna output, which includes residual interference and noise, it is difficult to achieve stable convergence of the weights when power of interference and/or noise is much larger than that of desired signal, that is, low desired signal-to-interference power ratio (DIR) and desired signal-to-noise power ratio (DNR), which are usual in a CDMA channel.

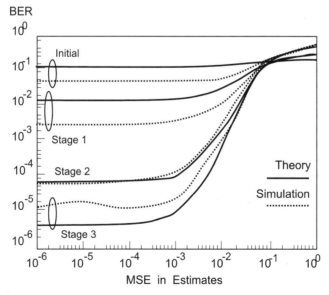

FIGURE 15.8 BER vs. MSE with $K = 64$ and $E_b/N_0 = 10$ dB for up to three canceller stages. $\bar{L} = 2$ and $N = 127$.

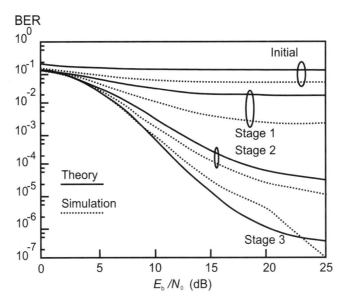

FIGURE 15.9 BER vs. E_b/N_0 with $K = 64$ and MSE $= 10^{-2}$.

Adaptive Array Antenna Using the Processing Gain To solve the above-mentioned problem, we proposed a method to more correctly update the weights using the inherent processing gain of a single-user receiver for DS/CDMA [34]. Figure 15.11 shows a DS/CDMA receiver with the proposed method of updating array. As the error signal for updating the weights, we can use the difference between the input and output

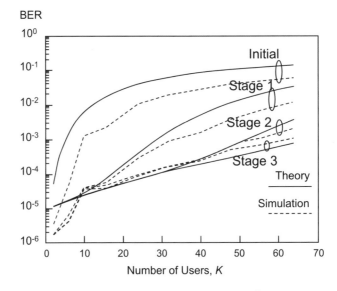

FIGURE 15.10 BER vs. K with $E_b/N_0 = 10$ dB and MSE $= 10^{-2}$ for up to three canceller stages. $\bar{L} = 2$ and $N = 127$.

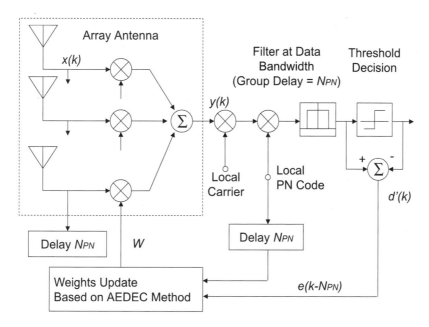

FIGURE 15.11 DS/CDMA receiver with the proposed adaptive array antenna using the processing gain.

signals of threshold decision. The input signal includes less interference and noise than the output signal of array even in case of low DNR and DIR, because these undesired signal can be suppressed by the processing gain, which is defined as a ratio of radio frequency (RF) bandwidth-to-baseband one.

However, convergence of the weights is disturbed by the delay in the controlling loop, which is due to despreading or demodulation in the correlator. If duration of a data bit equals to a period of spreading sequence, that is, N chips, the delay will be N. It is assumed that weights are updated chip by chip. To calculate the weights at instant k, we can use only the delayed error, which is obtained at instant $k - N$. Therefore, stable convergence of the weights cannot be guaranteed in a time-varying channel such as a fading channel.

To improve the convergence, we can introduce the same way as an AEDEC, that is, an automatic equalizer including a decoder of error-correcting code [35].

Frequency Performance of Array Antennas   In general, an adaptive array antenna is controlled so as to have maximum gain in direction of desired signal and null to that of interfering signals. It is called a spatial filter. On the other hand, since an array antenna such as least mean square (LMS) array can pass a signal possessing the same frequency component as the reference signal and reject that doing the different frequency one, it can also be considered as a temporal filter.

In an array antenna, there are different time delays among element antennas according to arrival angles of the signal waves. We can equivalently represent an array antenna as a model of tapped delay line, in which the delay intervals T_i of signal waves with arrival angles θ_i $(i = 0, 1, 2, \ldots, I)$ can be written by

$$T_i = \frac{L}{c}\sin(\theta_i), \tag{15.9}$$

where L is spatial interval among elements and c is a propagation speed of electric magnetic wave.

If arrival angles and received power of the signal waves are known, optimum weights of the array can be obtained by solving the Wiener–Hopf equation. Then the frequency performance of the array can be derived. Figures 15.12 and 15.13 show the frequency performance and the antenna pattern of array with two elements when there are two waves with arrival angles of $0°$ and $60°$.

TDL Array Antenna for Wideband Signals A tapped delay line (TDL) array antenna was proposed to reject wideband interference [33]. Figure 15.14 shows a structure of the TDL array antenna. Since TDL of each element antenna can more efficiently utilize correlation of received signals than a usual line array antenna, the frequency performance can be improved for wideband signals such as interfering SS signals.

Figure 15.15 shows the frequency performance of the TDL array antenna with two elements and five taps in an element. It is noted that Figure 15.8 can reject wider band interference than Figure 15.12.

TDL Adaptive Array Antenna Using the Processing Gain In particular, to reject co-channel interference with wideband in DS/CDMA when DIR and DNR are low, we can introduce the proposed updating method using the processing gain to the TDL array antenna. A DS/CDMA receiver with the TDL-adaptive array antenna using the processing gain can be constructed by combining Figures 15.11 and 15.14.

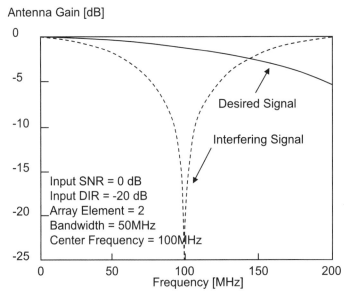

FIGURE 15.12 Temporal frequency performance of the array antenna with two elements.

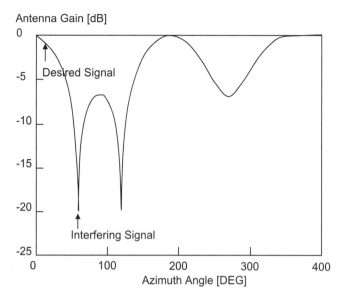

FIGURE 15.13 Antenna pattern of the array antenna with two elements.

System Evaluation Computer simulations illustrate performance of the proposed TDL-adaptive array antenna using the processing gain in comparison with conventional ones. In simulations, we assumed that there are desired and interfering SS signal waves with arrival angles of $20°$ and $60°$, respectively, a carrier frequency of 100 MHz, spreading sequences with a chip rate of 25 Mcps and a period of 40 ns, a data rate of 98 kbps, and a SS signal bandwidth of 50 MHz. The TDL array antenna has two elements and five taps in an element with a tap delay interval of 10 ns ($r = 2$).

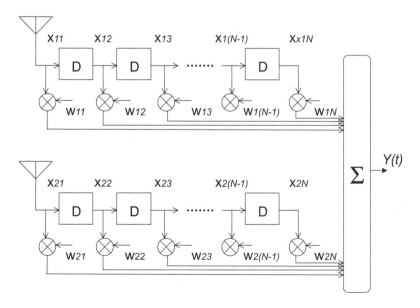

FIGURE 15.14 Structure of a tapped delay line (TDL) array antenna gain (dB).

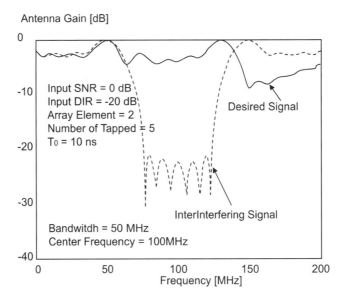

FIGURE 15.15   Temporal frequency performance of the TDL array antenna with two elements.

Figure 15.16 shows the output DIR corresponding to the bandwidth of SS signals. From this figure, we confirm that the proposed TDL array antenna can achieve the best DIR performance for wideband SS signals.

Figures 15.17 and 15.18 show BER performances vs. the input DNR and the input DIR, respectively. We can confirm that the proposed TDL-adaptive array antenna using the processing gain improves the BER performance in low DNR and DIR.

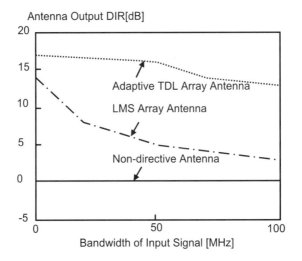

FIGURE 15.16   Output DIR corresponding to bandwidth of SS signals.

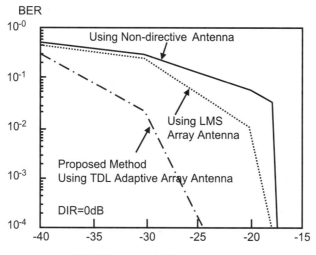

FIGURE 15.17 BER vs. input DNR.

15.4.3 Combination of Spatial and Temporal Filtering

If there is a high-level interfering SS signal from an undesired user with the same arrival angle as that of a desired user, an array antenna cannot suppress it. To solve this problem, a combination of an adaptive array antenna in the spatial domain and an adaptive interference canceller in the temporal domain was proposed [36]. The proposed system can suppress interfering SS signals, that is, co-channel interference, with arrival angles different from that of a desired user by using a null steering array antenna and eliminate by means of a canceller the residual interference and co-channel interference having an arrival angle the same as that of the desired SS signal.

Figure 15.19 shows the structure of the DS/CDMA system based on a combination of adaptive spatial and temporal filtering, which consists of array antenna and canceller using

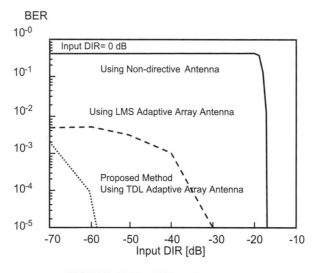

FIGURE 15.18 BER vs. input DIR.

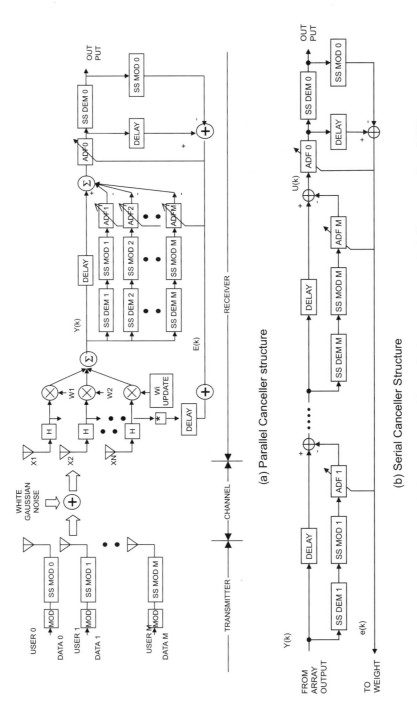

(a) Parallel Canceller structure

(b) Serial Canceller Structure

FIGURE 15.19 DS/CDMA system with an adaptive array antenna including a canceller of CCI.

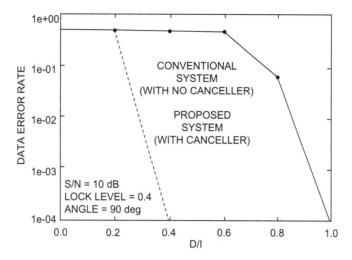

FIGURE 15.20   Demodulated BER vs. DIR.

adaptive digital filters (ADFs). Figure 15.20 shows BER of demodulated data as a function of the D/I ratio (DIR) for the same arrival angle. The proposed system can achieve stable demodulation and improve BER performance even in a heavy interference channel where a conventional array antenna system cannot achieve acquisition.

15.4.4   Combination of CCI Cancellation and Power Control

In the serial structure of the canceller shown in Figure 15.19(b), interfering SS signals or CCIs from multiple access users are canceled in the order of decreasing receiving power because it is easy to achieve acquisition, demodulation, and cancellation of an interfering SS signal having greater power and its cancellation makes it possible or easy to cancel other interfering SS signals [36, 37].

This discussion suggests that the performance of CCI cancellation or the capacity of CDMA depends upon receiving signal power distribution among multiple access users. In fact, when the receiving power distribution is exponential, the capacity can be improved to be 1.5 times as large as that in uniform power distribution.

Although transmission power control is considered as a key technology for increasing the capacity in a DS/CDMA cellular system, the receiving power of multiple access users is controlled so uniform as to solve the near–far problem. Errors in the power control results in severe degradation of the capacity [38]. However, if the power control is applied so as to make CCI cancellation more efficient, the power control errors can also be compensated by CCI cancellation.

15.5   CO-CHANNEL INTERFERENCE CANCELLATION FOR FH/CDMA

There is no near–far problem in an FH/CDMA system and its frequency diversity effect is useful to combat fading in personal, indoor and mobile radio communications. However, coincidence or collision of hopping frequencies of desired and undesired SS signals causes co-channel interference, namely "hit," which restricts the capacity of FH/CDMA.

In this section, several countermeasures against CCI in FH/CDMA are introduced.

15.5.1 Countermeasures for CCI

Coding Schemes An effective countermeasure against a hit or CCI in FH/CDMA is to introduce coding redundancy or an error-correcting code, so that if a given hop is hit, other hopefully more reliable received data symbols can be used to resolve the contaminated data symbol. This is also referred to as time diversity. Each data symbol is partitioned into several subsymbols. Each subsymbol is transmitted on a different hop using fast frequency hopping (FFH) or slow frequency hopping (SFH) with pseudorandom interleaving. The intention is that each subsymbol comprising a data symbol will have an independent chance of being hit. Since these subsymbols are distributed in time, this technique is termed time diversity. For FH/CDMA, to reduce average hit probability or to increase the capacity of FH/CDMA, coding redundancy should be designed for all multiple access users [2].

Multihopping Schemes To obtain more efficient diversity effect for FH/CDMA, two schemes of multihop-FFH in which several hopping frequencies are employed to convey the same information or datum in serial and parallel, that is, serial-hop-FFH and parallel-hop-FFH schemes, respectively, have been investigated [39]. The serial-hop-FFH scheme is defined as the FFH scheme in which several consecutive hopping frequencies are assigned to one data symbol, and the parallel-hop-FFH scheme is defined as the FFH scheme in which simultaneously several hopping frequencies are assigned to one data symbol. The serial-hop-FFH scheme is a conventional FFH scheme. Further, the single-hop-FFH scheme is defined as the scheme in which one hopping frequency is assigned to one data symbol. Multihopping schemes can improve the capacity of CDMA particularly in a fading channel.

Hybrid DS/FH Schemes Hybrid DS/FH schemes have been investigated to utilize advantages of both DS and FH, such as excellent resistance to AWGN, narrowband interference, and CCI in CDMA. In particular, even if there is a hit due to collision of several simultaneous accessing users' carrier frequencies in a given hop, DS despreading can distinguish between the hit users' signals in the hit frequency according to the processing gain of DS part. Optimization of bandwidth spread by DS and FH is interesting for a given bandwidth. Although acquisition for FH used to be difficult, acquisition for DS can be used to achieve that for FH in a hybrid DS/FH system [40, 41].

Spatial Filtering Schemes An adaptive array antenna or an adaptive beamformer is also available to suppress CCI in FH/CDMA in a similar way to CCI cancellation in DS/CDMA. In particular, we have proposed and investigated an algorithm for controlling weights of antenna elements that are derived from discrete Fourier transform (DFT) of spatially sampled signals in array of antenna elements [42, 43]. Since the weights of antenna elements are derived from the spatial frequency spectra instead of the temporal sampled signals unlike in LMS and constant modulus algorithm (CMA) algorithms, optimum weights can be quickly obtained independent of temporal sampling interval. The algorithm uses no reference signal.

15.5.2 Multilevel FSK/CDMA System Using Address Codes

The MFSK (multilevel frequency shift keying)/FH-CDMA system using the address code was proposed by Goodman et al. [25]. In the system using the address code as the hopping sequence, the carrier frequency in MFSK is hopped [44, 45]. Even when there is a hit or

CCI, the desired signal can be separated from the undesired signals by the decoding based on the address code. However, when many users' signals are overlapped or hit in CDMA, the desired signal cannot be always decoded correctly. As mentioned in Section 15.3.1, since this multiuser channel is considered as an OR channel, we can improve the capacity of CDMA by applying appropriate error-correcting coding and decoding [26, 27, 29, 46].

Coded MFSK/CDMA System   Figure 15.21 shows the system model of the coded MFSK/FH-CDMA system, and Figure 15.22 shows an example of time–frequency matrices that indicate the state of the frequency tones of the each chip in each part of Figure 15.21.

Modified Decoding Scheme for a Multiuser Receiver in the Coded MFSK/CDMA

In conventional decoding, if there are several rows having the largest number of entries in one time–frequency matrix decoded by the address code, then the received symbol, which corresponds to that matrix, must be considered as an erasure and decoded by the conventional error and erasure correction decoding.

The erasure is considered to be one among 2^K kinds of symbols with equal probability. In our modified decoding scheme, however, we use the information that indicates which symbols correspond to rows having the largest number of entries in the time–frequency matrix. Since we can reduce the number of possible symbols or candidates by the information, the error and erasure correcting capability can be improved and the user capacity can consequently increase.

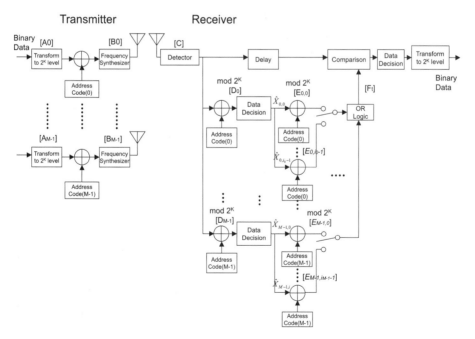

FIGURE 15.21   MFSK/FH-CDMA system using the proposed multiuser receiver based on CCI canceller.

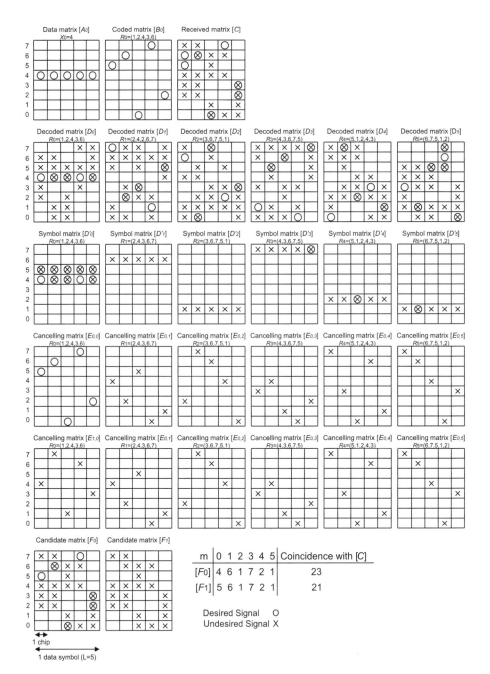

FIGURE 15.22   Example of decoding with CCI cancellation ($2^K = 8$, $L = 5$, $M = 6$; five users CDMA).

The modified decoding algorithm is as follows:

1. In the case where the number of erasures e with several rows having the largest number of entries in the decoded time–frequency matrix is less than $d_{min} - 1$, the error and erasure correction decoding scheme is performed in its usual manner. The error and erasure correcting capability of the Reed-Solomon (RS) code is expressed by

$$2s + e + 1 \leq d_{min} \tag{15.10}$$

where s and e denote the number of errors and erasures, respectively.

2. In the case where the number of erasures e is greater than $d_{min} - 1$, each information of symbols corresponding to rows having the largest number of entries of n decoded time–frequency matrices with one row and more than two rows having the largest number of entries are registered as candidates of the correct symbol. The decoding candidates are derived from $e_1 (\geq d_{min} - 1)$ decoded time–frequency matrices with one row and two rows having the largest number of entries. The check candidates are derived from $e_2 (= n - e_1)$ decoded time–frequency matrices with rows having the largest number of entries except e_1 decoded matrices containing the decoding candidates.

3. The symbols corresponding to e_2 decoded matrices containing the check candidates are considered as erasures, and the error and erasure correction decoding is performed by decoding each possible combination of the several decoding candidates. Consequently, we can obtain several candidates of the correct codeword.

4. The codeword closest to each possible combination of candidates of the correct symbol is selected as the correct codeword from several candidates of the correct codeword decoded by operation (15.3). If there are more than two codewords closest to the decoded codeword, we will select an arbitrary codeword among their codewords.

System Evaluation   Figure 15.23 shows the theoretically calculated user capacity normalized by the code rate versus the code rate for the modified decoding scheme in comparison with the usual decoding scheme without erasure, the conventional error and erasure correction decoding scheme, and the maximum-likelihood decoding scheme.

In the maximum-likelihood decoding scheme, the user capacity is theoretically calculated from the channel cutoff rate without restricting the utilized error-correcting code. Therefore, the user capacity of the maximum-likelihood decoding scheme using RS code is smaller than a bound of the user capacity derived from the channel cutoff rate because of the restriction of the length of RS code.

From Figure 15.23, it is noted that the user capacity in the proposed decoding scheme gives an extra 10 users at the code rate of 0.8 in comparison with that of the conventional error and erasure decoding scheme.

15.5.3 Multiuser Receiver Based on CCI Cancellation for a MFSK/CDMA System

As an optimum multiuser receiver for the MFSK/CDMA, we have proposed and investigated a multiuser receiver in which all users' data are correctly decided by canceling

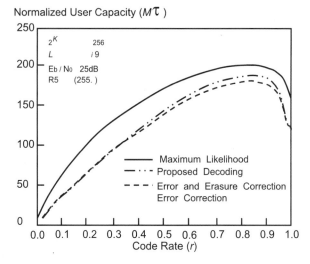

Normalized User Capacity ($M\tau$)

Maximum Likelihood
Proposed Decoding
Error and Erasure Correction
Error Correction

Code Rate (r)

FIGURE 15.23 Normalized user capacity vs. code rate in the coded MFSK/FH-CDMA system in a Rayleigh fading channel.

CCI to improve user capacity of CDMA [28, 47]. In decoding, we use the time–frequency matrices in which the desired and undesired signal tones are marked with OR logic. When in the decoding matrix there are not only a correct row corresponding to the desired user's data tone but also aliasing rows that consist of undesired users' signal tones and have no less entries than the correct row, the conventional decoding cannot correctly detect the desired user's data tone. However, since the proposed receiver can regenerate replicas of the undesired users' signal tones by using known address codes of every user, it can eliminate CCI by subtracting entries of the replicas from the decoding time–frequency matrix.

System Configuration Figure 15.21 shows the model of MFSK/FH-CDMA system with the proposed multiuser receiver, and Figure 15.22 illustrates the decoding and canceling CCI using the time–frequency matrices, which indicate the state of the frequency tones of the each chip in each part of Figure 15.21.

Multiuser Detection Based on CCI Cancellation The process of decoding and canceling CCI in MFSK/FH-CDMA system is described corresponding to Figures 15.21 and 15.22 as follows:

1. The matrices $[D_m]$ $(m = 0, 1, \ldots, M - 1)$ of Figure 15.22 show each user's time–frequency matrices decoded by each user's address code $\mathbf{R_m}$ $(m = 0, 1, \ldots, M - 1)$. $Z_{m,n,l}$ $(m = 0, 1, \ldots, M - 1)$ shows an entry of the vectors decoded by the address code or an entry of $[D_m]$ $(m = 0, 1, \ldots, M - 1)$. The subscript n denotes entries of nth user over the time–frequency matrix. When $n = m$, that shows entries of the desired user.

2. Let i_m $(1 \leq i_m \leq 2^K)$ be the number of rows having the largest number of entries in the time–frequency matrix decoded by the mth user's address code [e.g., there are two rows having five entries at $[D_0]$ in Figure 15.22, i.e., $i_0 = 2$, and $i_m = 1$ $(m = 1, \ldots, M - 1)$].

The symbols corresponding to the rows having the largest number of entries in Figure 15.22 $[D_m]$ $(m = 0, 1, \ldots, M - 1)$, are considered as estimates $\hat{X}_{m,j}$ $(j = 0, 1, \ldots, i_m - 1)$ of the correct data symbols of the mth user. $[D'_m]$ $(m = 0, 1, \ldots, M - 1)$ in Figure 15.22 shows the estimates.

3. Replicas of all users' transmitted vectors are regenerated by adding the estimates with the address codes, for every combination of estimates of all the user's data symbols.

   The replica $W_{m,j,l}$ regenerated by the estimate $\hat{X}_{m,j}$ of the mth user's data can be represented by $W_{m,j,l} = \hat{X}_{m,j} + R_{m,l}$ for $(m = 0, 1, \ldots, M - 1)$, $(l = 0, 1, \ldots, L - 1)$, $(j = 0, 1, \ldots, i_m - 1)$. $[E_{m,j}]$ in Figure 15.22 shows the replicas regenerated by the estimate $\hat{X}_{m,j}$. These replicas $\hat{W}_{m,j,l}$ $(m = 0, 1, \ldots, M - 1)$ are added in logical OR operation. Then, we can derive $\prod_{m-0}^{M-1} i_m$ candidates of the received matrices $[F_t](t = 0, 1, \ldots, \prod_{m=0}^{M-1} i_m - 1)$ in Figure 15.22.

4. If we select the $[F_t](t = 0, 1, \ldots, \prod_{m=0}^{M-1} i_m - 1)$, which has the most number of coincident entries with entries of the received matrix $[C]$, we can decode all users' data correctly.

System Evaluation Figure 15.24 shows BER versus the number of users derived by the theoretical analysis and computer simulation for the MFSK/FH-CDMA system without and with cancellation of CCI. Of course, if we jointly use the multi-user receiver based on CCI cancellation and the modified erasure decoding mentioned in Section 15.5.2 for coded MFSK/CDMA systems, optimum performance will be improved.

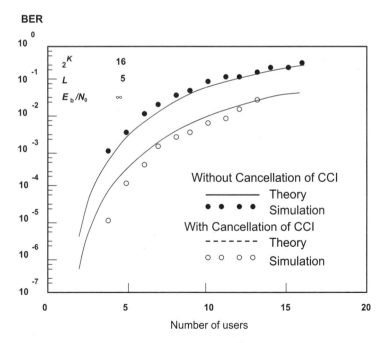

FIGURE 15.24 BER vs. number of users ($2^K = 16$, $L = 5$, $E_b/N_0 = 20$ dB).

15.6 CONCLUDING REMARKS

For DS/CDMA and FH/CDMA, optimum and suboptimum multiuser receivers based on CCI cancellation have been discussed. For wider commercial applications of SS techniques, that is, consumer and personal communications, improving transmission capacity of CDMA and combating with narrowband interference and fading are required compatibly. For this purpose, we should study total performance of capacity and complexity on practical implementation, relationship between regulation and security, common use of frequency bands with other systems, development of Application Specified IC (ASIC) for SS systems, and the like [48]. It would be worthwhile to study these subjects for future consumer and personal communications.

REFERENCES

1. Study Report on Standard for Wireless LAN, RCR July 1992.

2. M. K. Simon, J. K. Omura, R. A. Scholtz, and B. K. Levitt, *Spread Spectrum Communications*, Computer Science Press, 1985.

3. R. Kohno, "A Trend of Research and Development Concerning Low Power Radio Systems for Consumer Communications," *IEICE Tech. Rep.*, SSTA89-1, pp. 1–8, March 1989.

4. G. Marubayashi, M. Nakagawa, and R. Kohno, "Research and Development Activities on Spread Spectrum Communication Systems," *J. IEICE*, Vol. 73, No. 2, pp. 580–592, May 1989.

5. R. Schneiderman, "Spread Spectrum Gains Wireless Applications," *Microwaves & RF*, May 1992.

6. Federal Communications Commission (FCC) technical regulations 15.247.

7. K. S. Gilhousen et al., "On the Capacity of a Cellular CDMA System," *IEEE Trans. Veh. Tech.*, Vol. VT-40, No. 2, pp. 303–312, May 1991.

8. R. Kohno, "Interference Cancellation and Information Security in Spread Spectrum Systems," *IEICE Tech. Rep.*, SST91-5, pp. 29–36, April 1991.

9. R. Kohno, "Pseudo-Noise Sequences and Interference Cancellation Techniques for Spread Spectrum Systems," *IEICE Trans. Commun., Electronics, Inform. Sys.*, Vol. E74, No. 5, May 1991.

10. A. Salmasi and K. S. Gilhousen, "On the System Design Aspects of Code Division Multiple Access (CDMA) Applied to Digital Cellular and Personal Communications Networks." *Proc. IEEE VTS Conf.*, May 19–22, 1992.

11. J. S. Lehnert and M. B. Pursley, "Multipath Diversity Reception of Spread-Spectrum Multiple-Access Communications," *IEEE Trans. Commun.*, Vol. COM-35, pp. 1189–1198, Nov. 1987.

12. H. V. Poor and S. Verdu, "Single-User Detectors for Multiuser Channels," *IEEE Trans. Commun.*, Vol. COM-36, pp. 50–60, Jan. 1988.

13. R. Kohno, H. Imai, and M. Hatori, "Cancellation Techniques of Co-Channel Interference in Asynchronous Spread Spectrum Multiple Access," *IECE Tech. Rep.*, CS82-38, pp. 29–35, June 1982.

14. R. Kohno, H. Imai, and M. Hatori, "Soft Decision Receiver Using Viterbi Algorithm for Asynchronous CDMA," *Proc. IECE National Conf.*, 1322, p. 5–193, March 1983.

15. S. Verdu, "Minimum Probability of Error for Asynchronous Gaussian Multiple Access Channels," *IEEE Trans. Inform. Theory*, Vol. IT-32, pp. 85–96, Jan. 1986.

16. R. Kohno, H. Imai, and M. Hatori, "Cancellation Techniques of Co-Channel Interference in Asynchronous Spread Spectrum Multiple Access Systems," *Trans. IECE, Japan*, Vol. J66-A, No.

5, pp. 416–423, May 1983 (in Japanese). *Electronics & Communications in Japan*, Scripta, Vol. 66, No. 5, pp. 20–29, May 1983 (English version).

17. R. Kohno, H. Imai, M. Hatori, and S. Pasupathy, "An Adaptive Canceller of Co-channel Interference for Spread Spectrum Multiple Access Communication Networks in a Power Line," *IEEE J. Select. Areas Commun.*, Vol. SAC-8, pp. 691–699, May 1990.

18. T. Masamura, "Spread Spectrum Multiple Access System with Intrasystem Interference Cancellation," *IEICE Trans., Japan*, Vol. E71, No. 3, pp. 224–231, March 1988.

19. Z. Xie, R. T. Short, and C. T. Rushforth, "A Family of Suboptimum Detectors for Coherent Multiuser Communications," *IEEE J. Select. Areas Commun.*, Vol. SAC-8, pp. 683–690, May 1990.

20. M. K. Varanasi and B. Aazhang, "Multistage Detection in Asynchronous Code-Division Multiple-Access Communications," *IEEE Trans. Commun.*, Vol. COM-38, pp. 509–519, April 1990.

21. Y. C. Yoon, R. Kohno, and H. Imai, "A Spread-Spectrum Multi-Access System with a Cascade of Co-Channel Interference Cancellers for Multipath Fading Channels," *Proc. IEEE Int. Symp. Spead Spectrum Tech. & Appl.* (ISSSTA'92) pp. 87–90, Nov. 1992.

22. Y. C. Yoon, R. Kohno, and H. Imai, "Cascaded Co-Channel Interference Cancelling and Diversity Combining for Spread-Spectrum Multi-Access over Multipath Fading Channels," *IEICE Trans. Commun., Japan*, Vol. E76-B, No. 2, Feb. 1993.

23. Y. C. Yoon, R. Kohno, and H. Imai, "A Spread-Spectrum Multi-Access System with Co-Channel Interference Cancellation over Multipath Fading Channels," *IEEE J. Select. Areas Commun.*, Dec. 1993.

24. P. M. Grant, S. Mowbray, and R. D. Pringle, "Multipath and Co-Channel CDMA Interference Cancellation," *Proc. IEEE ISSSTA'92*, pp. 83–86, Nov. 1992.

25. D. J. Goodman, P. S. Henry, and V. K. Prabhu, "Frequency-hopped Multilevel FSK for Mobile Radio," *Bell Sys. Tech. J.*, Vol. 59, No. 7, pp. 1257–1275, Sept. 1980.

26. T. Kawahara and T. Matsumoto, "Optimum Rate Reed Solomon Codes for FFH/CDMA Mobile Radio," *IEE Elec. Letts.*, Vol. 27, No. 22, pp. 2066–2067, Oct. 1991.

27. T. Mabuchi, R. Kohno, and H. Imai, "An Advanced Decoding Scheme of a Coded MFSK/FH-CDMA System," *IEICE Tech. Rep.*, SST92-25, June 1992.

28. T. Mabuchi, R. Kohno, and H. Imai, "A Scheme of Cancelling Co-Channel Interference in MFSK/FH-CDMA System," *Proc. 15th Symp. Information Theory and Its Applications (SITA'92)*, Sept. 1992.

29. T. Kawahara and T. Matsumoto, "A DS/CDMA Mobile Radio System Featuring OR-Channel Characteristic," *Proc. IEICE National Conf.*, A-131, pp. 1–133, Sept. 1992.

30. D. V. Sarwate and M. B. Pursley, "Crosscorrelation Properties of Pseudorandom and Related Sequences," *Proc. IEEE*, Vol. 68, pp. 593–619, 1980.

31. H. Fukumasa, R. Kohno, and H. Imai, "Design of Pseudo-Noise Sequences with Good Odd and Even Correlation Properties," *Proc. IEEE ISSSTA'92*, pp. 139–142, Nov. 1992.

32. Y. C. Yoon, R. Kohno, and H. Imai, "Performance of a Cascade of Co-Channel Interference Cancellers for DS/CDMA in the Presence of Amplitude and Phase Estimation Errors," *IEICE Tech. Rep.*, SST92-65, pp. 31–36, Jan. 1993.

33. R. T. Compton, Jr., *Adaptive Antennas Concepts and Performance*, Prentice-Hall, 1988.

34. R. Kohno, H. Wang, and H. Imai, "Adaptive Array Antenna Combined with Tapped Delay Line Using Processing Gain for Spread-Spectrum CDMA Systems," *Proc. 3rd IEEE Int. Sym. Personal, Indoor and Mobile Radio Commun.*, pp. 634–638, Oct. 1992.

35. R. Kohno, H. Imai, and M. Hatori "Design of an Automatic Equalizer Including a Decoder of Error-Correcting Code," *IEEE Trans. Commun.*, Vol. COM-33, pp. 1142–1146, 1985.

36. R. Kohno, H. Imai, M. Hatori, and S. Pasupathy, "Combination of an Adaptive Array Antenna and a Canceller of Interference for Direct-Sequence Spread-Spectrum Multiple-Access System," *IEEE J. Selected Areas Commun.*, Vol. JSAC8, No. 4, pp. 675–682, May 1990.

37. A. J. Viterbi, "Very Low Rate Convolutional Codes for Maximum Theoretical Performance of Spread-Spectrum Multiple-Access Channels," *IEEE J. Select. Areas Commun.*, Vol. JSAC-8, pp. 641–649, May 1990.

38. E. Kudoh and T. Matsumoto, "Effects of Power Control Error on the System User Capacity of DS/CDMA Cellular Mobile Radios," *IEICE Trans. Commun., Japan*, Vol. E75-B, pp. 524–529, June 1992.

39. H. Yamamura, R. Kohno, and H. Imai, "Interference Cancellation and Capacity of a Cellular CDMA System Based on Multihop Fast Frequency Hopping," *Proc. IEEE Veh. Tech. Conf.*, pp. 71–76, May 1991

40. R. Kohno, H. Imai, M. Yamamoto, and A. Hoshikugi, "Remote Control Radio System Based on Hybrid DS/FH Spread Spectrum Technique," *Proc. SS Workshop, Montebello, Quebec*, pp. 21–23, May 1991.

41. A. Hoshikugi, M. Yamamoto, S. Ishi, R. Kohno, and H. Imai, "Implementation of an Industrial R/C System Using a Hybrid DS/FH Spectrum Techniques," *Proc. IEEE ISSSTA'92*, pp. 179–182, Nov. 1992.

42. T. Mabuchi, R. Kohno, and H. Imai, "A Note on Cancelling Cochannel Interference in Frequency Hopping Multiple Access (FH/CDMA)," *IEICE Tech. Rept.*, SSTA91-12, pp. 75–82, March 1991.

43. C. Yim, R. Kohno, and H. Imai, "An Adaptive Array Antenna Using DFT in Spatial Domain," *IEICE Trans. Commun., Japan*, Vol. J75-BII, No. 8, pp. 556–565, Aug. 1992.

44. U. Timor, "Multitone Frequency-Hopped MFSK System for Mobile Radio," *Bell Sys. Tech. J.*, Vol. 61, No. 8, pp. 3007–3017, Dec. 1982.

45. G. Einarsson, "Address Assignment for a Time–Frequency-Coded, Spread-Spectrum System," *Bell Sys. Tech. J.*, Vol. 59, No. 7, pp. 1241–1255, Sept. 1980.

46. T. Mabuchi, R. Kohno, and H. Imai, "Multihopping and Decoding of Error-Correcting Code for MFSK/FH-CDMA Systems," *Proc. IEEE ISSSTA'92*, pp. 199–202, Nov. 1992.

47. R. Kohno, T. Mabuchi, and H. Imai, "Multi-User Receiver Based on Co-Channel Interference Cancellation for Multilevel FSK/FH-Spread Spectrum Multiple Access," *Proc. IEEE ICC'93*, May 1993.

48. H. Imai, R. Kohno, and T. Matsumoto; "Information Security of Spread Spectrum Systems," *IEICE Trans. Commun.*, Vol. E74, No. 3, pp. 488–505, March 1991.

PART 5

WIRELESS SYSTEMS AND APPLICATIONS

CHAPTER 16

EDGE: Enhanced Data Rates for GSM and TDMA/136 Evolution

STEFAN JÄVERBRING

16.1 INTRODUCTION

Standardization of third-generation mobile communication systems is now rapidly progressing in all regions of the world. The International Telecommunication Union (ITU) has been developing recommendations for the third-generation systems IMT-2000 (International Mobile Telecommunications 2000) since the late 1980s. In line with the ITU work, various standardization bodies are now in the process of standardizing IMT-2000: European Telecommunications Standards Institute (ETSI) in Europe, Association of Radio Industries and Business (ARIB) in Japan, Telecommunications Industry Association (TIA) and T1P1 in the United States and TTA in South Korea [1].

In line with the efforts of the ITU to provide global recommendations for IMT-2000, a spectrum identification has been made, identifying parts of the 2-GHz band for IMT-2000 usage. Deploying IMT-2000-capable systems is, however, not limited to this spectrum band. This chapter describes the enhanced data rates for global evolution (EDGE) concept, a new time division multiple access (TDMA) radio access technology for increased data rates and efficient data access in both TDMA/136 (the North American digital TDMA standard) and Global Systems for Mobile Communications (GSM) systems. EDGE will provide third-generation capabilities in the existing (800, 900, 1800, and 1900 MHz) frequency bands.

16.2 BACKGROUND

GSM and TDMA/136 are two second-generation cellular standards with worldwide success. Although speech is still the main service in these systems, support for data communication over the radio interface is being rapidly improved. The current GSM standard provides data services with user bit rates up to 14.4 kbps for circuit-switched data and up to 22.8 kbps for packet data. Higher bit rates can be achieved with multislot

Wireless Communications in the 21st Century, Edited by Shafi, Ogose, and Hattori.
ISBN 0-471-155041-X   © 2002 by the IEEE.

operation, but since both high speed circuit switched data (HSCSD) and General Packet Radio Service (GPRS) are based on the original Gaussian-filtered minimum shift keying (GMSK) modulation, the increase of bit rates is modest [2–4]. For TDMA/136 evolution, similar standardization activities are ongoing. In IS 136+ the combination of multislot operation and the introduction of 8PSK based on the 30-kHz carrier bandwidth enables data rates approximately four times higher than today [5].

EDGE provides an evolutionary path from existing standards for delivering third-generation services in existing spectrum bands. The advantages of EDGE include fast availability, reuse of existing GSM and TDMA/136 infrastructure, as well as support for gradual introduction. For example, as a $\frac{1}{3}$ frequency reuse overlay to TDMA/136, EDGE can be deployed using as little as 600 kHz of total bandwidth. In GSM, EDGE can be introduced using a minimum of only one time slot per base station. EDGE was first proposed to ETSI as an evolution of GSM in the beginning of 1997. During 1997, a feasibility study was completed and approved by ETSI, making way for the currently ongoing standardization [6]. Although EDGE reuses the GSM carrier bandwidth and time slot structure, it is by no means restricted to use within GSM cellular systems. Instead it can be seen as a generic air interface for efficiently providing high bit rates, facilitating an evolution of existing cellular systems toward third-generation capabilities.

After evaluating a number of different proposals, EDGE was adopted by the Universal Wireless Communications Consortium (UWCC) in January 1998 as the outdoor component of 136 High Speed (136HS) to provide 384-kbps data services. One of the arguments in favor of this approach was leveraging the technology evolution for both GSM and TDMA/136 systems, also leading to opportunities for global roaming. Consequently, EDGE was included in the UWC-136 IMT-2000 proposal. UWC-136 was adopted by TR-45 in February 1998 and submitted by the U.S. delegation to ITU as a Radio Transmission Technology candidate for IMT-2000 [7]. Since then, EDGE development has been concurrently carried out in ETSI and UWCC to guarantee a high degree of synergy with both GSM and TDMA/136. The standardization roadmap for EDGE is based on two phases. In the first phase the emphasis has been placed on EGPRS (enhanced GPRS) and ECSD (enhanced circuit-switched data). Both were targeted in ETSI for standards release 1999 with products to follow shortly afterwards. The second phase of EDGE is concerned with the improvements for multimedia and real-time services.

In speech planned TDMA/136 or GSM network there is typically a distribution of user signal to interference ratio (SIR), where almost all users have an SIR above an operating point. Speech users do not normally gain from being above this threshold. The principle of EDGE is to utilize this excessive SIR to increase bit rates and spectral efficiency. This is accomplished by the use of higher order modulation (8PSK) in combination with a control mechanism for adapting the bit rate to the channel conditions. This control mechanism is called link quality control (LQC).

The EDGE concept and various aspects of its link and system performance have been described in the literature [8–16]. Although EDGE phase 1 supports both circuit-switched and packet-switched services, this chapter focuses on the packet-switched part, enhanced GPRS (EGPRS), which is based as much as possible on GPRS.

16.3 PHYSICAL LAYER

The EDGE air interface is based on the air interface of GSM. Higher order modulation, 8PSK, is introduced with as few changes of the parameters as possible.

16.3.1 TDMA Format and Modulation

The GSM carrier spacing is 200 kHz, and each carrier is divided into eight time slots, according to Figure 16.1. Within each time slot a burst is transmitted, consisting of payload symbols, training symbols, and tail symbols according to Figure 16.2. The symbol rate is $13/48$ MHz ≈ 271 kHz. The bursts in GSM are modulated with binary GMSK, and hence one symbol corresponds to one bit. Each burst contains 2×58 bits, and the gross bit rate is 23.2 kbps.

In EDGE linear 8PSK is introduced using the same burst format, thus giving $3 \times 2 \times 58 = 348$ payload bits per burst. The gross bit rate becomes 69.6 kbps, which is three times the gross bit rate of GSM. Since 8PSK is less robust than GMSK, EDGE adapts the modulation (GMSK or 8PSK) to the current radio and interference situation. As will be shown later, the amount of channel coding applied is also adapted to suit the channel conditions. The 8PSK symbol constellation is shown in Figure 16.3. Three bits are Gray mapped to one symbol.

To keep the 200-kHz carrier spacing the modulation is partial response, that is, intersymbol interference (ISI) is introduced on the transmitter side. The pulse shape used is a linearized GMSK pulse [12] (see Figure 16.4), which gives approximately the same spectrum and ISI in a receiver as normal GMSK.

However, nonideal power amplifiers will distort the spectrum more for 8PSK than for GMSK, and hence the spectrum requirements are slightly relaxed compared to the GMSK

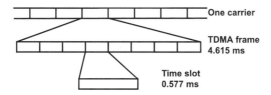

FIGURE 16.1 TDMA frame structure of GSM and EDGE.

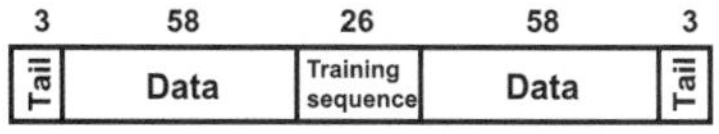

3	58	26	58	3
Tail	Data	Training sequence	Data	Tail

FIGURE 16.2 Burst format of GSM and EDGE.

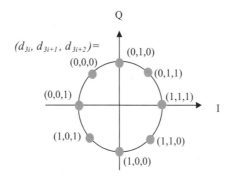

FIGURE 16.3 8PSK symbol constellation with Gray mapping of bits to symbols as in EDGE.

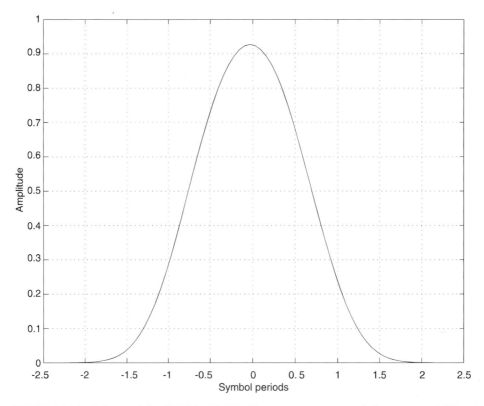

FIGURE 16.4 Pulse used for 8PSK in EDGE. The power spectrum and the amount of ISI are approximately the same as for GMSK.

requirements. GMSK is a constant-amplitude modulation, which enables efficient power amplifiers. 8PSK on the other hand has amplitude variations, the peak to average is 3.2 dB. For a given power amplifier this means that the maximum average output power for 8PSK is lower than the maximum average output power of GMSK. The difference is called average power decrease (APD) and could be in the order of 2 to 4 dB. Note, however, that this only is valid for the maximum power. If power control is used and the used power is less than the maximum possible, there should not be any difference between 8PSK and GMSK power.

To avoid zero crossings in the 8PSK symbol constellation, and thereby reducing the requirements on the power amplifier, a continuous rotation of $3\pi/8$ radians is applied. Actually any odd times $\pi/8$ rotation would have achieved the same thing, but for reasons explained below, $3\pi/8$ rotation gives better blind detection performance.

16.3.2 Blind Detection of Modulation

The eight binary training sequences of GSM (each 26 bits long) are reused for 8PSK in terms of symbols, that is, only two symbols of the eight are used during the training sequence, according to Figure 16.5.

Since EDGE utilizes both GMSK and 8PSK, the receiver needs to be able to handle both. In a radio environment with rapid quality changes, the modulation can be changed

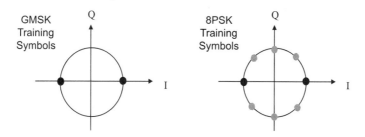

FIGURE 16.5   Training symbols for 8PSK.

dynamically and often. To signal such a change to the receiver in advance is undesired. Instead the receiver can exploit the fact that the same training sequences are used, but with a different constellation rotation. If the receiver makes a conventional linearized approximation of the GMSK modulation, the corresponding binary phase shift keying (BPSK) constellation will be rotated with $\pi/2$ radians, while the 8PSK constellation is rotated with $3\pi/8$ radians. This rotation factor can be identified during channel estimation, and thus the modulation format is detected by the receiver before the actual equalization. This is called blind detection of modulation, and the performance of this procedure depends on the cross correlation between a training sequence with the different rotations. For the specific GSM training sequences $3\pi/8$ rotation for 8PSK gives better performance than any other $k\pi/8$ rotation where k is odd.

16.3.3   Channel Coding

Although channel coding is a part of the physical layer, it is also related to the link layer for EDGE, and therefore it is described in the link layer section below. Basically the channel coding for EGPRS is based on punctured convolutional codes, where the puncturing is used to adapt the code rate to the channel quality. Enhanced circuit-switched data (ECSD) also utilizes Reed Solomon codes.

16.3.4   Physical Layer Performance

For GMSK it is common to use a full state equalizer, that is, a MLSE or a MAP receiver. However, for 8PSK this is unfeasible. Instead suboptimal receivers must be used. Since 8PSK is more sensitive to residual interference due to ISI not covered by the equalizer, the equalizing window for 8PSK needs to be larger than for GMSK, even if the symbol rate is the same. In Figure 16.6 the performance for the two different modulations, with an exemplary receiver for 8PSK, is shown. The receiver used in these simulations is low-complex and straightforward, and the performance of commercial receivers is expected to be significantly better.

16.4   LINK LAYER

The link layer contains automatic repeat request (ARQ) procedures and ways of adapting the bit rate to the channel quality, that is, link quality control (LQC). Both functions depend on the service type (e.g., packet-switched bearers or circuit-switched bearers) and

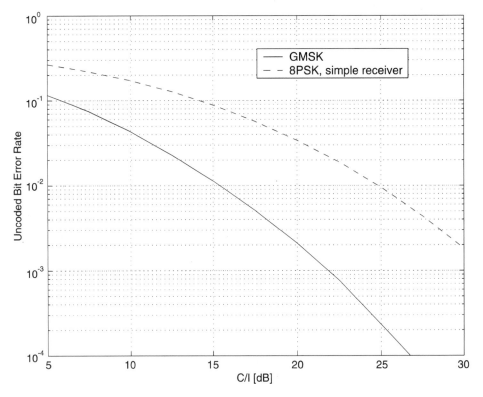

FIGURE 16.6 Performance for GMSK and 8PSK. The figures show the uncoded bit error rate for a low dispersive channel (typical urban profile).

are therefore described separately for these two cases. The channel coding is furthermore connected to the LQC, especially for EGPRS, and is therefore also described in the following sections.

16.4.1 Enhanced GPRS–EGPRS

EGPRS is a natural extension of GPRS, providing the packet switching of GPRS, but with higher data rates. Since the 8PSK modulation is more susceptible to noise and interference than GMSK, there is a need to adapt the transmission scheme used to the interference situation. This is essential for providing to each user the maximum throughput that the rapidly changing conditions allow at the moment. The LQC is also the main reason why the EDGE RLC protocol is somewhat different from the corresponding GPRS protocol.

EGPRS uses a combination of two methods: link adaptation (LA) and incremental redundancy (IR) for link quality. In short IR provides better performance than LA in most cases but is also more complex to implement, which is more elaborated in Eriksson [11]. The two methods and how they are used for EGPRS are described in the following sections.

Link Adaptation A pure LA scheme [18, 19] uses a set of type I hybrid ARQ schemes with different coding rates R and modulations, in the sequel called *modulation and coding schemes* (MCSs). A type I hybrid ARQ uses a forward error correcting (FEC) code to correct errors in blockwise encoded data, and additionally an ARQ mechanism to retransmit remaining erroneous blocks, detected by a frame check sequence (FCS). The channel quality is estimated continuously, and the MCS maximizing the link bit rate at the moment is chosen. Link adaptation in this way is introduced already in GPRS.

Incremental Redundancy In a pure IR scheme [20], a fixed type II hybrid ARQ scheme is used. The type II hybrid ARQ scheme first encodes a block of data with some low rate FEC code. Only a part of this codeword (a subblock) is transmitted initially, yielding some initial code rate R_1 (possibly, $R_1 = 1$). For erroneously decoded blocks, detected by an FCS, transmission of additional redundancy subblocks from the same codeword is requested, received, and combined with the first subblock, yielding a lower code rate R_{1+2}. This procedure is repeated until decoding succeeds, giving a stepwise increment of the amount of redundancy, or, equivalently, a decrement of the code rate $R_{1+\cdots+i}$.

Link Quality Control (LQC) for EGPRS A flexible LQC solution has been chosen for EGPRS, enabling pure LA, but also IR with different initial rates, and dynamic adaptation between all modes. The scheme enables a range of solutions with different trade-offs between complexity and performance [11]. The solution is as follows:

Nine MCSs are used, five using 8PSK and four using GMSK, each of which can be used in both LA and IR mode. The maximum bit rate (i.e., the bit rate after channel decoding without errors) ranges from 8.4 to 59.2 kbps. Some parameters for the MCSs are listed in Table 16.1.

For each MCS, an $R = \frac{1}{3}$ convolutionally encoded data block is divided into n subblocks (where n is either 2 or 3) by puncturing with n puncturing patterns, $P_1, \ldots, P_n$ (Figure 16.7). Initially, the subblock S_1 corresponding to P_1 is transmitted. On retransmission, one additional subblock S_i corresponding to P_i) is transmitted, where $i = 2, \ldots, n, 1, 2, \ldots$. Since each subblock for a given MCS is by itself a decodeable codeword, with the rate

TABLE 16.1 Parameters for MCS-1 to MCS-9 of the EGPRS LQC Scheme

Scheme	Modulation	Maximum Rate (kbps)	R_1	$R_1 + R_2$	$R_1 + R_2 + R_3$
MCS-9	8PSK	59.2	1.0	0.5	0.33
MCS-8		54.4	0.92	0.46	0.31[a]
MCS-7		44.8	0.76	0.38	0.25[a]
MCS-6		29.6	0.49	0.24[a]	—
MCS-5		22.4	0.37	0.19[a]	—
MCS-4	GMSK	16.8	1.0	0.5	0.33
MCS-3		14.8	0.85	0.42	0.28[a]
MCS-2		11.2	0.66	0.33	—
MCS-1		8.4	0.53	0.26[a]	—

[a] Code rates less than $\frac{1}{3}$ are obtained by repetition.

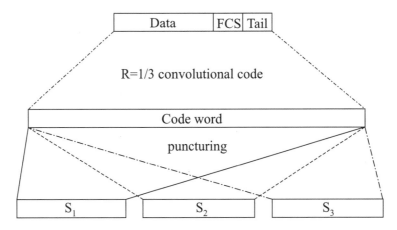

FIGURE 16.7   Encoding and subblock puncturing for the EGPRS LQC scheme.

$R_1 = \cdots = R_n$, the receiver can either discard or keep old subblocks when requesting retransmissions, thereby utilizing type I or type II hybrid ARQ.

Always altering in a cyclic manner among the subblocks S_i for an MCS enables the receiver to switch between combining and noncombining mode without notifying the transmitter. Thus, if the receiver temporarily enters noncombining mode due to lack of memory, IR operation will be possible as soon as memory is available again.

The network controls the choice of MCS in both uplink and downlink, based on the channel quality measured by the receivers. If IR combining is used in the receiver, this choice can be more aggressive, that is, less robust schemes can be used for a given channel quality.

The quality of the downlink is periodically reported to the network by the mobile. The short-term variations are typically faster than the reporting period. Therefore, there is a need to average the measures over time.

16.4.2   Enhanced Circuit-Switched Data

Enhanced circuit-switched data (ECSD) is a continuation of GSM's HSCSD. ECSD provides higher data rates per timeslot than HSCSD by utilizing the 8PSK modulation. In EDGE phase I, no extra ECSD service will be introduced compared to HSCSD, but by utilizing ECSD, the same data rates as in HSCSD could be achieved while using fewer time slots.

TABLE 16.2   Enhanced circuit-switched schemes

Scheme	Modulation	Radio Interference Rate (kbps)	Code Rate	Service Type[a]
ECSD TCS-1	8PSK	29.0	0.419	NT/T
ECSD TCS-2		32.0	0.462	T
ECSD TCS-3		43.5	0.629	NT

[a] T means transparent service and NT denotes nontransparent service.

The set of new coding schemes (Table 16.2) introduced in ECSD covers both transparent and nontransparent services. For the nontransparent services, the same ARQ mechanism as for HSCSD applies also for ECSD.

The new coding schemes introduced in ECSD all utilize the 8PSK modulation and make use of the same convolutional code polynomials as EGPRS. On top of that, the lowest ECSD code rates have been provided with a Reed Solomon code on top of the convolutional code. In a similar way as for GPRS and EGPRS, link adaptation is included in the ECSD concept to provide for usage of the most efficient coding scheme.

16.5 EGPRS PERFORMANCE

The downlink performance in a multiple cell network with dynamic packet traffic is evaluated by means of simulations. A standard three-sector frequency reuse pattern is used, using only three carriers in total. The time step of the simulator is 5 ms (corresponding to one burst) and the users produce packets according to a measurement-based Worldwide Web (WWW) traffic model. Queuing in the system is modeled. Finally, multipath fading is modeled on system level. More assumptions and details about the results can be found in Furuskär et al. [10]. Three different scenarios have been studied. First, EGPRS using the incremental redundancy and link adaptation mode is compared. Then, as a reference, a comparison to standard GPRS is also made.

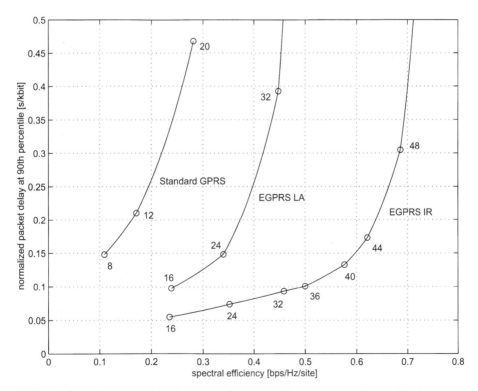

FIGURE 16.8 System capacity in terms of spectral efficiency vs. packet quality in terms of normalized delay (the offered load is also given as number of user per sector).

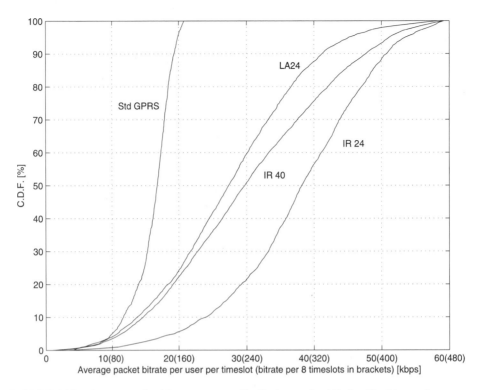

FIGURE 16.9   Average packet bit rate per user distributions at load limits. The bit rate figures are per time slot, supplemented by corresponding figures per eight time slots.

Used performance measures are normalized delay (i.e., the average packet delay in seconds per kilobit) and average packet bit rate (i.e., the average bit rate per user). Furthermore system load in terms of spectral efficiency (bps/Hz/site) is used. The more spectrum-efficient a system is, the higher system load is possible at a certain user performance (delay or throughput) or vice versa.

Figure 16.8 shows the spectral efficiency plotted against the 90th percentile of the normalized delay. It is seen that considerably higher spectral efficiencies are achieved using the incremental redundancy mode than when using the link adaptation mode. Assuming a delay requirement of 0.15 s/kbit at the 90th percentile (90% of the packets having a total delay of less than 0.15 kbps), a spectral efficiency of 0.60 bps/Hz/site (one site comprises three sectors) is obtained in the incremental redundancy mode. This corresponds to a 70% gain over the link adaptation mode achieving 0.35 bps/Hz/site. Even higher spectral efficiencies can be achieved if higher normalized delays can be accepted for the worst packets. At 0.4 kbps a spectral efficiency of 0.70 bps/Hz/site is reached, corresponding to a gain of 55% over the link adaptation mode. Compared to standard GPRS, the EGPRS spectral efficiency for the same delay requirement is approximately tripled.

The distribution of average packet bit rate per user per 1 (8) time slots is plotted in Figure 16.9. First, a case with offered loads that result in normalized delays just around 0.15 kbps is studied. Notice the significant increase in packet bit rate when EDGE is introduced compared to standard GPRS. For the link adaptation case, at the studied offered load (LA 24 curve, 24 being the number of users per sector), it is seen that approximately 5% of the users achieve a packet bit rate exceeding 48 kbps per time slot (384 kbps per 8 time slots), and that 84% of the users achieve a packet bit rate exceeding 18 kbps per time slot (144 kbps per 8 time slots). Using incremental redundancy, at the higher load limit of 40 users per sector, it is seen that (IR 40 curve), 10% of the users achieve a packet bit rate exceeding 48 kbps per time slot (384 kbps per 8 time slots), and that 86% of the users achieve a packet bit rate exceeding 18 kbps per time slot (144 kbps per 8 time slots).

It is also interesting to investigate how incremental redundancy affects the user data rates for the same offered load. The IR 24 curve shows the user data rates achieved at the load limit for the link adaptation case. It is seen that considerably higher rates are achieved: 48 (384) kbps is now reached by 20% of the users, whereas 95% reach 18 (144) kbps. Also notice the steeper cummulative distribution function (CDF) of the incremental redundancy operation, indicating a more fair system behavior.

16.6 CONCLUSIONS

EDGE is a common evolution of GSM and IS/136, providing third-generation services. Both packet-switched services (EGPRS) and circuit-switched services (ECSD) are provided. Depending on how the link layer protocol is used, the spectral efficiency of an EDGE packet data service can be twice or three times that of GPRS.

Acknowledgements The author wishes to thank Christer Edholm, Stefan Eriksson, Anders Furuskär, Sara Mazur, Frank Müller, and Håkan Olofsson for their large contributions to this chapter.

REFERENCES

1. E. Dahlman et al., "UMTS/IMT-2000 Based on Wideband CDMA," *IEEE Commun. Mag.*, 1998.
2. ETSI. TS 101 038 V5.0.1 (1997-04), "Digital Cellular Telecommunications System (Phase 2+); High Speed Circuit Switched Data (HSCSD)–Stage 2 (GSM 03.34)," version 0.4, 1997.
3. ETSI. TS 03 64 V5.1.0 (1997-11), "Digital Cellular Telecommunications System (Phase 2+); General Packet Radio Service (GPRS); Overall Description of the GPRS Radio Interface; Stage 2 (GSM 03.64 version 5.1.0)," version 11, 1997.
4. J. Cai and D. Goodman, "General Packet Radio Service in GSM," *IEEE Commun. Mag.*, Oct. 1997.
5. S. Labonte, "A Proposal for the Evolution of IS-136," *Proc. IEEE VTC'98*.
6. ETSI. Tdoc SMG2 95/97. "EDGE Feasibility Study, Work Item 184; Improved Data Rates through Optimised Modulation," version 0.3, Dec. 1997.
7. The UWCC-136 RTT Candidate Submission.
8. A. Furuskär, S. Mazur, F. Müller, and H. Olofsson, "EDGE, Enhanced Data Rates for GSM and TDMA/136 Evolution," *IEEE Commun. Mag.*, Vol. 6, No. 3, June 1999.
9. A. Furuskär, D. Bladsjö, S. Eriksson, M. Frodigh, S. Jäverbring, and H. Olofsson, "System Performance of the EDGE Concept for Enhanced Data Rates in GSM and TDMA/136," *Proc. IEEE WCNC'99*.

10. A. Furuskär, M. Höök, S. Jäverbring, H. Olofsson, and J. Sköld, "Capacity Evaluation of the EDGE Concept for Enhanced Data Rates in GSM and TDMA/136," *Proc. IEEE VTC'99 Spring*.

11. S. Eriksson, A. Furuskär, M. Höök, S. Jäverbring, H. Olofsson, and J. Sköld, "Comparison of Link Quality Control Strategies for Packet Data Services in EDGE," *Proc. IEEE VTC'99 Spring*.

12. P. Schramm, H. Andreasson, C. Edholm, N. Edvardsson, M. Höök, S. Jäverbring, F. Müller, and J. Sköld, "Radio Interface Performance of EDGE, a Proposal for Enhanced Data Rates in Existing Digital Cellular Systems," *Proc. IEEE VTC'98*.

13. A. Furuskär, M. Frodigh, H. Olofsson, and J. Sköld, "System Performance of EDGE, a Proposal for Enhanced Data Rates in Existing Digital Cellular Systems," *Proc. IEEE VTC'98*.

14. K. Zangi, A. Furuskär, and M. Höök, "EDGE: Enhanced Data Rates for Global Evolution of GSM and IS-136," *Proc. Multi-Dimensional Mobile Communications 1998 (MDMC'98)*.

15. A. Furuskär, M. Höök, C. Johansson, S. Jäverbring, and K. Zangi, "EDGE-Enhanced Data Rates for Global Evolution," *Proc. Nordic Radio Symposium 1998 (NRS'98)*.

16. H. Olofsson and A. Furuskär, "Aspects of Introducing EDGE in Existing GSM Networks," *Proc. IEEE ICUPC'98*.

17. P. Jung, "Laurent's Representation of Binary Digital Continuous Phase Modulated Signals with Modulation Index 1/2 Revisited," *IEEE Trans. Comm.*, Vol. COM-42, pp. 221–224, Feb./Mar./Apr. 1994.

18. J. Dunlop et al., "Estimation of the Performance of Link Adaptation in Mobile Radio," *Proc. IEEE VTC'95*.

19. A. Goldsmith and S. G. Chua, "Adaptive Coded Modulation for Fading Channels," *IEEE Trans. Comm.*, Vol. 46, No. 5, pp. 595–602, May 1998.

20. S. Kallel, "Analysis of a Type II Hybrid ARQ Scheme with Code Combining," *IEEE Trans. Comm.*, Vol. 38, No. 8, Aug. 1990.

CHAPTER 17

Continuing Evolution of CDMA into New and Improved Services

ANDREW J. VITERBI

17.1 COMMERCIAL CDMA: A BRIEF CONDENSED HISTORY

Over 20 years ago, this author began the main section of a tutorial article [1] on spread spectrum with the following paragraph:

The purpose and applicability of spread spectrum techniques is threefold:

- Interference suppression
- Energy density reduction
- Ranging or time delay measurement

Foremost among these is the suppression of interference which may be characterized as any combination of the following:

1. Other Users: intentional (hostile) or unintentional
2. Multiple Access: spectrum sharing by "coordinated" users
3. Multipath: self-jamming by delayed signal

The article was devoted primarily to the advantages of spread spectrum for military communications but did recognize commercial applications with the additional paragraphs:

Protection against in-band interference is usually called anti-jamming (A/J). This is the single most extensive application of spread spectrum communication. A similar application is that of multiple access by numerous users who share the same spectrum in a coordinated manner, in that each employs signaling characteristics or parameters (often referred to as codes) which are distinguishable from those of all other users. One reason for using this shared spectrum, so-called code-division multiple access (CDMA), is that by distinguishing signals in this way,

Wireless Communications in the 21st Century, Edited by Shafi, Ogose, and Hattori.
ISBN 0-471-155041-X © 2002 by the IEEE.

separation in the more common dimensions of frequency or time is not required, and hence the usual transmission tolerances need not be imposed on these parameters.

The third form of interference suppressed by spread spectrum techniques is the self interference caused by multipath in which delayed versions of the signal, arriving via alternate paths, interfere with the direct path transmissions.

The intervening two decades have witnessed how extensively the three applications of spread spectrum have permeated the commercial and consumer marketplace, well beyond the wildest expectations of this writer and his colleagues. First, the ranging and timing measurement techniques, which NASA had employed to track satellites and space vehicles since the fifties, were implemented in a network of 24 satellites known as the Global Positioning System (GPS). Today, tens of millions of GPS receivers have been put in service for position location in aircraft, automobiles, for personal and professional use, and shortly also for emergency location of individuals through their cellular phones. But it is the CDMA application that has had the most impact and promises to become the universal standard for wireless telephony and data transmission in the 21st century.

Spread spectrum is a generic term that refers to transmission by modulation of a carrier signal whose bandwidth greatly exceeds the rate of the information to be transmitted. This includes the use of frequency-hopped carriers as well as sinusoidal carrier signals whose center frequency is swept over the band, creating a so-called chirp signal. But by far the most common, as well as the most effective, form of spread spectrum is that known as direct sequence (DS) spread spectrum, wherein the carrier phase is shifted by $0°$ or $180°$ according to a pseudorandom sequence at time intervals (known as chips) inversely proportional to the desired spreading bandwidth. The information is then modulated on this pseudorandom carrier by modulating the carrier phase at the much lower information rate, but synchronous to a divisor of the chip rate. The theory of pseudorandom sequence generation, developed in 1953 [2], established that a simple linear feedback shift register can generate a sequence whose length, between repetitions, grows exponentially with the shift register length, which exhibits the three most important randomness properties of a "coin-flipping" sequence and consequently whose spectrum, with appropriate wave-shaping, is that of band limited white noise.

A principal requirement in spread spectrum transmission is that the receiver's direct sequence generator be synchronized with the transmitter's sequence, properly delayed by the propagation time. This is achieved by closed loop acquisition and tracking techniques [3, 4], which are fundamental to all digital communication systems but are particularly critical in synchronizing the spectrum-spreading pseudorandom sequence used at the transmitter with its replica at the receiver that performs the despreading operation. We shall return to synchronization issues in later sections as well as the related issue of phase coherence, which also relies on closed-loop acquisition and tracking concepts [5].

Satellite systems began to employ direct sequence spread spectrum (DS-CDMA) techniques primarily to allow small earth antennas, with correspondingly wide beams, to transmit uplink signals to a given satellite without unduly interfering with adjacent satellites within the beamwidth of the uplink. From such experience in the late 1980s on both commercial satellite CDMA systems used for two-way messaging in the transportation industry [6] and field trials for satellite mobile telephony [7], there arose the view that DS-CDMA might provide even more advantages in terrestrial cellular multiple access. The timing coincided with the cellular industry's plans to migrate to digital technology with the goal of improving spectral efficiency 10-fold over the analog cellular systems that had been in service since the early 1980s and that were beginning to exhibit spectral congestion. The European Telecommunications Standards Institute (ETSI) had in 1982 developed a pan-European standard based on time division multiple access (TDMA), which became known as the Global System for Mobile Communications (GSM). The corresponding North-American-based standards group created by the Telecommunication Industry Association (TIA) had also by 1988 selected TDMA for its digital standard, though the channelization, symbol rate, and other parameters differed significantly from the European version. This was identified as IS-54 and in a later revision as IS-136. Similarly the Japanese industry standards group, Association of Radio Industries and Businesses (ARIB), which operated under the auspices of Japan's Ministry of Post and Telecommunications, chose a third TDMA standard known as personal digital cellular (PDC), which shared many characteristics with the TIA version.

Given the seemingly settled state of cellular multiple access standards, the proposal in 1989 by QUALCOMM Incorporated to the North American service provider-led Cellular Telecommunication Industry Association (CTIA) that CDMA be also considered for standardization did not receive wide encouragement. Nevertheless, with the favorable attention and support of a few cellular providers led by the predecessor to Airtouch Cellular,* QUALCOMM proceeded to develop and demonstrate to the industry during November 1989 in San Diego an early prototype of a cellular mobile phone in transit between two base stations employing DS-CDMA technology with the characteristics and features described in the next section. This increased interest and reduced skepticism of CDMA in some quarters, especially after a second demonstration for NYNEX in Manhattan in early 1990. But it was not until an extensive field trial involving approximately 100 mobiles and 5 base stations was performed with the cooperation of several manufacturers and service providers in late 1991 that the CTIA agreed to consider standardization of CDMA. This initiated a series of industry colloquia after which the TIA convened a standards committee (TR45.5), which in mid-1993 completed the CDMA standard known as IS-95 [8]. In spite of this, there remained a small but committed and strident contingent of manufacturers and certain self appointed experts who denigrated[†] the capabilities of CDMA at every opportunity. This partly for competitive reasons and partly to attempt to exclude CDMA as a contender for the newly licensed personal communications services (PCS) band (1900 MHz) in North America. This last effort failed with the joint TIA/T1P1 standards group accepting all previous cellular standards as well as a few others that were never deployed.

Asian countries were the first to implement the IS-95 DS-CDMA standard with initial service in Hong Kong in the fall of 1995 and nationwide service in South Korea in 1996. Large-scale North American service began only in 1997 and primarily for the newly licensed PCS-band service providers, since the existing cellular-band (800 MHz) providers had a well-established analog service network and economic consideration dictated a slower conversion aimed principally at large usage customers. By 1997, five million subscribers were on CDMA networks and by 2001 penetration has grown to more than 100 million subscribers worldwide. All major countries of North and South America have large operating networks, usually with multiple service providers in each country. Similarly, in Asia and the Pacific region, multiple DS-CDMA (IS-95) networks operate in South Korea and Japan, rivaling the Americas for subscriber numbers, with service having begun also in China, Australia, Singapore, and numerous other countries of the hemisphere. In contrast,

---

*Then a subsidiary of Pacific Telesis, called PacTel Cellular.
†The criticisms and dire predictions about the capabilities and future of DS-CDMA ranged from claims of exaggerated expectations to utter failure and even outright fraud.

although some CDMA service exists in Eastern Europe and the Middle East and Africa, Western Europe currently remains totally committed to GSM and its TDMA evolution, as does most of that hemisphere. Nevertheless, as wireless telephony and particularly data transmission gains not only in numbers of subscribers but especially in minutes of use per subscriber, the analog and early digital networks are reaching saturation, with "network busy" signals and dropped calls becoming all too common. Partly as a result of this, an almost universal opinion has formed that DS-CDMA will provide such higher efficiency and service quality to make it the natural choice for a new high data rate digital service, categorized as "third-generation (3G) systems." In 1999, therefore, consensus was reached by all major standards organization on an almost common framework based solely on DS-CDMA technology. Although, as will be discussed in a later section, the European–Japanese structure differs in some details from the evolutionary North American approach, the fundamental concepts have been universally adopted and a worldwide standard based on DS-CDMA is approaching reality.

17.2   SYSTEM FEATURES OF CODE DIVISION MULTIPLE ACCESS

There are three primary requirements for any multiple access system, indeed for any digital communication system over a channel that varies with time. These are:

1. Channel *measurement*
2. Channel *control*
3. *Mitigation* of *interference* both from and to other users

The primary characteristic of CDMA is that each user's signal occupies a widespread bandwidth, so that mutual interference among users appears as wideband noise, which is relatively benign. Thus all users of all cells and sectors employ a common transmission bandwidth, as do all base stations in their transmissions to the users. This universal frequency reuse is a major contributor to the spectral efficiency of CDMA. The wide signal bandwidth facilitates channel measurement. But at the same time, the universal frequency reuse requires more stringent channel control and puts greater emphasis on interference mitigation, given the large number of sources of interference within the band.

The five most important parameters to be *measured* and *controlled* are:

1. Spreading sequence timing and its synchronization
2. Carrier frequency and phase tracking
3. Received signal power at user terminal and at base station
4. Channel characteristics—particularly relative delays and magnitudes of multipath components
5. Power received from neighboring base stations and neighboring sectors

The first two, which are critical to establishing any communication link are aided by the presence of an unmodulated pilot signal embedded in the base station transmission, and as we shall address below, from the user as well. Signal power measurements are necessary to control each user's transmitted and received power to prevent it from taking more than its equal share of channel resources and in this way also to minimize interference to other users (requirement 3 listed at the beginning of this section).

Measurement of multipath component magnitudes and delays affords the possibility of combining them constructively in an optimal manner by processing through a Rake receiver. This too is a form of channel control preventing the components from destructively combining occasionally to cause fades. Finally, measurement of power in pilots received from neighboring base stations and neighboring sectors affords the possibility to place the user's terminal in "soft handoff," meaning that the user terminal is communicating simultaneously with two or more base stations or sectors [9]; this converts an interferer into a collaborative signal component that can be combined constructively with the original signal by means of the same Rake processor used for multipath. All of these measurement and control features also serve to mitigate interference to other users by permitting the user terminals and base stations to transmit less power than would be needed if the channel parameters were not controlled and optimized. In addition, besides the above, three other processing features reduce the required transmitter power and thus contribute to interference mitigation:

1′. Forward error correcting (FEC) coding to reduce required link bit energy-to-interference density (E_b/N_0) for the desired link quality
2′. Voice activity detection to reduce transmitted data rate and hence transmitted power when voice activity is diminished
3′. Base station antenna sectorization

The first two reduce the power transmitted, while the third diverts it toward a subset of users, thus not interfering, or interfering less, with the remainder. Of course, all three of these processes can be applied in any digital multiple access system. However, only with DS-CDMA does the FEC coding redundancy not reduce data rate; only with CDMA can reduced voice activity autonomously permit a higher number of simultaneous users without additional channel control; only with CDMA does universal frequency reuse apply to adjacent sectors with tolerable interference between sectors.

A critically important feature of this DS-CDMA system concept is that the clocks of all base stations be synchronized. This minimizes the system complexity for acquisition and the parameter measurement for the original as well as neighboring base stations and both facilitates and speeds up the process of soft handoff. Typically, this is achieved by synchronizing each base station to GPS time, but as we shall discuss in a later section, base stations can be synchronized in other ways should the GPS system not be available or desired.

Finally, it should be noted that all eight processing features listed above are implemented in a single integrated circuit, which contains the digital voice compression [code-excited linear predictive (CELP) coding] algorithm as well. Over an eight-year period from 1991 to 1999 this technology has kept pace with Moore's law of integration, with a reduction of the application-specific integrated circuit (ASIC) size 10-fold, while increasing its processing capabilities 3-fold.

17.3 EARLY CDMA EVOLUTION FOR DATA SERVICES

Since the inception of cellular service, wireless has been almost exclusively dedicated to voice telephony. With the explosive growth of the Internet, the demand for mobile data has begun to be felt, and gradually service providers worldwide have started offering low data rate services such as stock quotes, weather and travel information, game scores, and news headlines and even wireless access to the Internet has begun. The CDMA base station architecture has from the beginning been oriented toward packet data. Thus it is readily enabled to support the above services, and in South Korea they began to be offered in 1997 at data rates up to 14.4 kbps. To "surf the Net," considerably higher data rates are required of course. This gave the impetus for the TIA's development in 1998 of the IS-95B standard [10], which provides for aggregating up to 8 channels in the forward (base station-to-subscriber) direction and up to 4 channels in the reverse direction to support data rates up to 115.2 and 57.6 kbps, respectively. (It is likely, however, that most providers will offer only 64 and 28.8 kbps initially). This standard was developed by making only minor changes to the basic (IS-95A) DS-CDMA standard that has been in force since 1993.

The ASIC processors, which implement IS-95A in both handsets and base stations, have been upgraded to implement IS-95B as well, and subscriber handsets are currently on the market with the data capabilities of IS-95B with backward compatibility to IS-95A.

17.4 IMPROVEMENT AND EVOLUTION TO CDMA2000

The evolution of CDMA continues with significant improvements and new capabilities. After 10 years of development and experimentation and 4 years of commercial deployment and operation, with a subscriber population in the tens of millions throughout the Americas and Asia, many lessons have been learned. These have led to improvements incorporated in a new evolutionary standard, designated as cdma2000. The most important of these refer back to the three fundamental requirements of Section 17.2: channel measurement, control, and interference mitigation. Under measurement, a major issue is phase tracking. For the forward direction (base station-to-subscriber), this is greatly facilitated by the pilot signal, whose only modulation is a spreading sequence, that significantly improves acquisition and time tracking of the sequence, as well as the tracking of the carrier phase. Moreover, this pilot is shared among all the subscribers being addressed by the base station in a given sector and thus represents a minor power overhead for each one, even though the pilot power is usually several times greater than each individual user's data signal power. In the reverse direction, on the other hand, since each user's transmitter would need to provide its own pilot, the overhead is much greater. Without phase measurement, noncoherent reception is possible, but with some degradation which is reduced if multiple orthogonal signal modulation is employed. This was the choice for the pilotless reverse channels in IS-95. Yet there remains a considerable advantage to phase-coherent reception, provided the pilot power can be kept well below the data-bearing signal power. This was not believed possible initially because the variability of the fading process was presumed to be too wideband. With considerable experience, it is now known that the fading bandwidth is usually an order of magnitude narrower than the data bandwidth for voice; and proportionately less for higher speed data. Thus the pilot tracking loop is narrow compared to the bit rate, allowing phase to be measured over many bit times and, since measurement accuracy depends on the pilot energy-to-noise, this permits the use of a much lower pilot signal power. The net result of coherent detection with binary phase-shift keying (BPSK) modulation, including the power allocation to each pilot per user, is a reduction of the required E_b/N_0 by about 2 dB.

A second improvement deals with control. Latency or delay in the control loop reduces its accuracy. For the reverse channels, which are many-to-one, accurate power control was deemed a necessity from the outset, so a tight power control loop kept their powers at the desired levels. The forward channel was not deemed to be as critical so a much looser control was applied that suffered from considerable delay. This often caused forward transmitted power to be a few decibels above the level desired, causing excessive interference to same cell and adjacent cell users. Furthermore, the forward transmitted power was not measured as accurately and not transmitted back as often, adding to the latency. All this has been greatly improved by making the forward channel power control similar to that of the reverse channels, consequently reducing the latency. Another forward link drawback was the excessive use of soft handoff. While on the reverse links soft handoff only means that more than one base station is receiving the subscriber's signal, on the forward link the information destined to the soft handoff user must be transmitted from two or more base stations or sectors. While presumably the original base station then transmits less power than it would otherwise need to, since transmission during soft handoff is from two or more base stations, the overall power and hence interference to other users is increased, reducing the number of subscribers that can be simultaneously supported. Changes in the handoff technique and parameter settings has reduced this interference while not sacrificing the benefits.

Finally, in the area of interference mitigation, code redundancy has been increased in both directions to improve performance. On the forward link, this was accomplished by employing quaternary phase-shift keying (QPSK) rather than BPSK modulation and in the reverse link by replacing the multiple orthogonal modulation by BPSK with coherent reception. One other feature of the new standard is the reduction of subscriber terminal power consumption, while on but not in use, by adopting a new signaling scheme which reduces the amount of processing required when the subscriber terminal is idle. This extends battery life severalfold.

To summarize, by changing the signal modulation and adding pilots to the reverse channel and with improved coding and minor power control latency improvement, the overall number of users supportable on the reverse link is nearly doubled. On the forward channel, major power control latency reduction, reduction in the percentage of soft handoff and improved coding reduces interference to the point that the number of supportable users is more than doubled. Thus these important evolutionary improvements, suggested by operational experience, afford the possibility of doubling in both directions. For two-way voice, it is unlikely that further significant improvement will occur, other than by the implementation of narrower beam (sectored), and possibly adaptive, antennas. For data, the new standard will provide transmission rates reliably up to 307.2 kbps in the forward and 153.6 kbps in the reverse direction with the same channelization and radio frequency (RF) components currently used for IS-95 voice. This standard, which is fully backward compatible with IS-95A and B, has recently been approved by vote of the TIA standards committees. It is commonly referred to as cdma2000-1X. The signal processing ASIC designs for IS-95 mentioned in the last section have been upgraded to include these improvements and their delivery and deployment began in 2000.

17.5   GENERATIONAL EVOLUTION AND EMPHASIS ON WIDER BANDWIDTHS

Much effort is being exerted in both technical and political committees and organizations toward evolution to a worldwide standard for both voice and high speed data interchange. The impetus seems to be the provision of data rates as high as 2 Mbits/s. These goals have been universally labeled as the third-generation (3G) wireless systems, and it is equally widely accepted that their implementation will occur in 5-MHz wide frequency bands. This might suggest categorizing as "first generation" those systems employing just 25 or 30 kHz (TACS, ETACS, AMPS, N-AMPS, D-AMPS, IS-136, PDC), as "second generation" those employing 200 kHz or 1.25 MHz (GSM and CDMA IS-95), and as "third generation" those employing 5 MHz or above. A less absurd definition might be according to technology. Thus analog frequency modulation (FM) begun in the early 1980s would be first generation, TDMA begun in the early 1990s would be second generation, and CDMA begun in the mid-1990s would be third generation. So in fact, we need not wait for a third generation; we are really just evolving it. And that we believe is the reality, particularly as we consider the two alternative approaches to wider bandwidths. The first is a direct extension of the evolutionary CDMA standard just described. Since the only difference is in the bandwidth occupied, the most natural evolutionary approach is to triple the direct sequence chip rate, speeding up the shift register generator clock from 1.2288 to 3.6864 MHz. This is the approach employed for the subscriber-to-base station reverse link. For the forward link the same approach can also be used. However, since base stations use spatial diversity transmission with dual antennas per sector, spatial and frequency diversity can be combined using a different technique. This involves keeping the chip rate at 1.2288 MHz, but repeating the signal three times (possibly with different coding) at three center frequencies separated by 1.2288 MHz. Then one antenna of the dual pair can be fed the middle frequency version of the trio and the other antenna the two end frequency versions. These two techniques have been labeled cdma2000-DS-3X and cdma2000-MC-3X, respectively.* Analyses and simulations of the two support the conclusion that there is virtually no performance advantage of either over the other.

With the tripled bandwidth, the data rate can be at least tripled, if only data is transmitted. Coexistence with voice traffic will reduce this multiplier. In Sections 17.7 and 17.8, we shall show that for a data-only service with proper packet-based traffic allocations, the rate can be more than tripled even in the current IS-95 carrier bandwidth.

17.6   ALTERNATE IMPLEMENTATION OF WIDER BAND CDMA

A second standard, generally called W-CDMA, has been recently formulated by manufacturers and service providers who for the most part had for the previous decade shunned CDMA in favor of the various TDMA standards (GSM, D-AMPS, and PDC). It involves a number of minor differences in parameters and one conceptual difference that has major implications for CDMA system implementation and performance. The minor parameter differences involve chip rate, frame rate, and pilot signal multiplexing. The major difference affects the critical synchronization functions. The chip rate chosen was 3.84 Mchips/s, a 4% difference from the evolutionary 3X chip rate of 3.6864 Mchips/s. The frame rate is 10 ms instead of 20 ms. The pilot signal multiplexing is similar on the reverse link and for the principal common pilot on the forward link, but individual pilots for forward channels are time-multiplexed rather than code multiplexed. All these are minor inconsequential changes introduced mainly to differentiate from the evolutionary system. The major conceptual difference is in the approach to network synchronization.

*3X refers to the tripled chip rate, whereas 1X referred to the current IS-95 chip rate.

As noted previously, in IS-95 and all its evolutionary descendents, including cdma2000 in all its manifestations, base stations are synchronized in time. Usually this is achieved by tracking GPS at each base station, but terrestrial means for synchronization are also available. This facilitates initial subscriber terminal acquisition as well as the identification and channel parameter measurement from newly encountered base stations. Effectively, once the first base station is acquired, all the others need not be searched because the network signaling information provides the relative delays of neighboring base stations. The alternate W-CDMA system assumes that each base station operates asynchronously from all others and possibly no attempt may be made to synchronize them. As a result, each time a new base station acquisition is attempted the subscriber terminal must initiate a new search. Furthermore, to facilitate this frequent reacquisition, the pilot modulation format is considerably more elaborate and the handset consumes more power due to the excess processing. In short, to avoid the necessity of synchronizing very infrequently hundreds of base stations at most (in a regional deployment), the alternate W-CDMA system requires resynchronization *of each mobile subscriber's handset frequently*, every time it wanders into a new base station's region of influence. The justification usually given for this is that GPS satellites may not always be available for commercial use (which would of course also incapacitate tens of millions of position location systems in use worldwide) or that underground base stations would have difficulty in connecting to a GPS receiver above ground. Even if these arguments are valid, alternate terrestrial means of base station synchronization have been developed and tested. One of the most effective employs the information already supplied by mobile users in IS-95, measuring new base station strengths and signaling this information via a "pilot strength measurement message," from which the relative distances between the new and the old base station can be determined quite accurately [11].

17.7   RE-EXAMINING THE GOAL: WIRELESS HIGH-SPEED DATA TRANSMISSION

The overriding experience and focus of most telecommunication service providers, whether wireless or wireline, has been and continues to be centered on voice telephony. Even for wireline transmission, where data traffic rivals voice, the bulk of the traffic reaches the consumer and even businesses via narrowband analog voice circuits. Yet standards bodies, numerous manufacturers, a few wireless service providers, and even some regulators in various parts of the world are trumpeting the remarkable benefits that will flow from the availability of wireless high-speed data transmission to and from the consumer. High speed refers to data rates on the order of 1 to 2 Mbps, an order of magnitude greater than what we might refer to as medium speed, 64 to 144 kbps, which are already available today or will be shortly based on the early evolution described in Section 17.4. These speeds are ample for such clearly identified uses as timely data retrieval, for stock quotes, for weather and sports scores, for recovering e-mail or voice mail messages and selected data from one's office or home computer, all via the display on one's portable phone or wireless personal digital assistant. To date, however, only one high-speed data requirement can be clearly identified, namely downloading data, messages, and files (including streaming audio and video) from the Internet or World Wide Web. The demand for such service is already manifested by the rapid growth of cable and digital subscriber line (DSL) service connections to the Internet in countries and regions in which they are available. These offer downloading speeds of a few megabits per second, with uploading reverse speeds typically an order of magnitude lower; hence, for example, the term asymmetric digital service line (ADSL) for one of the offering modes. Their value is for drastically reducing the latency of reception, which for ordinary analog phone modems is sometimes derisively called the "World Wide Wait".

Clearly wireless has an important role to play in this application, or more precisely two roles. The first is to provide an alternative to cable and DSL, for households and small businesses in regions not served by cable or where the local phone company does not find it economical to upgrade lines or provide modems for DSL and, even when one or both may be available, to provide an alternative that may be quicker and easier to install. A second and probably more important role is in serving the nomadic user who carries her or his portable laptop to remote locations. While this includes most business and professional people, it may include consumers as well, and the remote location need not be only hotel rooms, clients' offices, or airports; remote may apply to office meeting rooms or other rooms in the home or in the backyard, where Ethernet cable or high speed telephone lines are not available. Extrapolating on the identified high speed, low latency Internet usage, wireless service for nomadic or otherwise unserved populations seems to be a reasonable investment for a service provider. Investing in two-way high-speed links for such hypothetical applications as high resolution video telephony (which never materialized on wireline) would appear to be an expensive and risky venture. This, however, is in fact the proposition put forth by the strong proponents of "third-generation" wireless. Interestingly furthermore, the strongest proponents seem to be those whose current networks do not employ CDMA, so that for these manufacturers and service providers the approach is not evolutionary but involves total replacement, or reproduction, of infrastructure deployments. In addition, these wideband wireless systems would be intended to serve both voice and data users in the same bandwidth allocations, which sacrifices the capacity of both services. In the next section, we show instead that a high-speed wireless Internet data-only service can be efficiently implemented in the same bandwidth as current (IS-95) CDMA deployments and employing the same base stations and RF front ends, yet increasing data throughput manyfold by packet allocation techniques. This approach is designated "high data rate CDMA" (CDMA/HDR).

17.8 CDMA/HDR FOR HIGH-SPEED WIRELESS INTERNET ACCESS

Most data applications differ fundamentally from speech in two respects already noted, traffic asymmetry and tolerance to latency. The nature of data traffic is decidedly asymmetric. A much higher forward (or downlink) rate is required from the access point (base station) than that generated by the access terminal (user terminal) in the reverse (uplink) direction. Further, just as the fixed user, the nomadic user expects a response to her or his request that does not suffer from excessive latency. Our goal is again to satisfy these needs with an evolutionary approach that minimizes the time and cost for providing such capabilities in existing cellular infrastructure and with terminals that differ only at digital baseband from existing cellular and PCS handsets. In contrast two-way conversational speech requires strict adherence to symmetry for which latencies above 100 ms (which corresponds to about 1 Kbit of data for most speech vocoders) are intolerable. In contrast, high-speed data downlinked at 1 Mbps, for example, 100 ms represents 100 kbits or 12.5 Kbytes; furthermore latencies of 10 s are hardly noticeable and this corresponds to a record of 1.25 Mbytes. Thus smoothing over a variety of conditions, which is always advantageous for capacity, is easily accomplished.

All communication systems, wired as well as wireless, are greatly improved by a combination of techniques based on the three principles already noted in Section 7.2: channel measurement, channel control, interference suppression and mitigation. Our approach employs all three. First, on the basis of the received common pilot from each access point (or base station), each access terminal (subscriber terminal) can measure the received signal-to-noise-plus-interference ratio (SNR). The data rate that can be supported to each user is proportional to its received SNR. This may change continuously, especially for mobile users. Thus over each user's reverse (uplink) channel, the SNR or equivalently the supportable data rate value is transmitted to the base station. In fact, since typically two or more base stations may be simultaneously tracked, the user indicates the highest among its received SNRs and the identity of the base station from which it is receiving it, and this may need to be repeated frequently (possibly every slot*). In this way the downlink channel is controlled as well as measured. Further, by selecting only the best base station, in terms of SNR, to transmit to the user, interference to users of other base stations is reduced. Additionally, since data can tolerate considerably more delay then voice, error-correcting coding techniques that involve greater delay, specifically turbo codes [12], can be employed that will operate well at lower E_b/N_0 and hence lower SNR and higher interference levels.

Next, we show how unequal latency, for users of disparate SNR levels, can be used to increase throughput. Suppose we can separate users into N classes according to their SNR levels, and corresponding instantaneous rate levels supportable. Thus user class n can receive slots at rate R_n bps where $n = 1, 2, \ldots, N$, and suppose the relative frequency of user packets of class n is P_n. Suppose slots are assigned one at a time successively to each user class. Then the average rate, which we define as *throughput*, is

$$R_{\text{av}} = \sum_{n=1}^{N} P_n R_n \text{ bits/s}$$

This, of course, means that lower data rate (and SNR) users will have a proportionately higher latency. For if B bits are to be transmitted altogether for each class, the number of slots (and hence time) required for user class n will be B/R_n and hence the latency L_n is inversely proportional to R_n.

Suppose, on the other hand, that we require all users to have essentially the same latency[†] irrespective of the R_n they can support. Then as each user class is served, it will be allocated a number of slots inversely proportional to its rate. Let F_n be the number of slots allocated to class n, where $F_n = k/R_n$, k being a constant. In this case, the average rate or throughput is

$$R'_{\text{av}} = \frac{\sum_{n=1}^{N} P_n R_n F_n}{\sum_{n=1}^{N} P_n F_n} = \frac{1}{\sum_{n=1}^{N} P_n/R_n} \text{ bps}$$

---

*In speech-oriented CDMA, voice frames are 20 ms long. In the next section, we shall establish corresponding lengths for data, which will be called slots. Multiplying R_{av} by slots/second yields throughput in bits/second.
†This is the case for voice. The only difference is that in speech transmitter power levels are controlled to equalize received power, while for data, time, in terms of frames, is controlled to equalize energies.

However, the latency of all user classes will be the same (assuming the total number of bits to be large and thus ignoring edge effects).

To assess the cost in throughput for equalizing latency, consider the extreme case of only two user classes, each equally probable ($P_1 = P_2 = \frac{1}{2}$) but capable of supporting very disparate rates $R_1 = 16$ kbps, $R_2 = 64R_1 = 1024$ kbps. Then in the first case,

$$R_{av} = \tfrac{1}{2}(R_1 + R_2) = 520 \text{ kbps}$$

but $L_1/L_2 = R_2/R_1 = 64$. In the second case, $L_1 = L_2$, but

$$R'_{av} = \frac{2}{(1/R_1) + (1/R_2)} = 31.51 \text{ kbps}$$

To see that there is a more rational allocation that is less "unfair" than a latency ratio of 64, and still achieves a better throughput than R'_{av}, consider a compromise that guarantees that the highest latency is no more than, for example, 8 times the lowest latency. Then in the second case, we would assign 8 slots to class 1 for every slot assigned to class 2. The result would be $L_1/L_2 = 8$ as required and

$$R''_{av} = \frac{P_1 R_1 8F + P_2 R_2 F}{P_1 8F + P_2 F} = \frac{8R_1 + R_2}{9} = 128 \text{ kbps}$$

It can be shown [13] that for the general case of N classes and latency ratio L_{max}/L_{min}, the maximum achievable throughput, denoted by C, is

$$C = \frac{\displaystyle\sum_{n=1}^{n_0} P_n + \sum_{n=n_0+1}^{N} P_n(L_{min}/L_{max})}{\displaystyle\sum_{n=1}^{n_0} P_n/R_n + \sum_{n=n_0+1}^{N} (P_n/R_n)(L_{min}/L_{max})} \text{ bps}$$

where $R_1 < R_2, \ldots, R_N$ and n_0 is such that $R_n \le C$ for all $n \le n_0$, while $R_n > C$ for all $n > n_0$.

Surprisingly, with this maximizing strategy, each user's latency is either L_{max} (for those for which $R_n < C$) or L_{min} (for those for which $R_n > C$). To determine the maximum throughput, it is necessary to have a histogram of the achievable rates for users of the wireless network in question. This will be discussed in the next section. Also, as we shall find there, practical numerology considerations may require us to deviate from this strict bimodal latency allocation, although the ratio L_{max}/L_{min} will remain as the principal constraint.

17.9 IMPLEMENTATION OF CDMA/HDR

In the last section we discussed the key factors and parameters of a wireless system designed to optimize the transport of packet data. In the following we describe such a system design, which leverages in many ways the lessons learned from the development and operation of CDMA IS-95 networks, but makes no compromises in optimizing the air interface for data services. Furthermore, a compelling economic argument can be made for a design that can reuse large portions (to be exact all but the baseband signal processing elements) of components and designs already implemented in IS-95 products, both in the access terminals (AT) and access points (AP).

Due to the highly asymmetric nature of the service offered, we will focus most of our attention on the downlink. In the IS-95 downlink, a multitude of low data rate channels are multiplexed together (with transmissions made orthogonal in the code domain) and share the available base station transmitted power with some form of power control. This is an optimal choice for many low rate channels sharing a common bandwidth. The situation becomes less optimal when a low number of high rate users share the channel. The inefficiencies increase further when the same bandwidth is shared between low rate voice and high rate data users, as their requirements are vastly different as already discussed. It should be noted that increasing the bandwidth available for transmission cannot help in this regard if the data rate of the users is increased proportionally as well.

Therefore a first fundamental design choice is to separate the services, that is, low rate data (voice being the primary service in this category) from high rate data services, by using possibly adjacent but nonoverlapping spectrum allocations. In short, a better system is one that uses an IS-95 or cdma2000-1X RF carrier to carry voice and a separate CDMA/HDR RF carrier to deliver high rate packet bursts.

With a dedicated RF carrier, the HDR downlink takes on a different form than that of the IS-95 designs. As shown in Figure 17.1, the downlink packet transmissions are time multiplexed and transmitted at the full power available to the AP but with data rates and slot lengths that vary according to the user channel conditions. Furthermore, when users' queues are empty, the only transmissions from the AP are those of short pilot bursts and periodic transmissions of control information, effectively eliminating interference from idling sectors.

The pilot bursts provide the access terminals with means to estimate accurately and rapidly the channel conditions. Among other parameters the access terminal estimates the received E_c/N_t of all resolvable multipath components and forms a prediction of the effective received* SNR. The value of the SNR is then mapped to a value representing the maximum data rate that such an SNR can support for a given level of error performance. This channel state information, in the form of a data rate request, is then fed back to the AP via the reverse link data rate request channel (DRC) and updated as fast as every 1.67 ms, as shown in Figure 17.2. The reverse link data request is a 4-bit value that maps the predicted SNR into one of the data rate modes of Table 17.1. In addition, the access terminal requests transmission from only one sector (that with the highest received SNR) among those comprising the active set. Here the definition of active set is identical to that of IS-95 systems, but unlike IS-95, only one sector transmits to any specific access terminal at any given time. The main coding and modulation parameters are summarized in Table 17.1.

The forward error correcting (FEC) scheme employs serial concatenated coding and iterative decoding [12, 14]. Table 17.2 summarizes the main parameters of the three code structures used in the system.

---

*E_c represents the received signal energy density which is the energy in each chip, and N_t represents the total nonorthogonal single-sided noise density. N_t comprises intercell interference, thermal noise, and possibly nonorthogonal intracell interference.

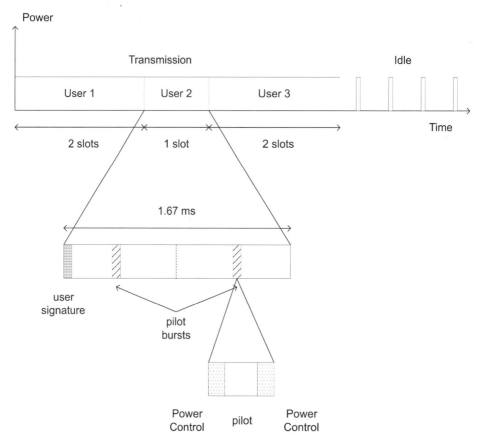

FIGURE 17.1   Access point transmission diagram.

Following the encoder, these traditional signal processing steps are applied: symbol repetition is performed on the lower data rate modes; scrambling, channel interleaving, and the appropriate modulation are applied to obtain a constant modulation rate of 1.2288 MHz for all modes. The in-phase and quadrature channels are then each demultiplexed into 16 streams each at 76.8 kHz and 16-ary orthogonal covers are applied to each stream. The

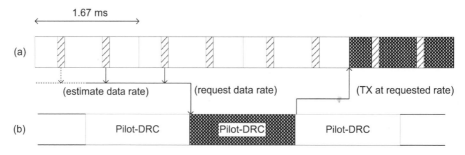

FIGURE 17.2   Channel estimation and data request channel timing diagram: (a) access terminal receive and (b) access terminal transmit.

TABLE 17.1

Data Rate (Kbps)	Packet Length (bytes)	FEC rate (bits/sym)	Modulation
76.8	128	$\frac{1}{4}$	QPSK
102.6	128	$\frac{1}{4}$	QPSK
153.6	128	$\frac{1}{4}$	QPSK
204.8	128	$\frac{1}{4}$	QPSK
307.2	128	$\frac{1}{4}$	QPSK
614.4	128	$\frac{1}{4}$	QPSK
921.6	192	$\frac{3}{8}$	QPSK
1228.8	256	$\frac{1}{2}$	QPSK
1843.2	384	$\frac{1}{2}$	8PSK
2457.6	512	$\frac{1}{2}$	16QAM

TABLE 17.2

Concatenated Code Rate	Rate $\frac{1}{4}$	Rate $\frac{3}{8}$	Rate $\frac{1}{2}$
Outer convolutional code rate (R_{CO})	$\frac{1}{2}$	$\frac{1}{2}$	$\frac{1}{2}$
Inner convolutional code rate (R_{CI})	$\frac{1}{2}$	$\frac{1}{2}$	$\frac{1}{2}$
Outer code puncture rate (R_{PO})	—	—	$\frac{3}{4}$
Inner code puncture rate (R_{PI})	—	$\frac{4}{6}$	$\frac{4}{6}$
Effective outer code rate (R_O)	$\frac{1}{2}$	$\frac{1}{2}$	$\frac{2}{3}$
Effective inner code rate (R_I)	$\frac{1}{2}$	$\frac{3}{4}$	$\frac{3}{4}$
Effective number of tail bits (L)	5	5	5

resulting signal, obtained by adding the 16 data streams, is then spread by quadrature pseudorandom sequences, band limited and up-converted. The resulting RF signal has the same characteristics of an IS-95 signal, thus allowing the reuse of all analog and RF designs developed for IS-95 base stations, including the power amplifiers, and the receiver designs for subscriber terminals.

Table 17.3 summarizes the SNR required to achieve a 1% packet error rate. Note that at the lower rates this corresponds to E_b/N_t of 2.0 to 2.5 dB, a result of using iterative decoding techniques on serial concatenated codes, while for the two highest rates, E_b/N_t increases considerably because 8PSK and 16QAM modulations are employed.

At this point we are able to estimate the maximum achievable throughput per sector as discussed in the previous section. Figure 17.3 shows a graph of the cumulative distribution function of the SNR for a typical embedded sector of a large three-sector network deployed with a frequency reuse of one. In particular, the SNR values are those of the best serving sector and representative of a uniform distribution of users across the coverage area. From the results of Figure 17.3 and the knowledge of the SNR required to support a given data rate, it is straightforward to derive the histogram of data rates achievable in such an embedded sector. The result is shown in Figure 17.4 where the SNRs used in the calculation are those of Table 17.3 with an additional 2 dB of margin to account for various losses. Finally, Figure 17.5 shows the realized throughput per sector per 1.25 MHz of bandwidth versus the parameter $L_{\max}/L_{\min}$. Note that the throughput is doubled for a

TABLE 17.3

Data Rate (kbps)	E_c/N_t (dB)	E_b/N_t (dB)
76.8	−9.5	2.5
102.6	−8.5	2.3
153.6	−6.5	2.5
204.8	−5.7	2.1
307.2	−4.0	2.0
614.4	−1.0	2.0
921.6	1.3	3.0
1228.8	3.0	3.0
1843.2	7.2	5.4
2457.6	9.5	6.5

latency ratio $L_{max}/L_{min} = 8$. The overall throughput of data with variable latency is better than three times the total throughput of voice. As just shown, a factor of 2 is the result of the variable latency. The remainder is a combination of turbo coding, again with decoding latency greater than that tolerable by voice, and the reduction of intersector and intercell interference. If voice traffic were to be mixed with data, the advantage for data would be much reduced and voice with its intolerance for latency would not be better served. This leads to the obvious conclusion that, for efficiency, voice and data are best segregated to separate bandwidth allocations.

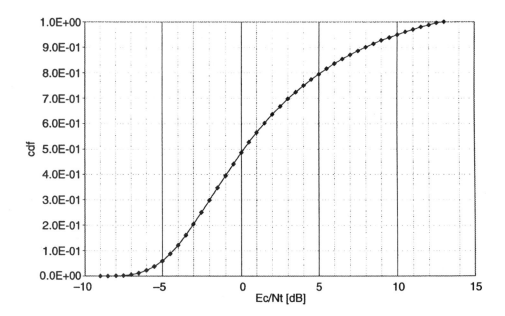

FIGURE 17.3 E_c/N_t distribution for a typical embedded sector in a three-sector network with universal frequency reuse in each cell.

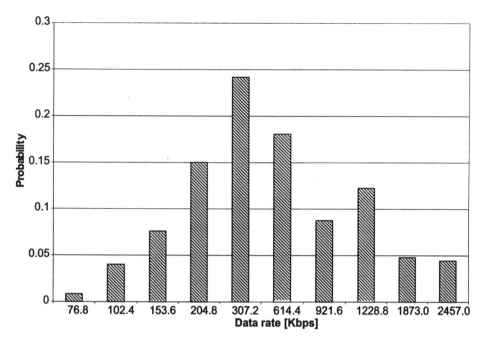

FIGURE 17.4 Data rate histogram.

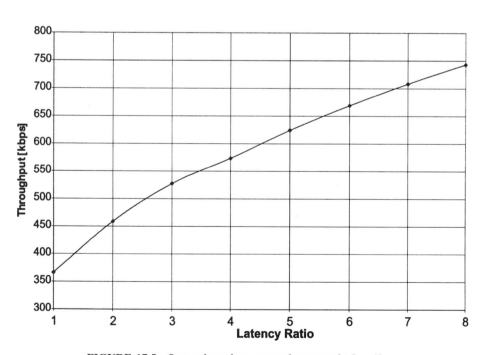

FIGURE 17.5 Sector throughput versus latency ratio L_{max}/L_{min}.

17.10 SUMMARY AND CONCLUDING REMARKS

CDMA is a proven technology now used worldwide by more than 100 million users. To make such headway as a latecomer in a TDMA-technology-dominated world, it has had to demonstrate its superior capacity, coverage, and voice quality characteristics. These are in part a consequence of the more effective approach to satisfying basic multiple access requirements provided by spread spectrum techniques. Improvements and enhancements of this proven technology are producing higher voice user capacity and medium-speed data services. For high-speed data services, the industry-favored approach seems to be to triple the spread bandwidth occupancy of each CDMA RF carrier. This, however, does not increase the voice-user capacity much beyond what is expected as a consequence of the bandwidth expansion. Furthermore, even though higher speed data services are possible, the plan for supporting both voice and data users on the same carrier will limit both capacity and maximum data speeds. An alternate approach, based on optimized packet allocation and variable latency service for data-only traffic provides for better than three times the throughput on the current (1X) RF carrier bandwidth, with peak data rates exceeding 2 Mbps. Voice users can then use a separate 1X or possibly 3X carrier more efficiently without the encumbrance of high-speed data users competing for the same carrier resources.

Acknowledgement The author is grateful to R. Padovani and N. Sindhushayana for providing much of the material in Sections 17.8 and 17.9.

REFERENCES

1. A. J. Viterbi, "Spread Spectrum Communications—Myths and Realities," *IEEE Commun. Mag.*, pp. 11–18, May 1979.

2. S. W. Golomb, "Sequences with Randomness Properties," Glenn L. Martin Co., Baltimore, MD. *Terminal Prog. Rep.*, Contract No. 639498, June, 1955; see also *Shift Register Sequences*, Holden-Day, San Francisco, CA, 1967.

3. J. J. Spilker, Jr., "Delay Lock Tracking of Binary Signals," *IEEE Trans. Space Electronics Telemetry*, Vol. SET-9, pp. 1–8, March 1963.

4. A. J. Viterbi, "Phase-Locked Loop Dynamics in the Presence of Noise by Fokker-Planck Techniques," *Proc. IEEE*, Vol. 51, pp. 1737–1753, Dec. 1963.

5. A. J. Viterbi, *Principles of Coherent Communications*, McGraw-Hill, New York, 1966.

6. I. M. Jacobs et al., "The Application of a Novel Two-Way Mobile Satellite Communications and Vehicle Tracking System to the Transportation Industry," *IEEE Trans. Veh. Tech.*, Vol. 40, pp. 57–63, Feb. 1991.

7. K. S. Gilhousen, I. M. Jacobs, R. Padovani, and L. A. Weaver, "Increased Capacity Using CDMA for Mobile Satellite Communication," *IEEE Trans. Select. Areas Commun.*, Vol 8 , pp. 503–514, May 1990.

8. TIA/EIA/IS-95 Interim Standard "Mobile Station-Base Station Compatibility Standard for Dual-Mode Wideband Spread Spectrum Cellular System," Telecommunication Industry Association, Arlington, VA, July 1993.

9. A. J. Viterbi, A. M. Viterbi, K. S. Gilhousen, and E. Zehavi, "Soft Handoff Extends CDMA Cell Coverage and Increases Reverse Link Capacity," *IEEE J. Selected Areas in Commun.*, Vol. 12, pp. 1281–1288, Oct. 1994.

10. TIA/EIA IS-95B, "Mobile Station Compatibility Standard for Dual-Mode Spread Spectrum System," Telecommunications Industry Association, Arlington, VA, March 1999.

11. C. E. Wheatley III, "Self-Synchronizing a CDMA Cellular Network," *Microwave J.*, Vol. 42, pp. 320–328, May 1999.

12. C. Berrou, A. Glavieux, and P. Thitimajshima, "Near Shannon Limit Error Correcting Coding and Decoding: Turbo Codes," *Proc. Of International Conf. Comm.*, Geneva, Switzerland, May 23–26, pp. 1064–1070, 1993.

13. P. Bender, P. Black, M. Grob, R. Padovani, N. Sindhushayana, and A. J. Viterbi, "CDMA/HDR: A Bandwidth Efficient High Speed Wireless Data Service for Nomadic Users," *IEEE Commun. Mag.*, 2000.

14. S. Benedetto, G. Montorsi, D. Divsalar, and F. Pollara, "Serial Concatenation of Interleaved Codes: Performance Analysis, Design and Iterative Decoding," *JPL TDA Progress Report* 42–126, August 15, 1996; also *IEEE Trans. Inform. Theory*, Vol. 44, pp. 909–926, May 1998.

CHAPTER 18

WCDMA Radio Access Technology for Third-Generation Mobile Communication

ERIK DAHLMAN, FREDRIK OVESJÖ, PER BEMING, CHRISTIAAN ROOBOL, MAGNUS PERSSON, JENS KNUTSSON, and JOAKIM SORELIUS

18.1 INTRODUCTION

Standardization of third-generation mobile communication systems (UMTS/IMT-2000) has been rapidly progressing on a global basis. Third-generation systems extend the services provided by current second-generation systems (GSM, PDC, IS-136, and IS-95) with high-rate data capabilities, in the first phase supporting data rates up to 384 kbps with wide-area coverage and up to 2 Mbps in indoor and small-cell outdoor environments. A main application for these high-rate data services will be wireless packet access, especially wireless access to the Internet.

This chapter presents the background and current technical status of the UMTS radio access concept, also known as UMTS terrestrial radio access (UTRA). The focus is on the UTRA Frequency Division Duplex (FDD) mode, which is based on wideband code division multiple access (WCDMA) technology.

The chapter is outlined as follows: In Section 18.2, a short history of the WCDMA technology is given. Section 18.3 presents an overview of the UMTS/IMT-2000 system architecture. A discussion on the WCDMA radio protocols is given in Section 18.4 followed by a detailed description of the physical layer in Section 18.5 and a discussion on radio resource management in Section 18.6. Finally in Section 18.7 we discuss some technologies that can be used to improve the WCDMA performance.

18.2 BACKGROUND TO WCDMA

Work on WCDMA as an access technology for third-generation mobile communication begun in the late 1980s. In the RACE/CODIT project [1] a WCDMA concept fulfilling the requirements on third-generation mobile communication was first developed. The

Wireless Communications in the 21st Century, Edited by Shafi, Ogose, and Hattori.
ISBN 0-471-155041-X   © 2002 by the IEEE.

WCDMA technology was then further refined into the FMA2 (FRAMES Multiple Access 2) concept developed within the FRAMES project [2–4]. In December 1996, the FMA2 concept was submitted to the European Telecommunications Standards Institute (ETSI) as a candidate technology for UTRA. In ETSI, the FMA2 proposal was first merged with other WCDMA proposals into the so-called Alpha concept and in January 1998 WCDMA, was chosen by ETSI as the technology for the UTRA FDD mode.

In parallel to the WCDMA activities in Europe, extensive work on WCDMA for third generation mobile communication was also carried out in Japan, first by NTT DoCoMo and later within the Association of Radio Industries and Business (ARIB) [5]. Due to extensive co-operation between European and Japanese companies, the WCDMA concepts developed by ARIB and ETSI were very similar.

Additional WCDMA concepts were also developed in Korea and North America. These WCDMA concepts were very much in line with the WCDMA concepts of Europe and Japan.

In January 1999, the work on WCDMA in Europe, Japan, North America, and Korea was merged in the global third-generation partnership project (3GPP). The task of 3GPP was to develop a worldwide third-generation specification based on WCDMA.

In parallel with the work on WCDMA, work on an evolution of second-generation CDMA (IS-95) into a third-generation system was carried out in North America, first by the Telecommunications Industry Association (TIA) and later on by the third-generation partnership project 2 (3GPP2). This work led to the development of the cdma2000 concept [6]. The cdma2000 concept consisted of two modes, one based on a direct-spread wideband CDMA downlink (3.75 MHz) and one based on a multicarrier downlink (multiple 1.25-MHz narrowband carriers).

In July 1999, after extensive discussions between operators in the so-called Operators Harmonization Group (OHG) and major manufacturers, a merge between WCDMA and cdma2000 into a single harmonized CDMA-based third-generation concept was unanimously agreed upon. According to the OHG agreement, the harmonized third-generation CDMA standard should consist of three modes:

- A direct-spread wideband CDMA FDD mode based on the UTRA FDD mode (WCDMA) to be developed by 3GPP
- A multicarrier CDMA FDD mode based on the cdma2000 multicarrier mode to be developed by 3GPP2
- A time division duplex (TDD) mode based on the UTRA TDD mode to be developed by 3GPP

The OHG agreement also stated that all three modes of the harmonized third generation CDMA standard should be able to interface to both GSM and ANSI-41 core networks.

18.3 UMTS/IMT-2000 SYSTEM OVERVIEW

18.3.1 UMTS/IMT-2000 System Architecture

The overall system architecture of UMTS/IMT-2000 as defined by 3GPP is illustrated in Figure 18.1. It includes user equipment (UE), universal terrestrial radio access network (UTRAN), and core network. The architecture also includes two general interfaces: the Iu

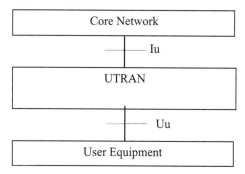

FIGURE 18.1   UMTS/IMT-2000 overall system architecture as defined by 3GPP.

interface between UTRAN and the core network, and the Uu interface (the "radio interface") between UTRAN and the UE. The functional division between UTRAN and the core network is such that UTRAN handles all radio-specific procedures whereas the core network handles the service-specific procedures, including mobility management and call control.

18.3.2   Radio Access Bearer and Radio Bearer

The core network requests UTRAN to provide radio access bearers (RABs) to the core network. A RAB includes two subparts: a radio bearer over the radio interface and an Iu bearer over the Iu interface. A RAB is characterized by means of quality-of-service (QoS) parameters. UTRAN is then responsible for selecting the settings of the radio-bearer and Iu bearer parameters so that the QoS is fulfilled. How these settings are selected is not standardized. Hence, it is up to the operator to select the way of providing RABs.

18.3.3   UTRAN Internal Architecture

Figure 18.2 shows the UTRAN internal architecture. It consists of two types of nodes. The radio network controller (RNC) controls several cells, similar to a GSM base station controller (BSC). Node B is a logical node handling the transmission and reception of one or more cells, similar to a GSM base tranceiver station (BTS). It is the RNC that terminates the Iu interface, whereas to Node B one or several cells are connected. Within UTRAN, two interfaces are defined: the Iur interface connects two RNC while the Iub interface connects an RNC to a Node B.

18.3.4   Radio Interface Protocol Architecture

Figure 18.3 shows the WCDMA radio interface protocol stack. Layer 1 comprises the WCDMA physical layer. Layer 2 comprises the medium access control (MAC) and radio link control (RLC) sublayers. The network layer (layer 3) of the control plane is split into the radio resource control (RRC) sublayer and the mobility management (MM) and connection management (CM) sublayers. CM and MM terminate in the core network and are thus not shown in Figure 18.3.

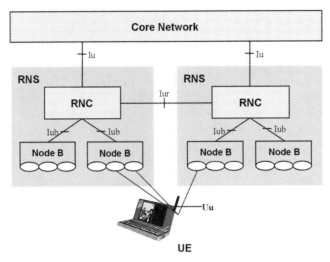

FIGURE 18.2 UTRAN internal architecture.

Transport Channels—L1 Services The physical layer offers information transfer services to the MAC layer. These services are denoted transport channels and are characterized by the way in which data is transmitted over the radio interface. The following transport channels are defined:

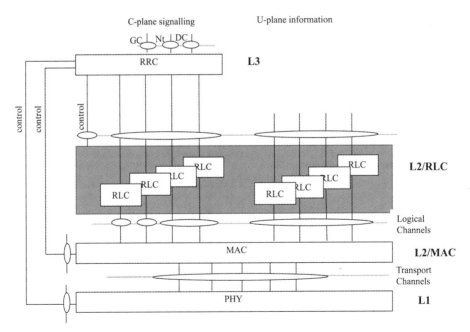

FIGURE 18.3 WCDMA radio interface protocol stack. Only protocols terminating in UTRAN are shown.

Broadcast Channel   BCH is a fixed rate point-to-multipoint channel used to convey system information in the whole coverage area of the cell.

Paging Channel   PCH is a point-to-multipoint channel used to reach a UE in the whole coverage area of the cell. The UE can perform efficient sleep mode when camping on this transport channel.

Forward Access Channel   FACH is a point-to-multipoint channel used to convey data to one or more UEs. In some cases, the FACH may be transmitted over only a part of a cell, using beam forming.

Random Access Channel   RACH is used by the UE to transmit short user data packets and control packets, for instance to initiate packet transfer on the dedicated channels.

Common Packet Channel   CPCH is used by a UE to transmit longer user data packets and control packets. The CPCH is a contention-based channel, similar to RACH.

Dedicated Channel   DCH is a point-to-point bidirectional channel used to convey data from/to a UE. The DCH may be transmitted over only a part of a cell, using beam forming. Furthermore, DCH can be used in soft handover.

Downlink Shared Channel   DSCH is a point-to-multipoint channel with dedicated characteristics. It is always associated with a DCH that, by means of layer 1 signaling, conveys the information on whether the UE should receive the DSCH or not.

High-speed Downlink Shared Channel   HS-DSCH is an enhanced downlink shared channel to be included in WCDMA release 5. MHS-DSCH supports high-speed downlink packet access with peak data rates above 10 Mbps.

A transport block is the basic unit by which data is exchanged over transport channels, that is, between the physical layer and the MAC layer. On each transport channel, a number of transport blocks arrive to or are delivered from the physical layer every transmission time interval (TTI); see Figure 18.4. The TTI is transport-channel specific from the set TTI $\in \{10, 20, 40, 80 \text{ ms}\}$. For HS-DSCH, the TTI will be 2 ms. The number of transport blocks as well as the size of each transport block may vary between each TTI.

Logical Channels—MAC Services   The MAC layer offers information-transfer services of specific information to the RLC layer. These services are denoted logical channels and are characterized by the kind of information they transfer. The following logical channels are defined:

Broadcast Control Channel   BCCH is used to transfer system information.

Paging Control Channel   PCCH is used to transfer paging information.

Common Control Channel   CCCH is used to transfer information when the UE does not have a unique MAC identity in the current cell. This channel is used when accessing the system after power on.

Dedicated Control Channel   DCCH is used when MAC has a unique identity in the current cell or when a DCH is used.

Common Traffic Channel   CTCH is used for multicast services such as short message service (SMS) cell broadcast.

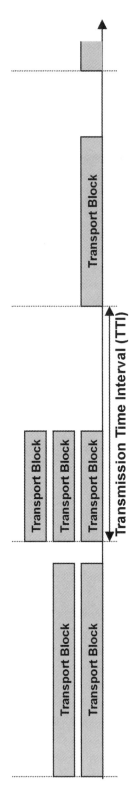

FIGURE 18.4 Transport-channel structure.

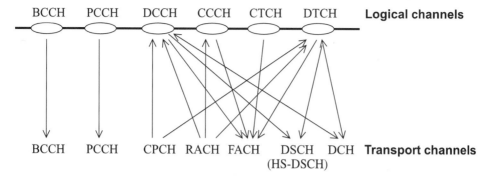

FIGURE 18.5   Logical channels mapped onto transport channels, as seen from the UTRAN side.

Dedicated Traffic Channel   DTCH is used for user data traffic. The channel can only be used if MAC has a unique identity in the current cell or if a DCH is used.

Mappings of Logical Channels onto Transport Channels   The possible mappings of logical channels onto transport channels are shown in Figure 18.5. From the figure it can be seen that DCCH and DTCH can be mapped onto both common and dedicated transport channels. This enables transport-channel-type switching, which means that a logical channel can switch from a dedicated transport channel to a common transport channel and vice versa.

RLC Services   The RLC layer offers information transfer to higher layers by means of acknowledged mode (AM), unacknowledged mode (UM), and transparent mode (TM) operation. The transparent mode does not add any overhead whereas the unacknowledged mode does. Furthermore, the unacknowledged mode provides error detection and hence it does not deliver erroneous service data units (SDUs). The acknowledged mode provides an even more reliable transfer since it includes retransmissions.

Radio Bearers   A radio bearer is provided by means of the RLC/MAC/L1 combinations giving the requested QoS required for the RAB. It should be noted that the same QoS can be provided with different settings of the RLC/MAC/L1 parameters. In Section 18.4 we will further discuss how to control the settings.

18.4 WCDMA RADIO PROTOCOL OPERATION

In Figure 18.6 the relation between the different radio protocols is depicted. The figure illustrates the central role of the RRC. The RRC controls all the radio resources, allocates resources, and configures all the radio protocols. Furthermore, RRC handles the mobility of the UEs having an RRC connection, that is, handover and cell updates.

If a radio access bearer is to be set up, the RRC will allocate the corresponding radio resources to the UE. In short, the RRC will allocate transport channel with an associated set of transport formats (TF), the so-called transport format set (TFS). A transport format describes the way by which data is transmitted on a transport channel. More specific, a transport format includes information about for example, the channel coding, the

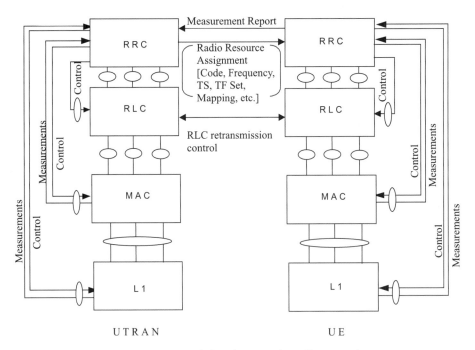

FIGURE 18.6 Relations between the radio protocols.

transmission time interval (TTI), the number of bits in each transport block, and the number of transport blocks per TTI. The TFS together with the configuration of the RLC protocol are the tools with which a certain QoS of a RAB is supported by UTRAN. The MAC can change the TF every TTI provided that the new TF is within the TFS allocated by the RRC. This feature is useful for services with time-varying rate. If, at a certain time, the source data rate is low, the MAC will choose a TF corresponding to that data rate. If at a later instance the rate increases, the MAC will choose a TF with a higher data rate, provided that this TF is within the TFS.

In case the QoS cannot be maintained with the allocated TFS, the RRC needs to reallocate a new TFS and perhaps even reconfigure the RLC protocol. It is clear that this is a more complex and time-consuming process than just changing TF within the allocated TFS.

In case two RABs are to be established for one UE, it is possible to do this by means of two different transport channels. The RRC allocates for each transport channel a TFS within which the MAC may choose the TF. In general, the MAC cannot use all possible combinations of transport formats. Instead, the RRC allocates a transport format combination set (TFCS). The TFCS is the allowed set of combinations of transport formats, that is, MAC can only use combinations of transport formats that are within the TFCS. The TFCS prevents that the MAC will allocate transport formats to the different transport channels without taking into account the total output power. The used transport format combination is signaled over layer 1 using the so-called transport format combination indicator (TFCI).

It can be seen from Figure 18.6 that there is only RRC and RLC peer-to-peer signaling. The MAC has no peer-to-peer signaling. The main task of the MAC is to schedule and prioritize data transmission between users and also between services for one user. In addition, the MAC has a multiplex function. Data originating from different radio access bearers with the same QoS requirement may be multiplexed on MAC level and thus share the same transport channel.

The RLC protocol has three modes, the transparent mode, the unacknowledged mode, and the acknowledged mode. In the transparent mode the only possible function of the RLC is segmentation of higher layer protocol data units (PDU) into smaller RLC PDUs to be transmitted over the radio interface. However, in the transparent mode the RLC does not add any overhead to the RLC PDU, so segmentation needs to be done in a predefined manner, or the receiver will not know how to reassemble the PDUs.

In the unacknowledged mode, the RLC performs segmentation of higher layer PDUs into smaller RLC PDUs as well. In this mode the RLC includes a header including a sequence number in the RLC PDUs. Consequently, segmentation does not need to be done in a predefined manner, since the reassembly can now be done with help of information in the header. Due to the sequence number the RLC can also detect errors in the unacknowledged mode. If there are missing PDUs, the receiver RLC will immediately detect that, due to the fact that there are missing sequence numbers. However, the RLC will not request for retransmission of missing PDUs in unacknowledged mode.

In the acknowledged mode, the RLC will also perform segmentation and reassembly and additionally it will perform error correction by means of retransmissions. The RLC specification is such that many types of ARQ mechanisms are possible to support. Which type of mechanism to be used is determined by the RRC. Based on a certain type of service to be supported, the RRC will configure the RLC. The RLC can operate such that it polls for status information on a regular basis, but it can also operate such that it polls only when the sender needs to move its transmission window. It can be configured such that the receiver side will immediately send status information as soon as it detects an error. It is possible for the receiver to quickly detect failed retransmissions and failed transmissions of status information. Exactly how an RLC entity is configured depends on the type of service that uses that entity.

18.5 WCDMA PHYSICAL LAYER

18.5.1 Transport-Channel Coding and Multiplexing

Figure 18.7 illustrates how channel coding and interleaving is applied to each transport channel.

First, a cyclic redundancy check (CRC) is added to each transport block. The CRC allows for detection of errors in the received transport blocks. The error detection can, for example, be used for:

- Uplink soft handover combining
- Threshold setting for closed-loop power control
- General indication of erroneous transport blocks to higher layers, for example, for error concealment for speech codecs.

As shown in Figure 18.7, the CRC can be of different length depending on the error detection requirements for the transport channel. The CRC length is part of the transport format.

Second, channel coding for forward error correction (FEC) is applied to each transport channel. Two different coding schemes can be applied:

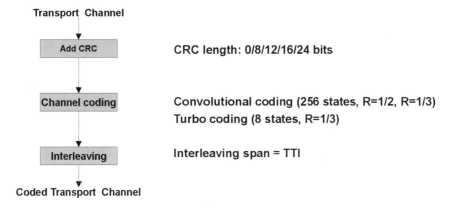

FIGURE 18.7 Coding of transport channels.

- Convolutional coding
- Turbo coding

What coding scheme to use for a transport channel is determined by the RRC and is part of the transport format. Typically turbo coding is used for higher-rate services or services with high-quality requirements, that is, low error rates.

Finally, block interleaving with an interleaving span of one TTI is applied to each coded transport channel.

Figure 18.8 illustrates how two or more coded transport channels with, in general, different TTI and different channel coding, are multiplexed together and jointly mapped to a physical channel. The quality/rate matching applies puncturing and/or unequal repetition to the coded transport channels in a coordinated way to ensure that the total bit rate

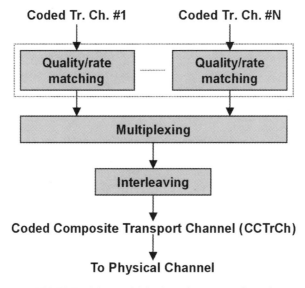

FIGURE 18.8 Multiplexing of transport channels.

after multiplexing matches that of the physical channel. The rate matching is applied in such a way that the quality of the different transport channels is approximately balanced. As an example, more puncturing may be applied to a coded transport channel with lower quality requirements. After multiplexing of the transport channels, block interleaving is applied over one 10-ms radio frame.

The result of the multiplexing is a coded composite transport channel (CCTrCh) that is mapped to a physical channel. If the total bit rate of the CCTrCh is too high to fit into one physical channel, mapping can be done to several parallel physical channels, so-called multicode transmission.

18.5.2 Physical Channel Structure

Figure 18.9 illustrates the physical channel structure of WCDMA. Transport channels as described in Section 18.3 are mapped to different physical channel types. There are also a number of physical channels to which no transport channels are mapped, that is, physical channels that carry information generated internally within the physical layer. Below, some of the physical channels are described in more detail.

Dedicated Physical Channel The dedicated physical channel (DPCH) carries the DCH transport channel. Each DPCH consists of two parts, a dedicated physical data channel (DPDCH) and a dedicated physical control channel (DPCCH). The DPDCH part carries the actual higher layer data, that is, the DCH information, while the DPCCH part carries information generated internally within the physical layer. This physical-layer information consists of predefined pilot bits that can be used, for example, for channel amplitude and phase estimation and bits for transmit power control (TPC), TFCI, and feedback information (FBI). The feedback information is, for example, used in case of feedback transmit diversity, see Section 18.7.

Figure 18.10 illustrates the structure of the uplink DPCH. Each 10-ms DPCH frame consists of 15 slots, where each slot corresponds to one power control period. The DPDCH and DPCCH are transmitted in parallel within each slot. The number of DPDCH bits per slot is variable in the range of 10 to 640 bits, depending on the DPDCH spreading factor

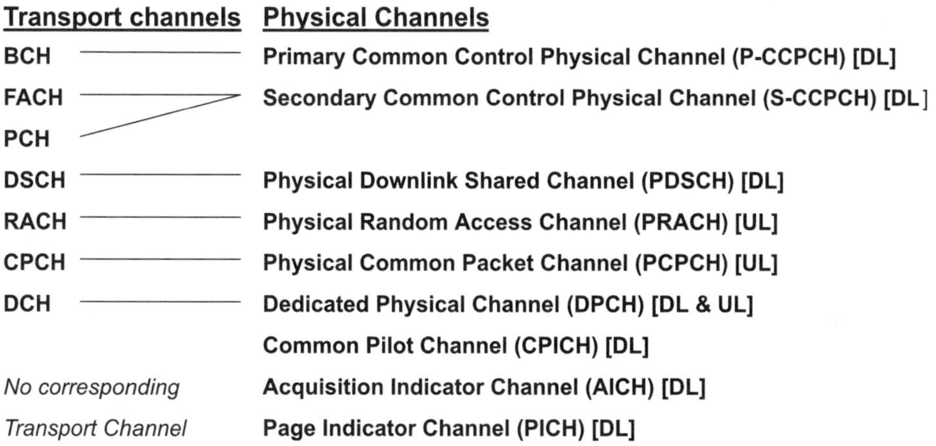

Transport channels   **Physical Channels**

BCH ——————— Primary Common Control Physical Channel (P-CCPCH) [DL]

FACH ——————— Secondary Common Control Physical Channel (S-CCPCH) [DL]

PCH

DSCH ——————— Physical Downlink Shared Channel (PDSCH) [DL]

RACH ——————— Physical Random Access Channel (PRACH) [UL]

CPCH ——————— Physical Common Packet Channel (PCPCH) [UL]

DCH ——————— Dedicated Physical Channel (DPCH) [DL & UL]

Common Pilot Channel (CPICH) [DL]

No corresponding   Acquisition Indicator Channel (AICH) [DL]

Transport Channel   Page Indicator Channel (PICH) [DL]

FIGURE 18.9 Physical channel structure (release 4).

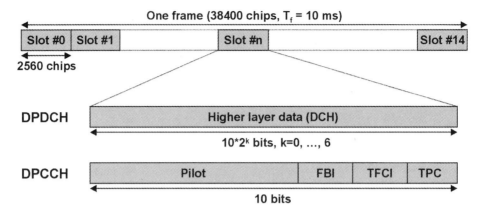

FIGURE 18.10   Structure of uplink DPCH.

(SF ranging from 256 to 4). The maximum DPDCH bit rate is thus 960 kbps. There are 10 DPCCH bits per slot (fixed spreading factor SF = 256). The exact format for the DPCCH (number of bits for pilot, TPC, TFCI, and FBI) varies and depends on if the TFCI and FBI fields are needed or not. The pilot and TPC fields are always present on the uplink DPCCH.

Figure 18.11 illustrates the structure of the downlink DPCH. For the downlink, the DPDCH and DPCCH parts are time-multiplexed within each slot. The number of DPCH bits per slot is variable in the range 10 to 1280 bits, depending on the spreading factor (SF ranging from 512 to 4. The maximum downlink DPCH bit rate is thus 1920 kbps including layer 1 information.

The exact format for the downlink DPCH varies and depends on the spreading factor and if TFCI is needed or not. The data, pilot, and TPC fields are always present on the downlink DPCH.

Primary Common Control Physical Channel   The primary common control physical channel (P-CCPCH) carries the BCH transport channel. The structure of the P-CCPCH is similar to that of the downlink DPCH except that the P-CCPCH only includes the data field, that is, the pilot, TPC, and TFCI fields are not present on the P-CCPCH. The P-CCPCH carries 20 bits per slot (fixed spreading factor SF = 256).

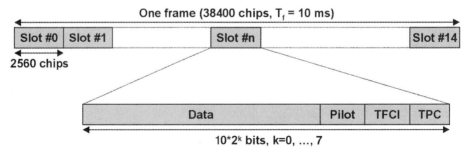

FIGURE 18.11   Structure of downlink DPCH.

Secondary Common Control Physical Channel The secondary common control physical channel (S-CCPCH) carries the PCH and FACH transport channels. The structure of the S-CCPCH is basically the same as that of the downlink DPCH except that the S-CCPCH does not include the TPC and FBI fields, but only data, pilot, and TFCI. Furthermore, both the pilot and TFCI fields are optional. If pilot bits are not included in the S-CCPCH, the common pilot channel (see below) should be used for channel estimation. The number of bits per S-CCPCH slot is variable in the range of 20 to 1280 bits (SF ranging from 256 to 4).

Common Pilot Channel The common pilot channel (CPICH) has the same structure as the primary P-CCPCH, except that the data is replaced by a predefined bit sequence of length 20 bits. The CPICH can be used as phase reference for coherent detection of other downlink channels. In case of transmit diversity, see Section 18.7, the CPICH is transmitted from both antennas but with different orthogonal bit sequences. In this way, the UE can estimate the channel from each of the antennas separately.

18.5.3 Spreading and Scrambling

Spreading and scrambling is the process by which the bit stream of a physical channel is multiplied by a user-specific code and spread to the chip rate.

Downlink Spreading and Scrambling Figure 18.12 illustrates the spreading and scrambling of downlink physical channels in WCDMA. A physical channel is first serial-to-parallel converted to a pair of bit sequences. These two sequences are then spread to the chip rate (3.84 Mcps) by the same user-specific channelization code c_{ch} and transformed into a complex signal $I + jQ$. Finally, the signal $I + jQ$ is scrambled (multiplexed) by a cell-specific complex scrambling code of length 38,400 chips.

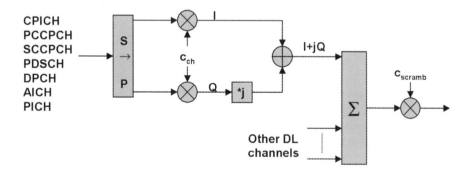

c_{ch}: **channelization code**

c_{scramb}: **scrambling code**

FIGURE 18.12 Downlink spreading and scrambling.

If the downlink CCTrCh bit rate exceeds the maximum bit rate of a downlink DPCH (1920 kbps), multiple downlink DPCH spread by different channelization codes can be transmitted in parallel to one UE. A maximum of three downlink DPCH can be transmitted in parallel corresponding to a maximum theoretical channel bit rate of 5.76 Mbps to one UE.

The channelization code c_{ch} is from the set of orthogonal variable spreading factor (OVSF) codes [7]. This set of codes has the property that orthogonality between transmitted signals is preserved even if they have different channel bit rates and thus are spread with different spreading factors.

There is only a limited number of OVSF codes available, for example, the number of OVSF codes is limited to 256 if all downlink channels use a spreading factor of 256 and less than that if some downlink channels use a lower spreading factor. This means that, in some cases, the downlink capacity may be code limited, that is, there are no channelization codes available although the interference limit has not yet been reached. To avoid such code-limited capacity, multiple downlink scrambling codes can be used in one cell. For each primary scrambling code there can be up to 16 additional secondary scrambling codes in each cell. For each of the secondary scrambling codes, the entire set of channelization codes can be reused, making the risk for code-limited capacity negligible. The disadvantage with multiple scrambling codes is that the orthogonality between transmitted signals will be lost if they are scrambled by different codes. However, as long as multiple scrambling codes are only used in the case of code-limited capacity, this is not a serious drawback.

If multiple scrambling codes are used, the CPICH and the P-CCPCH should always be scrambled by the primary scrambling code. All other downlink channels could be scrambled by either the primary scrambling code or a secondary scrambling code.

There are a total of 512 primary scrambling codes defined. These are divided into 64 groups with 8 codes in each group. This is done to simplify the cell search procedure; see Section 18.5.4. The large number of scrambling codes implies that basically no downlink code planning is needed.

Uplink Spreading and Scrambling   Figure 18.13 illustrates the spreading and scrambling of the uplink DPCH. The DPDCH and DPCCH parts are spread to the chip rate by two different channelization codes and are then combined into a complex signal $I + jQ$. Note that the powers of the I and Q branches are normally different due to the different bit rates of the DPDCH and DPCCH.

Finally, the signal $I + jQ$ is scrambled with a UE specific complex scrambling code of length 38,400 chips (one frame). As an alternative, scrambling can be done with a scrambling code of length 256 chip, so-called short-code scrambling. The use of short-code scrambling allows for easier implementation of some advanced receiver structures (multiuser detectors, etc.) in the base station. The network decides if the UE should use long (38,400 chips) or short uplink scrambling codes.

Similar to the downlink, if the uplink CCTrCh bit rate exceeds the maximum bit rate of an uplink DPDCH (960 kbps), multiple DPDCH can be transmitted in parallel from one UE. Multiple DPDCH can be transmitted on either the I branch or the Q branch and signals transmitted on the same branch should use different channelization codes. As one channelization code is used by the DPCCH, a maximum of six DPDCH can thus be transmitted in parallel corresponding to a maximum theoretical channel bit rate of 5.76 Mbps from one UE.

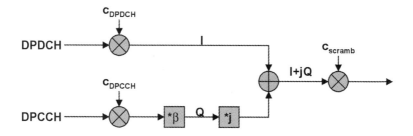

c_{DPDCH}: **DPDCH channelization code**
c_{DPCCH}: **DPCCH channelization code**
c_{scramb}: **scrambling code**

β: **relative DPCCH power (P_{DPCCH}/P_{DPDCH})**

FIGURE 18.13   Uplink spreading and scrambling.

18.5.4 Physical Layer Procedures

This section discusses some essential WCDMA physical layer procedures.

Cell Search   One of the most important and most critical aspects of a CDMA-based cellular system is the process by which the mobile terminal carries out cell search. There are different types of cell search:

- Initial cell search, that is, the process by which a mobile terminal finds the "best" cell at power-on.
- Idle-mode cell search, that is, the process by which a mobile terminal in idle mode searches for neighbor cells. If a neighbor cell that is sufficiently "better" than the current cell is found, the mobile terminal should, for example, use this cell to listen to page messages.
- Active-mode cell search, that is, the process by which a mobile terminal in active mode searches for neighbor cells. If a neighbor cell that is sufficiently "good" compared to the current cell is found, the mobile terminal should add this cell to the active set.

The third case, that is, active-mode cell search, is by far the most critical. If a neighbor cell is not properly added to the active set, there may be excess interference with a loss of capacity as a consequence. There are thus tight time constraints for active-mode cell search. However, the process is simplified by the fact that some kind of *a priori* information is normally available to the mobile terminal. This information is typically in form of a neighbor list that contains information about the primary scrambling code of potential neighbor cells and possibly also *a priori* timing information.

The WCDMA cell search process differs from that of second-generation CDMA systems in that it should be possible to deploy WCDMA systems without tight intercell synchronization. In second-generation CDMA (IS-95), all cells are synchronized to each other with an accuracy of a few microseconds. This simplifies cell search as the timing of a neighbor cell is then approximately known *a priori*. Consequently, the mobile terminal must only search for a neighbor cell within a limited time window. However, tight intercell synchronization implies that a common accurate timing reference must be available to all cells. Typically in IS-95 systems this is ensured by deploying receivers for the Global Positioning System (GPS) at each cell site. However, for WCDMA, where indoor deployment has been seen as an important application, a requirement on accurate intercell synchronization was not acceptable.

WCDMA cell search is based on the following three signals broadcast by every cell; see Figure 18.14:

- The primary search code (PSC) is a specific chip sequence of length 256 chips, transmitted at a predefined position in every slot interval (2560 chips). The PSC is the same for every cell and every slot interval and has been chosen to have very good aperiodic autocorrelation properties.
- The secondary search code (SSC) is also a chip sequence of length 256 chips and is transmitted in parallel with the PSC. In contrast to the PSC, the SSC is in general different between different cells and also different between slot intervals. The 15 consecutive SSC in a frame define a codeword of length 15. A total of 64 such SSC code words have been defined, and they specify to which scrambling-code group the primary scrambling code of the cell belongs; compare Section 18.5.3.
- The CPICH, see Section 18.5.2.

The cell search procedure is typically carried out in three steps:

1. Find slot timing. This can be done by matched filtering of the received signal with a filter matched to the PSC.
2. Find frame timing and scrambling-code group. This can be done by correlating the received signal, at the timing found from step 1, using the set of valid SSC codewords. Frame timing can be found because the SSC codewords are shift-unique, that is, a shift of one SSC codeword does not coincide with a different shift of the same codeword or an arbitrary shift of any of the other codewords.
3. Find primary scrambling code. This can be done by correlating the received signal, at the frame timing found from step 2, with the CPICH corresponding to all codes in the scrambling-code group found from step 2.

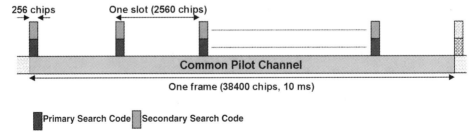

FIGURE 18.14   Signals used for WCDMA cell search.

When cell search is completed, the UE has identified the primary scrambling code and the frame timing of the found cell. It can then read the BCCH to get further information, for example, full cell identity and system frame number.

A significant effort has been carried out to optimize and evaluate the performance of the WCDMA cell search performance, see, for example, [8]. An example of the cell search performance is illustrated in Figure 18.15.

Although WCDMA cell search does not require the use of tight intercell synchronization, synchronization may still be beneficial from a cell search complexity point of view. In

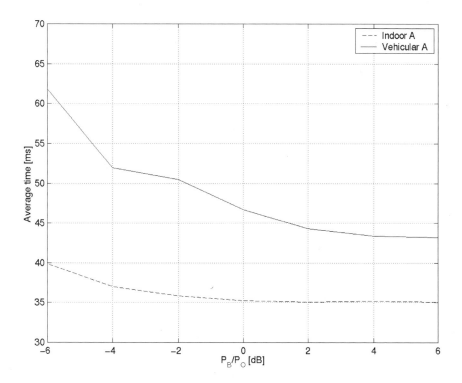

Assumptions/notations

- P_A: Own-cell power
- P_B: Target-cell power
- P_0: Other-cell interference
- P_B/P_A = -6 dB (neighbor cell search)
- P_B/P_0 = [-6 dB, 6 dB]
- PSC+SSC power (%of total cell power):
  - 10% (peak)
  - 1% (average)
- Channel models (from ITU, see [13])
  - Indoor A, 3 km/h
  - Vehicular A, 3 km/h

FIGURE 18.15   WCDMA cell search performance (from [8]).

many cases, intercell synchronization based on GPS reception is relatively straightforward, for example, in large-cell outdoor deployments. For WCDMA, four levels of accuracy for intercell synchronization have been defined:

- Level 1 (accuracy ≤ 10 µs). The cell search can be based on an extended step 3, that is, steps 1 and 2 are not needed.
- Level 2 (accuracy ≤ 256 chips). Matched filtering of step 1 can be carried out over only a part of a slot interval.
- Level 3 (accuracy ≤ 2560 chips). Search for frame timing in step 2 is not needed.
- Level 4 (accuracy > 2560 chips). Full cell search needed.

For each of the three first levels, a reduction in cell search complexity can be achieved. Consequently, although WCDMA supports fully asynchronous intercell operation, it also includes the features needed to take advantage of different degrees of intercell synchronization.

Interfrequency Cell Search and Compressed Mode WCDMA supports a frequency reuse of one, that is, neighbor cells normally use the same carrier frequency. This means that intrafrequency cell search, that is, searching for a cell that is using the same carrier frequency as the current cell, is the normal case. Consequently, cell search does not require a special RF receiver in the UE. Instead a single RF receiver can be shared between cell search and normal reception.

With so-called compressed mode (see Figure 18.16), the UE can also carry out interfrequency cell search, that is, search for a cell that uses a different carrier frequency than the current cell, without the need for a second receiver. In compressed mode, the spreading factor is decreased by a factor of 2 in some frames. The channel bits normally transmitted during an entire 10-ms frame can then be transmitted in a 5-ms fraction of a frame. During the remaining part of the frame, the UE receiver is idle and can be used for interfrequency cell search. To compensate for the reduced reception time, the instanta-

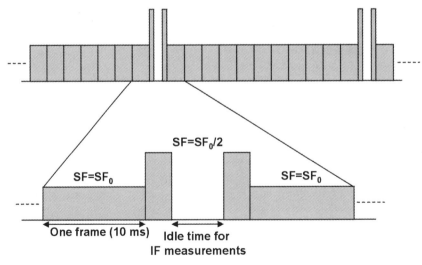

FIGURE 18.16 Compressed mode operation.

neous transmit power should be increased by approximately 3 dB during the compressed frames.

WCDMA Random-Access Procedure   The WCDMA random-access procedure is the process by which a UE accesses the network when it has not been allocated any dedicated channel. Compared to uplink dedicated-channel transmission, random-access transmission is subject to the following constraints:

- The network does not know the exact timing of the random-access transmission. The random-access transmission must therefore be preceded by the transmission of a predefined chip sequence (a preamble) to which the network can synchronize.
- In case of dedicated-channel transmission, the closed-loop power control ensures that the uplink signal is received with approximately the correct power. For random-access transmission, the UE can only determine the required transmit power from an estimate of the uplink path loss (open-loop power control). Furthermore, the uplink path loss can only be estimated from an estimate of the downlink path loss. For several reasons, the open-loop power estimate will deviate from the correct transmit power, for example, due to errors in the downlink path-loss estimate and nonreciprocity between uplink and downlink path loss. To avoid that the random-access transmission is received with too high power, the UE should therefore start the random-access transmission with a sufficiently low power and apply some kind of power ramping until the correct power is reached.

WCDMA uses preamble-based power ramping [9] that allows for fast adjustment of the transmit power to the correct level. It consists of the following steps; see also Figure 18.17:

- The UE transmits a preamble with an initial transmit power based on the open-loop estimate but modified by a predefined negative power offset.
- After transmission of the preamble, the UE checks for the reception of an acquisition indicator (AI) on the downlink acquisition indicator channel (AICH). The AI is

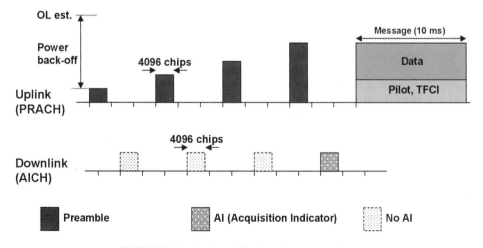

FIGURE 18.17   Preamble-based power ramping.

transmitted on the reception of a sufficiently strong preamble. If no AI is detected, the UE retransmits the preamble with a predefined increase in the transmit power. If a positive AI is received, the UE transmits the random-access message.

As shown in [9], preamble-based power ramping allows for faster access, compared to power ramping where the entire random-access message is retransmitted for each power-ramping step.

18.6   WCDMA RADIO RESOURCE MANAGEMENT

During operation of the system, there are a number of radio access bearers with corresponding QoS requirements established in the radio network. The overall goal of the radio network is to fulfill all negotiated QoS contracts and on the other hand ensure high spectrum efficiency. This is accomplished by a set of radio resource management functions, including power control, handover, admission control, packet scheduling, and congestion control.

18.6.1   Power Control on DPCH

In contrast to second generation CDMA (IS-95), WCDMA employs closed-loop power control on dedicated channels not only on the uplink but also on the downlink. The basic power control rate is 1500 Hz with a step size that can be varied in the range 1 to 3 dB, for example, to adapt to UE speed and operating environment. Signal-to-interference ratio (SIR) based power control is used, that is, the receiver compares the estimated received SIR with a SIR target and sends a corresponding increase power or decrease power command. The closed-loop power control simultaneously increases or decreases the power of all physical channels on one connection.

The SIR target is controlled by an outer power control loop. The outer loop measures the link quality, typically the transport block error rate and adjusts the SIR targets accordingly. Ensuring that the lowest possible SIR target is used at all times maximizes capacity. In addition, the outer loop can control the relative power of different physical channels belonging to the same connection. As an example, the DPDCH and DPCCH power difference can be varied by the outer loop to compensate for the variations in DPDCH performance between different environments.

18.6.2   Handover

The normal handover in WCDMA is soft intrafrequency handover, where the UE is simultaneously connected to two or more cells that use the same carrier frequency. The UE continuously searches for new soft handover candidates, using the cell search technique described in Section 18.5.4. The neighbor list informs the UE in which order to search for the different scrambling codes, and it can also limit the search to a subset of all available codes.

A special case of soft handover is the softer handover, where the user equipment is connected to two cells belonging to the same Node B. Instead of doing the uplink combining in the RNC, as is the case for normal soft handover, softer handover combining can be done within Node B. This allows for more efficient uplink combining, for example, maximum-ratio combining.

In WCDMA, soft and softer handover use relative handover thresholds. By doing so, it is guaranteed that all connected cells contribute to the received signal quality and the radio network planning task is simplified.

Further, as softer handover can employ more efficient combining in the uplink and has lower network transmission load, the handover margin for softer handover will typically be larger than for soft handover. It is also possible to have service- and load-dependent handover thresholds. Even though much of the handover functionality reside in the user equipment, the network can still put a veto on the user equipment's suggestion of cells to connect to.

As already mentioned, the normal handover in WCDMA is soft intrafrequency handover. However, interfrequency handover between cells using different carrier frequencies may be needed in some cases:

- Hot-spot scenarios, where a cell uses more carriers than the surrounding cells
- Hierarchical cell structures, where different cell layers use different carrier frequencies
- Handover to second-generation systems (GSM, PDC, IS-136, and IS-95)

As described in Section 18.5.4, WCDMA supports interfrequency cell search using compressed-mode transmission.

18.6.3 Admission Control

Admitting a new call will always increase the interference level in the system. This will reduce the cell coverage, so-called cell breathing, with a risk for dropped calls as a consequence. To secure the cell coverage and avoid dropped calls when the load increases, the admission control will limit the interference, see Figure 18.18. The basic strategy is to protect ongoing calls, by denying a new user access to the system if the system load is already high, since dropping is assumed to be more annoying than blocking.

In a highly loaded system, the interference increase may cause the system to enter an unstable state and may lead to call dropping. Hence, in addition to securing cell coverage, the admission control is used to achieve high capacity and still maintain system stability [10].

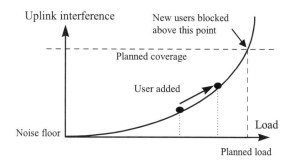

FIGURE 18.18 Uplink interference as function of traffic load. The admission control guarantees the coverage.

In WCDMA, admission control is required in both links since the system is capable of serving different services. Furthermore, different services demand different capacity as well as different quality. Hence, service-dependent admission control thresholds should be employed. These service-dependent thresholds should preferably depend on load estimates, for instance, the received power level at the base station as an uplink load estimate and the total transmitted power from a base station as a downlink load estimate [11]. Since the received power level as well as the transmitted power level may change rapidly, event-driven measuring and signaling are preferred. The measurement values are obtained at the base station, where the admission decision should be made, unless global information is required. Arrivals of high bit rate users, particularly the ones that require a large amount of resources in the downlink may demand global information to make an efficient admission decision.

18.6.4 Packet Scheduling

Packet scheduling exploits the fact that different services have different quality requirements. The basic strategy is to fill out the load gap between the load generated by real time traffic and the load that the system is planned to handle. Thus, the system will be operating at a high load as long as there is packet data traffic in the system. As a result, a high spectrum efficiency is achieved.

Further, by viewing the capacity results in [12], it can be seen that the highest spectrum efficiency is achieved for high bit rate packet transmissions. This is due to the fact that WCDMA has been optimized for that kind of services.

18.6.5 Congestion Control

Even though an efficient admission control algorithm and an efficient scheduling procedure are employed, overload situations may still occur. When reaching an overload situation, the output powers are rapidly increased due to the closed-loop power control until one or several transmitters are using their maximum output power. The connections unable to achieve their required quality are considered useless and are only adding interference to the system. This is of course an unacceptable behavior. Hence, a procedure to remove the congestion is needed. The congestion problem is particularly severe in the uplink, where the high interference levels may propagate in the system. The impact of the high uplink interference level, due to overload, may be limited by integrating the uplink power control with the uplink congestion control procedure. This is achieved by slightly degrading the quality of the users in the overloaded cell during the time it takes to resolve the congestion. The congestion control consists of several steps:

- Lowering the bit rate of one or several services that are insensitive to increased delays. This is the most preferred method.
- Performing interfrequency handovers to another carrier in the same cell or to a carrier in a overlapping cell.
- Removing one or several connections.

The congestion control is activated once the congestion threshold is exceeded. Thus both the uplink and the downlink thresholds correspond to a certain load. This means that the same measurements as in the admission control are used. However, to detect overload, these measurements have to be updated continuously since the considered values vary very rapidly when overload situation occurs. To make an efficient decision regarding which connections to deal with, that is, minimizing the number of altered connections, the congestion control algorithm is likely to require global information. This information is obtained by event-driven signaling, triggered by the occurrence of overload. Once the connections to alter are identified, the required signaling is typically the same as for altering bit rates, performing an interfrequency handover or call termination.

18.7 PERFORMANCE-ENHANCING TECHNOLOGIES

In this section some technologies that can be used to enhance the performance of WCDMA will be briefly discussed.

18.7.1 Transmit Diversity

Radio channels suffer from fading due to multipath propagation. Such multipath fading leads to large variations in the received signal-to-interference ratio (SIR) and to increased error rates.

To combat fading, different means for diversity can be applied. Due to the large bandwidth (≈ 5 MHz), WCDMA has inherent frequency diversity. Furthermore, WCDMA uses efficient channel coding combined with interleaving to achieve time diversity. However, in some cases, especially in indoor and small-cell outdoor environments, the coherence bandwidth of the radio channel can often be of the same order or larger than the signal bandwidth in which case there is no or very little frequency diversity. Furthermore, the mobile terminal is, especially in these environments, often stationary or only slowly moving. This will, to a large extent, remove the efficiency of time diversity.

Receiver diversity, that is, simultaneous reception with two (or more) antennas spaced sufficiently far apart is a well-known technique to combat fading. Uplink receiver diversity is extensively used in second-generation systems and is expected to be part of normal deployment also for WCDMA. However, for the downlink, receiver diversity is less attractive. Downlink receiver diversity implies that two complete RF chains need to be implemented in the mobile terminal. Although the cost, size, and power consumption associated with this may be acceptable for high-end terminals, it is clearly not desirable for low-end terminals, for example, voice-only terminals.

At low Doppler spread, that is, low mobile-terminal speed, the closed-loop transmit power control of WCDMA can track and compensate for the fading, with an almost constant received SIR as a consequence. However, this is achieved at the expense of an increase in the average transmit power ("excess transmit power"). This leads to higher interference levels and reduced system performance. Thus power control does not remove the negative effect of fading.

To combat fading on the downlink, different methods for downlink transmit diversity, that is, downlink transmission from two antennas, are therefore supported by WCDMA.

Open-Loop Transmit Diversity Open-loop transmit diversity are transmit diversity schemes that do no rely on feedback information from the mobile terminal. There are different schemes for open-loop transmit diversity; see Figure 18.19:

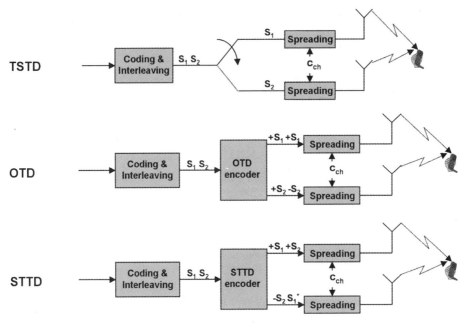

FIGURE 18.19 Different transmit diversity schemes.

- Time-switched transmit diversity (TSTD)
- Orthogonal transmit diversity (OTD)
- Space–time transmit diversity (STTD)

With TSTD, coded symbols are alternatively transmitted from two antennas. As coding and interleaving is carried out before the antenna switching, there will be a diversity effect that improves the performance after channel decoding. TSTD benefits from its simplicity. However, it suffers from the fact that power balancing between the two base station antennas cannot be guaranteed. TSTD thus imposes higher power amplifier requirements, compared to other transmit diversity schemes.

With OTD, odd and even symbols are transmitted in parallel on each antenna. The performance of OTD is basically identical to that of TSTD. However, in contrast to TSTD, OTD guarantees power balancing between the two antennas.

STTD, is very similar to OTD but with an additional coding between the two antennas. This gives additional diversity and better performance, compared to TSTD and OTD. STTD also gives a gain for noncoded symbols, such as power control commands. Furthermore, STTD also avoids the power-balancing problems of TSTD. Because of these advantages, STTD has been chosen as the open-loop transmit diversity scheme for WCDMA.

Feedback Transmit Diversity Open-loop transmit diversity does not utilize any information about the downlink channel conditions. In contrast, feedback transmit diversity uses feedback information from the mobile terminal to adjust weight factors for the two antennas, see Figure 18.20. Feedback transmit diversity allows for further improvement of downlink performance, compared to open-loop transmit diversity.

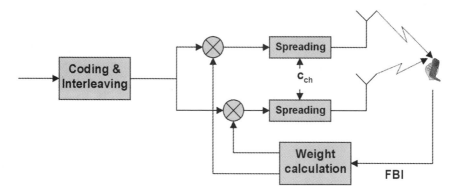

FIGURE 18.20   Feedback transmit diversity.

It should be noted that feedback transmit diversity is only applicable to dedicated physical channels. For common physical channels, such as the primary and secondary CCPCH, STTD is the only option for transmit diversity.

Performance Gain with Transmit Diversity   The performance gain with open-loop (STTD) and feedback transmit diversity is illustrated in Table 18.1. At low speed, the closed-loop power control keeps the received SIR almost constant. Consequently, transmit diversity does not give any significant gain in terms of required received E_b/N_0. However, there is a significant reduction in the excess transmit power, that is, the use of transmit diversity gives a gain in overall system capacity. The gain is especially large in case of feedback transmit diversity as, at low speed, the feedback loop can accurately track the channel variations.

At higher speed, the closed-loop power control cannot accurately track the fading. Consequently, there is not much excess transmit power to reduce. On the other hand, despite the time diversity at high speed, there is still a noticeable gain in required received E_b/N_0 using STTD. On the other hand, the feedback loop is too slow to accurately track the channel variations, that is, there is no gain with feedback mode transmit diversity.

Finally, for a channel with significant time dispersion (vehicular A), there is basically no gain with transmit diversity as the multipath propagation provides sufficient diversity.

TABLE 18.1.   Performance Gain with Transmit Diversity[a]

	Required E_b/N_0 at receiver (dB)		Transmit excess power (dB)	
	STTD	FB TX div.	STTD	FB TX div.
Indoor 3 km/h	0.2	0.2	2.6	4.6
Indoor 30 km/h	0.8	−0.3	1.0	3.4
Indoor 120 km/h	1.0	−0.3	0.2	0.7
Vehicular 240 km	0.3	−0.4	0.1	0.4

[a]Channel models according to [13].

18.7.2 Adaptive Antennas

Another interesting technique to increase the performance and capacity of a WCDMA system is to deploy an array of two or more antennas at the cell site. Such an installation is often called an adaptive antenna system. The various transmit diversity techniques discussed above all rely on the assumption that the antennas are placed sufficiently far apart so that the multipath fading associated with each antenna are sufficiently uncorrelated, thereby giving rise to a diversity gain. An adaptive antenna system uses the spatial dimension in a different way.

Instead of placing the antenna elements far apart, they are placed sufficiently close to each other so that there is a unique correspondence between a phase shift between the antennas and a certain angle of arrival (AOA) of the incoming wave front. As an example, for an array of uniformly spaced sensors, the mapping between phase shift and AOA is unique if the antenna elements are not placed more than half a wavelength apart. It is this directivity of the adaptive antenna that is used to increase link quality and system capacity. Indeed, the deployment of an antenna array allows for space–time processing of the incoming signal. As opposed to time only processing such as in the classical Rake receiver, a space–time processing algorithm can exploit both the spatial and temporal dimensions simultaneously to increase the performance of the system.

The adaptive antenna concept is somewhat different in up- and downlink. In the uplink, the outputs of the different antenna elements are phase shifted and coherently combined to increase the received signal strength coming from a desired direction. Note that the entire cell is received, that is, the spatial processing merely forms a virtual beam toward the desired user. No information is lost in the process, and separate beams are formed for the remaining users. Since each user experiences an increased link quality, their transmit powers can be decreased, which in turn leads to reduced interference that can be traded for capacity in the system. In the downlink, the beam is formed for each user prior to transmission. This means that only the part of the cell corresponding to the direction of the desired user is illuminated. This leads to a reduction in the transmitted power, which again can be traded for increased capacity. Common channels, such as the RACH in the uplink and the BCH in the downlink, must of course be received and transmitted from the entire cell. As explained above, this causes no particular problem in the uplink since the entire cell is received by the adaptive antenna. In the downlink, however, the common channels must be transmitted over the entire cell. This is done by designing the adaptive antenna in such a way that the transmission with no phase shifts yields a total antenna diagram that resembles that of a conventional sector antenna.

One important signal-processing step in the algorithms for adaptive antennas is to determine the phase shifts corresponding to the AOA of the desired user. For this estimation, the pilot bits transmitted within the physical channels are used. The reason not to use the common pilot channel, which is transmitted over the entire cell, is that the pilot and the data must experience the same propagation channel for the estimation to be reliable.

The estimation step can be performed in at least two fundamentally different ways. In a parametric approach, the AOAs of the multipaths (fingers) of the desired user are determined using a parameterized model of the signal environment, and the uplink beams are then steered in the determined direction. In a nonparametric approach, the AOAs of the desired user are not estimated based on a structured model. Instead, only the spatial channels (or spatial signatures) are determined, giving rise to an estimated array weight vector for each finger. The main difference between the two classes of methods, from an implementation point of view, is that the high-performance parametric approaches require accurate calibration of the antenna array and coherency in the receiver chain. The nonparametric approaches do not suffer from this potential problem but have in general lower performance than the parametric methods.

Acknowledgements The authors would like to thank Magnus Sundelin, Henrik Olofsson, Maria Edvardsson, and Leo Hedlund for providing performance results presented in this chapter.

REFERENCES

1. A. Baier, U.-C. Fiebig, W. Granzow, W. Koch, P. Teder, and J. Thielecke, "Design Study for a CDMA-Based Third Generation Mobile Radio System," *IEEE J. Select. Areas Commun.*, Vol 12, May 1994.

2. H. Holma, F. Ovesjö, E. Dahlman, M. Latva-aho, and A. Toskala, "Physical Layer of FRAMES Mode 2—Wideband CDMA," *Proc. 48th IEEE Veh. Tech. Conf., VTC'98*, Ottawa, Canada, May 1998.

3. E. Dahlman, A. Toskala, and M. Latva-aho, "FRAMES FMA2, a Wideband-CDMA Air-Interface for UMTS," *Proc. 2nd CDMA Int. Conf., CIC'97*, Seoul, South Korea, October 1997.

4. E. Nikula, A. Toskala, E. Dahlman, L. Girard, and A. Klein, "FRAMES Multiple Access for UMTS and IMT-2000," *IEEE Pers. Commun.*, Vol. 5, No. 2, April 1998.

5. S. Onoe, K. Ohno, K. Yamagata, and T. Nakamura, "Wideband-CDMA Radio Control Techniques for Third-Generation Mobile Communication Systems," *Proc. 47th IEEE Veh. Tech. Conf., VTC'97*, Phoenix, Arizona, May 1997.

6. TIA/EIA/IS-2000-2-A, "Physical Layer Standard for cdma2000 Spread Spectrum System," TIA/EIA, 2000.

7. F. Adachi, M. Sawahashi, and K. Okawa, "Tree-Structures Generation of Orthogonal Spreading Codes with Different Lengths for Forward Link of DS-CDMA Mobile Radio," *Elect. Lett.*, Vol. 33, January 1997.

8. H. Olofsson, M. Sundelin, M. Edvardsson, and E. Dahlman, "Cell Search Performance in UTRA," *Proc. VTC 1999—Fall*, Amsterdam, The Netherlands, September 1999.

9. H. Olofsson, M. Edvardsson, and G. Frank, "Performance Evaluation of Different Random Access Power Ramping Proposals for the WCDMA System," *Proc. PIMRC '99*, Osaka, Japan, September 1999.

10. J. Knutsson, P. Butovitsch, M. Persson, and R. D. Yates, "Evaluation of Admission Control Algorithms for CDMA Systems in a Manhattan Environment," *Proc. 2nd CDMA Int. Conf., CIC'97*, Seoul, South Korea, October 1997.

11. J. Knutsson, P. Butovitsch, M. Persson, and R. D. Yates, "Downlink Admission Strategies for CDMA Systems in a Manhattan Environment," *Proc. 48th IEEE Veh. Tech. Conf., VTC'98*, Ottawa, Canada, May 1998.

12. G. Brismark, H. Olofsson, J. Knutsson, M. Gustafsson, and E. Dahlman, "UMTS/IMT2000 Based on Wideband CDMA," *Proc. Multi-Dimensional Mobile Communications, MDMC'98*, Menlo Park, California, September 1998.

13. "Guidelines for Evaluation of Radio Transmission Technologies for IMT-2000/FPLMTS," IYU-R M.1225 (REVAL).

14. S. Parkvall, E. Dahlman, P. Frenger, P. Beming, and M. Persson, "The High Speed Packet Data Evolution of WCDMA," *Proc. PIMRC'2001*, San Diego, California, September 2001.

CHAPTER 19

New Systems for Personal Communications via Satellite

J. V. EVANS*

19.1 INTRODUCTION

While telegraph cables laid across the Atlantic in the 1860s provided a means of linking the United States and Europe, it was not until the 1950s that cables capable of supporting voice circuits were developed and laid down. These employed coaxial cables (similar to those used to connect a TV set to its antenna) in which a wire runs inside a surrounding shield. The undersea phone cables consisted of several of these coaxial cables bundled together and encased in a steel wire armored jacket. Transmitted through each coaxial cable was a radio wave onto which a number of voice signals were modulated. However, loss (or attenuation) of the radio wave down the cable required that amplifiers be installed roughly every 5 km to reboost the signals.

In this era, many small nations had no direct communications access with the rest of the world, as all their traffic had to transit through other countries (and the cost of phone calls was commensurately high).

This then was the state of the technology when, in 1962, the U.S. Congress passed the Communications Satellite Act, which was signed into law by President Kennedy. The act called for the establishment of a single global telecommunications satellite system for the benefit of all humankind. Within the United States, the Communications Satellite Corporation (COMSAT) was incorporated to carry out the intent of the act and began raising money on the stock market. With the help of the State Department, 10 other countries were persuaded to join in the endeavor and, in 1964, they formed a treaty organization called the International Telecommunications Satellite Organization (INTEL-SAT) for this purpose. INTELSAT launched its first communications satellite *Early Bird* (later renamed INTELSAT I) over the Atlantic Ocean in 1965. This satellite could support

*J. V. Evans was with COMSAT Corporation. He is now retired.

Wireless Communications in the 21st Century, Edited by Shafi, Ogose, and Hattori.
ISBN 0-471-155041-X © 2002 by the IEEE.

240 voice circuits (almost equal to the 300 circuits carried by all the cables at the time) but could also carry television, which the undersea cables could not.

By 1969 INTELSAT had established satellite communications over each of the three major ocean regions (Atlantic, Indian, and Pacific). Over the years since, its membership has grown (to 141 countries today) and it has launched nine generations of ever more capable satellites. At one time these carried about 70% of the voice traffic between countries.

One of the ironies of the Satellite Act was that it caused Bell Laboratories, which had built the first experimental communications satellite (Telstar, launched in 1962), to cease working on satellite technology and develop fiber-optic technology instead. Today, fiber optics is the preferred medium for carrying telecommunications traffic. A large number of undersea cables have been laid east–west across the Atlantic, and east–west and north–south in the Pacific, and plans exist for cables through the Indian Ocean. INTELSAT satellites continue to be used to link countries that have no access to fiber-optic cables but are increasingly being used to carry television and data rather than voice traffic.

Regional and domestic satellite systems were established in the 1970s, serving Europe, the Middle East, and the United States. Some of these systems initially carried voice traffic, but later they were used almost exclusively for distributing television to cable head ends. One of the unique features of a geostationary earth orbit (GEO) satellite is that approximately one-quarter of the earth's surface lies within its field of view. Thus, it is very efficient to use such satellites to deliver the same information to a large number of users. Accordingly, all of the TV-cable systems in the United States receive their programming via satellite, and a number of states are linking schools via satellite for "distance learning."

Another growing use for satellite systems is in supporting private networks for banks, chain stores, and other geographically dispersed institutions. The technology employed goes under the acronym VSAT for "very small aperture terminal" and achieves its low cost by linking a number of small terminals to a single larger hub terminal (which may be located at the corporation's headquarters).

All of these services are for corporate users who require reliable communications and find that satellite systems best suit their needs. There is now, however, a growing effort to exploit satellite technology to provide services to consumers by providing so-called personal communications service (PCS). This chapter provides a review of some of these projects and discusses some of the technology involved.

Before describing these new services it is worthwhile mentioning the one existing service now delivering communications to consumers. This is DBS (direct broadcast service) or DTH (direct-to-the-home) distribution of television. In the United States the distribution of television to cable heads ends via satellite created a cottage industry of receive-only terminals—backyard dishes 2 to 3 m in diameter used to "eavesdrop" on these broadcasts. The number of such installations is now thought to be around 2 million.

To reach a larger subscriber base, it was widely accepted that a small-size receiving antenna is necessary: one that can readily be mounted on the side of a house, for example. Hughes was the first in the United States to approach this market. In 1994 it launched a high-power (approximately 120 W per transponder), 16-transponder satellite capable of beaming over 100 digitally compressed TV channels to viewers, who receive signals with a 40-cm-diameter antenna and set-top box converter that initially cost $700. Presently, DTH TV has almost 10 million U.S. viewers. DTH television has enjoyed even more solid growth in Europe, due, in part, to an earlier start and, in part, to the poorer market penetration of cable systems. In all, it is estimated that there are currently about 30 million

European subscribers to DTH pay-TV. Still larger markets are believed to exist in the Asia-Pacific region, where several single-country projects are under way. According to some forecasts, the DTH market will grow to 100 million subscribers worldwide by 2010, with as many as one-third in the Asia-Pacific region.

19.2 MOBILE SATELLITE SERVICES

We are presently witnessing the deployment of a second consumer market for communications satellite services: the provision of cellular phone service via satellite.

While satellites are uniquely suited for broadcast applications, they are also well suited to providing service to mobile users. (Indeed, this is true of all wireless systems.) Mobile satellite communications for commercial (i.e., nonmilitary) users had its beginning in 1976, when COMSAT purchased three Marisat satellites to provide communications for the U.S. Navy at ultra high frequency (UHF) and to commence a commercial service for mariners at L-band. (Table 19.1 gives the radio frequency bands that are set aside for satellite communications and are referred to in this discussion.) This led to creation of the International Maritime Satellite Organization (Inmarsat) in 1979. Inmarsat, a treaty organization similar to INTELSAT grew to 81 members before voting to become a private corporation in 1999. It was authorized to provide communications to ships, aircraft, and (although not in the United States) land mobile users. Initially, Inmarsat leased its satellite capacity from COMSAT and from the European Space Agency MARECS satellites. Subsequently, Inmarsat's capacity was provided by INTELSAT, which added maritime packages to some of its V-series satellites.

In 1991, Inmarsat deployed four Inmarsat-2 satellites constructed to its own specifications, and subsequently launched a fleet of five Inmarsat-3s. These later satellites, like their predecessors, have a global coverage beam, but in addition have five "spot" beams that allow communication (voice and fax) via a small terminal the size of a laptop computer. This service is made possible through the use of digital voice compression, in which the speaker's voice signals are sampled, analyzed, and represented by a low-rate digital bit stream that drives a speech synthesizer at the receiver. Digital speech compression algorithms have been developed in recent years that provide reasonably good-quality speech (i.e., the speaker is recognizable and quality is equally good for male and female speakers) with rates as low as 4.8 kbps. The processing power of silicon chips has made the technology both practical and low cost.

TABLE 19.1 Frequency Bands Assigned to Communications Satellite Services

Satellite Band	Earth-to-Space Frequencies	Space-to-Earth Frequencies	
L	1626.5–1660.5 MHz	1525–1559 MHz	(Inmarsat)
	1621.35–1625.5 MHz	1621.35–1625.5 MHz	(Iridium)
	1610.0–1618.25 MHz	1980–2110 MHz	(Globalstar)
S	2170–2200 MHz	1980–2110 MHz	(ICO)
C	5.850–6.425 GHz	3.6–4.2 GHz	
Ku	12.75–13.25 GHz	10.7–12.75 GHz	
	13.75–14.8 GHz	17.3–17.7 GHz	
Ka	27.5–30.0 GHz	17.7–21.2 GHz	
Q/V	47.2– 50.2 GHz	39.5–42.5 GHz	

Compressing the voice signal in this fashion makes it possible to transmit it in a channel that is only 5-kHz wide and thereby achieve the sensitivity necessary to communicate with a satellite in geostationary orbit. Several regional mobile satellite communications systems have now been built that provide similar services. These exist in the United States (American Mobile Satellite Corporation), Canada (Telesat Mobile, Inc.), Australia (Optus), and Japan (NTT Mobile Communications Network, Inc.) operating at either L- or S-band (Table 19.1). In addition, Mexico and India are in process of implementing systems. All of these regional systems, however, are intended to serve terminals that are mounted in vehicles or are transportable, and this has greatly limited their adoption. By contrast, the growth of cellular phone service worldwide has been nothing short of spectacular, with almost 200 million subscribers at the end of the decade for all forms of personal communications (both digital and analog). In 1992, this enormous growth spurred Motorola to proceed with a bold plan to create a global personal satellite system called Iridium, employing 66 satellites in low earth orbit (LEO). Other proposals for LEO systems followed, causing Inmarsat to consider what type of personal communications system it might launch. Guided to some extent by design studies performed by TRW, Inmarsat adopted a system employing satellites placed in 6-h orbits at 10,000-km altitude (i.e., above the Van Allen radiation belts). This system is now being built by an affiliate company called ICO-Global.

In providing service to handheld terminals, designers are handicapped by severe limitations. The terminal's antenna is likely to be too small to yield any significant gain, and the amount of power to be radiated must be kept low (below 0.5 to 1 W) to preserve battery life and for possible medical reasons. Thus, the entire burden of completing the link is placed on the satellite and can be achieved only if the satellite employs very narrow spot beams that illuminate regions of the earth of the order of 100 to 200 km in diameter. A large number of these beams must then be formed to obtain coverage of a reasonably sized service area.

Figure 19.1 shows the amount of the earth's surface visible from LEO, intermediate circular orbit (ICO), and GEO. The higher the satellite altitude, the fewer satellites are

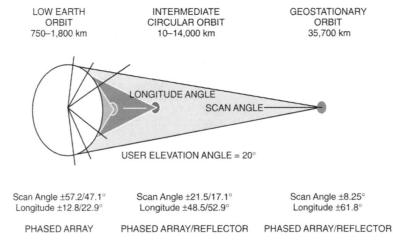

LOW EARTH ORBIT 750–1,800 km   INTERMEDIATE CIRCULAR ORBIT 10–14,000 km   GEOSTATIONARY ORBIT 35,700 km

LONGITUDE ANGLE   SCAN ANGLE

USER ELEVATION ANGLE = 20°

Scan Angle ±57.2/47.1°   Scan Angle ±21.5/17.1°   Scan Angle ±8.25°
Longitude ±12.8/22.9°   Longitude ±48.5/52.9°   Longitude ±61.8°

PHASED ARRAY   PHASED ARRAY/REFLECTOR   PHASED ARRAY/REFLECTOR

FIGURE 19.1 Relative amounts of earth coverage afforded by satellites in LEO, shown here as the closest satellite position; ICO, shown as the next closest satellite to the earth; and GEO, shown as the most distant satellite location.

required to provide global coverage. However, to preserve the link margin on the handheld-to-satellite link, the spot size of the beam on the earth's surface must be kept small. This requires that larger satellite antennas be employed the further out the satellites are placed, and each antenna must form a greater number of spot beams to maintain overall coverage. Thus, LEO satellites tend to be smaller, lighter, and cheaper than ICO spacecraft, which in turn are likely to be less expensive than GEO spacecraft. This offsets to some extent the benefit of developing systems with fewer satellites.

19.3 PROPOSED GLOBAL SATELLITE PHONE SYSTEM DESIGNS

Studies performed by the proponents of global satellite telephone systems have identified four potential markets:

1. *International Business Travelers*   Primarily business travelers from the developed world traveling to less-developed countries.
2. *National Roamers*   Primarily business travelers who need mobile communications in their own countries but who travel beyond the reach of terrestrial cellular systems.
3. *National Rural Fixed Service*   An extension of the national fixed services to regions where they are presently unobtainable.
4. *Government Agencies*   Law enforcement, fire, public safety, and other services.

The designs of the announced global satellite cellular phone systems differ considerably, reflecting different assumptions regarding the business to be attracted from these four segments. While a large number of systems have been proposed, only the five that appear to have some prospect of being fielded are discussed here. Four are U.S.-based systems that have received licenses from the U.S. Federal Communications Commission (FCC), and the fifth is the ICO spin-off from Inmarsat.

Table 19.2 lists some of the salient features of these five systems, none of which intends to employ geostationary orbits. All of the U.S. systems except Iridium operate at S-band for the subscriber-terminal-to-satellite link and L-band for the satellite-to-subscriber link. ICO operates entirely at S-band, and Iridium entirely at L-band. The feeder links from the satellite to the gateway earth station, which provide access to the public switched network, may be at C-, Ku-, or Ka-band.

19.3.1 Iridium

From a technical standpoint the Iridium system is the most ambitious of the five. The system was purchased and is being operated by a separate company (Iridium, Inc.), which has secured investment from many parts of the world. The design employs 66 satellites placed in circular polar orbits at 780-km altitude. The satellites are deployed into six equispaced orbital planes, with 11 satellites equally separated around each plane. Satellites in adjacent planes are staggered with respect to each other to maximize their coverage at the equator, where a user may be required to access a satellite that is as low as $10°$ above the horizon.

Users employ small handsets operating in frequency division multiplexed/time division multiple access (FDM/TDMA) fashion to access the satellite at L-band. Eight users share 45-ms transmit and 45-ms receive frames in channels that have a bandwidth of 31.5 kHz

TABLE 19.2 Planned Global Satellite PCS Systems[a]

Parameter	Iridium	Globalstar	ICO-Global	Ellipso	Aires (ECCO)
Company	Motorola	Loral/Qualcom	ICO-Global	Mobile Communication Holdings, Inc.	Constellation Communications, Inc.
No. of active satellites	66	48	10	17	46
Orbit planes	Six circular polar (86.5°)	Six circular inclined (52°)	Two circular inclined (45°)	Two elliptical inclined (116.6°) One circular equatorial	Seven circular inclinedOne circular equatorial
Orbit altitude (km)	780	1414	10,355	N.A. 8060 equatorial	? 2000 equatorial
Satellites per orbit plane	11	8	5	Five in each elliptical orbit Seven in equatorial orbit	Five in each inclined orbit Eleven in equatorial orbit
Beams per satellite	48	16	163	61	1
Reported cost ($B)	4.7	2.5	4.6	0.56	1.15

[a]Some of these satellite systems are now experiencing financial difficulties and may or may not survive.

and are spaced 41.67 kHz apart. Users are synchronized so that they all transmit and all receive in the same time windows alternately. This approach is necessary because (three) phased-array antennas are used for both transmitting and receiving. Figure 19.2(a) is a sketch of the satellite, and Figure 19.2(b) shows the 48 spot beams formed at L-band projected onto the earth at the equator.

The Iridium system requires onboard processing to demodulate each arriving TDMA burst and retransmit it to its next destination. This can be to the ground if a gateway earth station is in view or, failing that, to one of the four nearest satellites: the one ahead or behind in the same orbital plane, or the nearest in either orbital plane to the east or west.

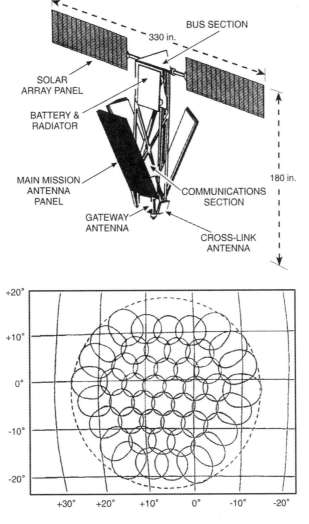

FIGURE 19.2   (a) Sketch of the Iridium system satellite. The 48 spot beams (see b) are formed by the three phased-array antennas cantilevered from the three sides of the spacecraft (Courtesy of Motorola Corporation). (b) Service (L-band) spot beams formed by an Iridium satellite over the equator.

These satellite crosslinks operate at 1.3-cm wavelength (Ka-band), while the links to the gateway earth stations are at 1.5-cm wavelength.

The use of crosslinks greatly complicates the design of the system, but allows global service to be provided with a small number (11) of gateway earth stations. To properly route traffic, each satellite must carry a set of stored routing tables from which new routing instructions are called every 2.5 minutes.

The onboard processor was constructed using very large scale integrated circuits designed specifically for the project. It includes 512 demodulators that, via the signaling channel, can cause the arriving handheld bursts to be centered in frequency and time. The observed Doppler shift of these arriving bursts is routed to the intended destination gateway earth station to determine the user's location. Service is then provided (or denied) based on country-by-country service agreements. Each satellite is capable of handling as many as 1100 simultaneous calls.

The services provided include voice (at 2.4 kbps encoding), data at 2.4 kbps, and high-penetration paging that affords 11 dB more power than the regular signal. The design, however, already provides a link margin of about 16 dB, which is higher than any of the competing systems. This is because Motorola required that the handheld unit be usable from inside a vehicle (such as a taxi)—a capability dictated by their business plan, which depended heavily on serving international business travelers.

One of the complicating aspects of the Iridium system is the need to hand off a subscriber from beam to beam as a satellite flies by. Since a typical satellite pass takes less than 9 min and the average international call duration is about 7 min, there is also a need to hand off some calls to the next satellite to appear above the horizon. The handoff will be into one of the adjacent orbits, and hence in a somewhat different direction from the first, raising the possibility of the call being dropped if buildings block the view. A dropped call rate of about 10% has been reported in the press.

The Iridium satellites are station-kept using onboard propulsion, to overcome atmospheric drag and have sufficient fuel for an 8-year life. Four telemetry, tracking, and control facilities were built to manage satellite operations (in Hawaii, at Yellowknife and Iqualvit in Canada, and at Eider in Iceland), as well as a separate engineering facility to diagnose problems that may arise (such as the failure of a crosslink). The Master Control Facility is in Lansdowne, Virginia, with a backup in Rome, Italy.

Motorola succeeded in launching all of its satellites (without a single launch failure) by the spring of 1998 and began offering service in November of that year. Unfortunately, it failed to quickly attract a large number of subscribers owing in part to the high cost of the terminal ($3000) and the service ($6 to $8 per minute). Other factors include the fact that, in contrast to the time that the project was begun, cellular service is now widely available in many countries and manufacturers have developed multimode phones capable of operating with different standards worldwide. In 1999 Iridium was forced to file for bankruptcy protection and presently continues to offer service (to about 30,000 customers). Its long-term future must, however, be very much in doubt.

19.3.2 Globalstar

The Globalstar system is being purchased by a limited partnership in which Loral and Qualcomm of the United States are principal partners. The satellites were designed by Loral, while Qualcomm developed much of the ground segment. Unlike the Iridium system, which offers a true *global* service, Globalstar's business plan calls for launching the space segment and franchising its use to partners in different countries. Over 90 such relationships have been established.

The Globalstar system employs 48 satellites organized in 8 planes of 6 satellites each. The satellite orbits are circular, at 1414 km and 52° inclination. The use of an inclined orbit concentrates the available satellite capacity at lower latitudes where the largest populations exist; little or no coverage is provided beyond ±70° latitude. As can be seen in the figure, two or more satellites are visible (above 10° elevation) between 25° and 50° latitude at all times, and from the equator to 60° latitude 80% of the time. Like the Iridium satellites, the Globalstar spacecraft are three-axis-stabilized, with a mission life of 7.5 years (minimum).

Because the Globalstar system does not employ satellite crosslinks, a subscriber can gain access to the system only when a satellite in view can also be seen by a gateway earth station. Typically this means that service areas are within 1000 miles of each gateway earth station. To achieve truly global coverage would require the construction of more than 200 earth stations, which seems unlikely to happen. Thus, Globalstar is more likely to serve national roamers than international business travelers.

In contrast to Iridium, each Globalstar satellite covers a comparable area of the earth's surface with only 16 spot beams. This, together with sharing of the receive channels on board the satellite by many more users, reduces the available link margins to about 3 to 6 dB, although for a small number of users this can be increased to 11 dB. Access to and from the satellite is at L- and S-band, respectively, using code division multiple access (CDMA) in channels that are 1.25 MHz in bandwidth. Voice is encoded at a rate in the range of 1 to 9 kbps, depending on speaker activity. The satellites employ simple "bent pipe" transponders, with the feeder links at C-band.

Since all 16 beams of all of the 48 satellites are always active, each satellite that is in view of a subscriber will pick up the subscriber's signal and retransmit it in its feeder link. Thus, by tracking the several satellites that are in view of a given gateway earth station, two channels can be kept open to the subscriber. The channel providing the stronger signal can then be selected for connection to the public switched network. This feature should mitigate blockage by buildings and provide an automatic "soft" handoff from satellite to satellite.

As of November 1999, Globalstar had 44 of its 48 satellites in orbit and is planning to begin offering a limited service in regions that surround nine of the earth stations that have been completed. It hopes to avoid the fate of Iridium by offering lower cost handsets (~$1000 each) and service ($1 to $2 per minute). Since the business model is very different from Iridium, in that the regional service providers were required to invest by building an earth station, and therefore have an economic incentive to sell terminals, there remains optimism that Globalstar will be a commercial success.

19.3.3 ICO

ICO-Global is a spin-off from Inmarsat, which initially owned 15% of the corporation. The remainder was owned by Inmarsat signatories and by Hughes, the builder of the spacecraft. The ownership structure is in the process of changing, however, as discussed below.

ICO-Global has chosen an intermediate circular orbit for its system (10,355-km altitude), with 10 satellites arranged 5 in each of 2 inclined circular orbits. (Actually, 12 satellites are to be launched in order to provide a spare in each orbital plane.) The inclination of the orbits is 45°—making it the lowest of the three systems employing circular inclined orbits. This reduces the coverage at high latitudes, but allows for the fewest satellites.

To improve the link margins on the ICO satellites, Inmarsat chose a design that employs 163 spot beams. Routing signals to the correct spot beam becomes extraordinarily difficult to accomplish with analog filters and instead will be performed using a digital filter bank (which performs a fast Fourier transform on the signals arriving from the gateway earth station). Therefore, to access a given spot beam, the gateway earth station must transmit at a particular frequency.

Given that a large digital signal processor is required onboard the ICO satellites, it was decided to form the 163 beams by phase combining the signals digitally rather than in a microwave matrix. While this further complicates the onboard processor, it is expected to greatly simplify the checkout and calibration of the antenna systems.

ICO differs from Globalstar in that a TDMA scheme has been adopted for the service links, with six subscribers multiplexed into channels 25.2 kHz in width at a bit rate of 36 kbps. In this respect, ICO more closely follows the approach being adopted for Iridium. A disadvantage of this access scheme is that a soft handoff (e.g., from beam to beam) is not automatic, and it is more difficult to exploit dual-satellite visibility. One method being considered would send a burst via an alternate satellite (say) every fifth burst. By noting the strengths of the regular and alternate bursts, the subscriber terminal could determine which satellite affords the best path to the gateway earth station at a given moment and could adjust its own burst time and frequency to select that satellite.

The ICO satellites are being built by Hughes Space and Communications Division, and the ground segment by a team consisting of NEC, Ericsson, and Hughes Network Systems Division. ICO had hoped its system operational in the 2000 to 2001 time frame, but Iridium's financial difficulties came at a time when ICO was in the midst of raising more capital, and doomed its efforts to do so. It therefore has also sought bankruptcy protection.

Inmarsat presently offers a wireless internet access service using a beefed-up M terminal that provides data access at 64 kbps. In this regard, it is important to note that considerable work has been done within the framework of ITU-R to develop standards (IMT-2000) for a third-generation terrestrial wireless system. The new system standards include provision for high-speed data (2 Mbps) delivery as well as provision of service by LEO, ICO, or GEO satellites. The existence of these standards, together with the fielding of the third-generation cellular systems, could create an opportunity for a satellite service provider to successfully enter the mobile data market.

19.3.4 Ellipso and Aires

Ellipso, to be built by Mobile Communications Holdings Inc. (MCHI) of Washington, D.C., and the Aires system proposed by Constellation Communications Inc. (CCI) based in Reston, Virginia, both received licenses from the FCC in July 1997, that is, long after Iridium and Globalstar. Both systems aim to be low-cost providers of telephony service to developing countries and plan to initiate service with satellites in circular orbits above the earth's equator. The financial difficulties encountered by Iridium have made it virtually impossible for these projects to proceed at the present time, and they are likely to go forward only if and when it becomes apparent that Globalstar and/or ICO are financially successful.

19.3.5 Regional Mobile PCS Systems

In addition to the five global satellite PCS systems discussed above there are a number of projects underway to provide a similar service within a given region, employing geostationary satellites. Table 19.3 lists six such projects. Figure 19.3 shows the regions of the earth that might be covered by them. It can be seen that most are intended for parts of Asia, and, as such, have been thrust into doubt (or delayed) by the financial crisis in that part of the world. Of the ones listed in Table 19.3 it is believed that only Thuraya and ACeS are presently proceeding according to plan. The American Mobile Satellite Corporation appears to have abandoned plans for a second generation system (AMSC II), and the Euro-African Satellite Telecommunications (EAST) Company appears to have concluded that it will be unable to raise sufficient financing to proceed with its system. The U.S. government, meanwhile, has canceled Hughes's license to export the Asia Pacific Mobile Telecommunications (APMT) satellite to China owing to heightened fears concerning technology transfer. This will likely force APMT to seek a European or Japanese manufacturer for its satellite. The Asia Cellular Satellite System's (ACeS) first satellite (*Garuda*) is due to be launched soon, and the Thuraya satellite is expected to be launched in the year 2000.

The chief technical difficulty in fielding a geostationary satellite to provide PCS systems is the need to keep the power flux density about the same as that achieved with a LEO or ICO systems. That is, the size of the spots illuminated on the ground must be kept constant, and hence the angular extent of the beam must be reduced. This can be achieved at these long wavelengths only through the use of very large (10 to 12-m-diameter) antennas, and Figure 19.4(a) and 19.4(b) illustrate the designs being employed by Lockheed Martin and Hughes. Such large antennas must be unfurled in space; this represents a technically risky step since it is difficult to test the deployment in a manner that reproduces the absence of gravity encountered in orbit. To achieve adequate regional coverage a large number (>200) of beams must be employed so that onboard processing (similar to that planned for ICO) must be employed to steer the signals uplinked from the gateway earth station(s) into the correct downlink beams.

The economics of these systems is such that they could offer service at a lower cost to their subscribers than any of the global systems, including Globalstar. They require, however, a greater degree of user cooperation, (i.e., finding a clear view of the satellite) and this may limit their appeal. These systems could be employed to provide fixed rural telephony via suitably constructed pay phones, but the cooperation of a government entity (such as the post office) would seem necessary to make this a paying proposition because of the obvious collection problem.

19.3.6 Technical Challenges Facing PCS System Providers

The principal challenge for satellite communications systems that provide land-mobile service is poor propagation. Signals can arrive at the subscriber terminal via multiple paths, but fortunately the high elevation of the source makes the time dispersion of the signals much less of a problem than is encountered in terrestrial cellular systems. The chief difficulty instead is shadowing or blockage of the direct ray. These effects have been studied by a number of groups for different settings, such as tree-lined roads and urban environments. The results show that systems operating with ICO satellites should enjoy an advantage, as the satellites are generally visible at higher elevations and cross the sky more

TABLE 19.3 Some Parameters of Six Regional Mobile PCS Systems Reported in the Trade Press

	Agrani	AMSC II	APMT	East	Garuda	Thuraya
Operator (country of HQ)	Afro-Asian Satellite Communcations Ltd. (Mauritius)	American Mobile Satellite Corporation	Asia-Pacific Mobile Telecommunications (Singapore)	Euro-African Satellite Telecommunications Ltd.	ACeS (Indonesia)	Al-Thuraya Satellite Communications Co. (UAE)
Total cost (in millions)	$800	SEE TEXT	$640		$800	$1100
First satellite launch date	2000		mid-2000		end 1999	May 2000
Beginning of commercial service	2001		late-2000		early 2000	Sept. 2000
Coverage	54 countries in Southern Asia, parts of W. Asia and Africa		22 countries in Asia Pacific and Southern Asia	SEE TEXT	23 countries in Southern Asia and Asia Pacific	five countries in Southern Asia
Orbital position	N/A	—	95°E and 125°E		123°E and 80.5°E	24.5°E
Prime contractor	Lockheed Martin	Hughes	tbd	Matra Marconi	Lockheed Martin	Hughes S&C
Number of spotbeams/satellite	N/A	—	230		N/A	250–300
Service data rates	N/A	—	2.4–9.6 kbps	—	N/A	2.4–9.6 kbps
Voice circuits/satellite	16,000	—	16,000	—	3000	13,750

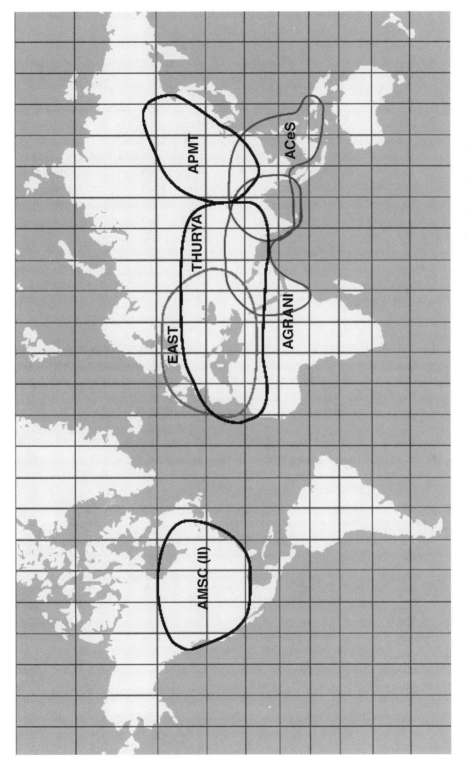

FIGURE 19.3 Proposed coverage areas of the regional mobile PCS systems listed in Table 19.3.

slowly. Dual-satellite visibility, where it can be exploited, should also be valuable in mitigating blockage.

Handoff from beam to beam is necessary in all of the global systems and can be likened to the situation arising in a terrestrial cellular system, only in reverse: The subscriber is essentially stationary and the cells are moving past. Since, by definition, a subscriber who has successfully placed a call is in a known beam, it is possible to manage a beam-to-beam handoff relatively straightforwardly. More difficult is a satellite-to-satellite handoff, which is likely to occur for any LEO system. CDMA systems can allow this to happen in a soft fashion (albeit at some loss of capacity), since the transfer need not occur until the subscriber's signal is being received in the following satellite. This is more difficult to accomplish in FDMA/TDMA systems since the Doppler shifts (relative to the channel width) are quite large and in different directions for the arriving and departing satellites. Also, call transfers must be handled carefully to avoid missing a portion of the conversation or having the terrestrial system erroneously detect a call hang-up and take down the call.

Some form of network control is required for all of the systems described here. Call setup usually entails a signaling channel, which the subscriber unit uses initially to request a call. This is particularly true for TDMA systems, which must be told what burst time to employ. All of the systems require that the call be handled in a different channel from the signaling channel.

Network control for the Iridium system is particularly challenging, since essentially the largest portion of the network resides in the satellites and their crosslinks. To avoid congestion, these links must be designed so that each crosslink can handle all of the service traffic from a given satellite. In the event that a crosslink fails, new routing tables must be delivered to all of the satellites in the vicinity of the failed link. Linking between satellites that are in ascending and descending planes is particularly difficult and may require that packets be routed around the globe in the opposite direction. Finally, there is a need to monitor the number of times a packet has been routed via a node and to drop any when this exceeds a certain value, lest the system become clogged with undeliverable traffic.

The regional satellite systems are conceptually simpler since beam-to-beam handover is not required, and for stationary users the liklihood of a dropped call is quite low. Here the largest technical risk would appear to be in deploying the very large antennas required (Figure 19.4) and constructing these in such a way that temperature distortions (which would move the beam positions) are minimized.

19.4 DATA AND MULTIMEDIA SERVICES

Deregulation in the United States and overseas is speeding delivery of new services to the marketplace and fueling the investment of enormous amounts of capital in new facilities. Central to all of this activity are two developments that are affecting the planning and introduction of new satellite systems. The first is the tremendous growth worldwide in the use of cellular and PCS systems, as has already been discussed. The second is the explosion in the use of the Internet. These, together with the distribution of multimedia, are now seen as a basis for new satellite systems that will operate at Ka-band, Q/V-band or (in non-geostationary orbits) at Ku-band (Table 19.1).

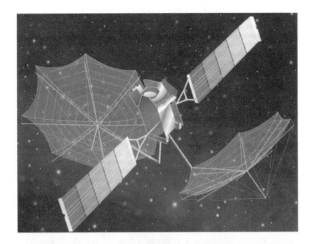

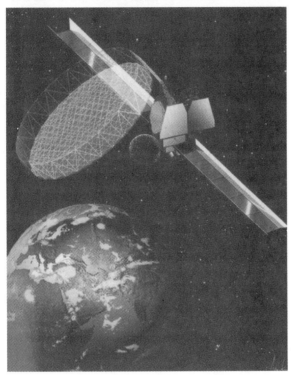

FIGURE 19.4 Artists sketches of the two geostationary satellites that will provide regional PCS service. (a) Lockheed Martin built "Garuda" satellite of the ACeS system and (b) Hughes design for the Thuraya satellite.

The choice of Ka-band is driven largely by the absence of a suitable alternative. As satellites providing domestic or regional fixed satellite services were deployed, agreement had to be reached on their separation along the orbital arc to avoid interference between systems due to an earth station with a small antenna illuminating satellites on either side of the intended one. Initially set at 3°, this spacing is now set at 2° around most of the orbital arc.

The absence of suitable C-band orbital slots drove satellite constructors to build satellites operating at Ku-band (Table 19.1), and most INTELSAT satellites are now built with transponders operating in both C- and Ku-bands. Powerful DTH TV broadcasting satellites all operate at Ku-band. These trends have made it almost impossible to secure an orbital location in which a satellite can operate at C- or Ku-band without interfering with its neighbors. This has spurred interest in operating at Ka-band and the still higher Q/V-band.

Until recently, the use of Ka-band was confined to experimental satellites launched by the United States, Italy, and Japan. This is because, unlike at C-band, rain greatly attenuates Ka-band signals (and to some lesser extent, Ku-band signals), making this a difficult band in which to provide satellite services. A group of private U.S. investors initially proposed a Ka-band satellite system, known as The Calling[sm] Network, that would provide global, low-cost telephone capability. This system was later redesigned to provide users with wideband data distribution, and was renamed Teledesic. As redesigned, the system was to employ 840 low-altitude satellites, each of which could relay to its 8 nearest neighbors.

Despite the very ambitious nature of this proposal, the Teledesic organization was successful in lobbying the 1995 World Administrative Radio Conference for Ka-band frequency assignments. This caused the FCC to proceed with a rule-making offering other applicants the opportunity to seek Ka-band spectrum and orbital locations. In all, 13 applications were submitted (in addition to the one from Teledesic)—all of them for GEO satellite systems. Of these, 7 (including Teledesic) proposed to offer global or near-global service. AT&T subsequently withdrew its filing for a 12-satellite global system, and in May 1997, the FCC approved the applications of the 12 remaining Ka-band geostationary fixed satellite systems (Table 19.4). Three U.S. systems that appear to be proceeding are reviewed below.

TABLE 19.4   U.S. Licensed Ka-Band Satellite Systems[a]

System	Orbit	Coverage	No. of Satellites	Capacity (Gb/s)	ISL Capacity	Onboard Switching	Capital Inv. ($B)
Astrolink	GEO	Global	9	9.6	1 Gbps	FPS	4.0
Cyberstar	GEO	Limited global	3	4.9	1 Gbps	BBS	1.05
Echostar	GEO	United States	2	5.8	120 MHz	BBS	0.34
Galaxy/ Spaceway	GEO	Global	20	4.4	1 Gb/s	BBS	5.1
GE*Star	GEO	Limited Global	9	4.7	None	None	4.0
KaStar	GEO	United States	2	7.5	155 Mbps	FPS	0.645
Millennium	GEO	United States/ Americas	4	5.2	1 Gbps	FPS	2.3
Morning Star	GEO	Limited global	4	0.5	None	None	0.823
NetSat 28	GEO	CONUS	1	772.0	None	Optical SW	0.25
Orion	GEO	US/IOR	3	2.9	TBD	FPS	0.725
PanAmSat	GEO	AOR	2	1.2	None	None	0.409
Teledesic	LEO	Global	840	13.3	1 Gbps	FPS	9.0
VisionStar	GEO	CONUS	1	1.9	None	None	0.208

[a]FPS, fast packet switching; BBS, baseband switching.

Worldwide there are now believed to be more than 50 proposed Ka-band projects requiring approximately 270 geostationary orbit locations. Most of these proposals appear to be for national or regional systems, and not a great deal has been published about them so far. Of these non-U.S. Ka-band systems, one advanced by Alenia of Italy appears to have the best prospect of proceeding and we include it in the review presented here.

The FCC opened a second Notice of Proposed Rule Making (NPRM) in 1997 for others to file for Ka-band system and a third for systems operating in the Q/V-bands. Some 8 additional Ka-band systems were filed for and 16 at Q/V-band, none of which have so far been licensed.

At the 1997 World Administrative Radio Conference France was able to secure for Alcatel Alsthom authorization to build a low-earth orbiting satellite system operating at the same frequency band as the existing geostationary Ku-band satellites. This system is called SkyBridge. A U.S. subsidiary of SkyBridge LCC subsequently filed for a U.S. license with the FCC, prompting a fourth NPRM for new satellite systems. This one closed in January 1999 with some six (including SkyBridge) applications. None of these systems have so far been licensed, but SkyBridge seems likely to proceed without a U.S. license, and hence we include it in our review.

Table 19.5 gives some of the system parameters of the three U.S. Ka-band systems discussed below as taken from their respective FCC filings. However, all three are undergoing further design study and refinement, so that these should, at best, be taken as a guide. Fundamental to each of the designs, however, is the need to access small (inexpensive) terminals in a manner that overcomes rain fading and achieves ≥95.5% availability. This entails the use of multiple narrow spot beams to achieve high effective isotropic radiated power (EIRP).

19.4.1 Astrolink

The Astrolink system proposed by Lockheed Martin will employ nine GEO satellites in five orbit locations, as illustrated in Figure 19.7. The total capacity of each satellite is about 7.7 Gbps. By placing two satellites at the same orbit location and operating them with orthogonal polarization, a total capacity of about 15.4 Gbps is achieved over the Americas, Europe, and Asia. Crosslinks between satellites operating at 60 GHz provide a means of routing traffic around the globe, and each link has a capacity of 1 Gbps.

In its original design, Lockheed planned to use a system of 64 hopping beams to provide global coverage. By adjusting the dwell time, a match could be achieved to the volume of traffic at each beam location. In a subsequent redesign 80 fixed uplink beams are formed using an active phased array that supports users. The beams are assigned a bandwidth adequate to carry traffic in the range of 20 to 330 Mbps. These channels are subjected to a frequency division (demultiplexing) analysis to select out each carrier, which is then demodulated and decoded. A packet switch routes packets to the appropriate downlink channel, where they are multiplexed with other packets destined for the same beam (or the intersatellite link). The packets are then reencoded and modulated onto the carrier. Up to 52 downlink beams can either "stare" or "hop" to deliver traffic to each user in the 80 uplink beams. Up to 16 of these beams are dual polarized and capable of delivering twice as much traffic to the regions they serve. Both the up- and downlink beams are 0.8° in diameter.

Astrolink proposes to equip users with terminals that employ antennas of 65, 85, and 120 cm in diameter, which can be operated at power levels in the range 0.25 to 10 W, at

TABLE 19.5 System Parameters of Three Licensed U.S. Ka-Band Systems[a]

System	No. of Beams	Number of Transponders per Satellite	Transponder Bandwidth (MHz)	TWTA Power (W)	Beam Size	EIRP (dBW)	Satellite Power (W)	Dry Mass (kg)	Satellite Life (yr)
Astrolink	192 (64 HBs) + 3 FB + 1 SB	68	125	56	$1°$	56	10,500	2185	12
Galaxy/Spaceway	48	48 (Ka) 24 (Ku)	125 (Ka) 36 (Ku)	20 & 60 (Ka) 75 (Ku)	$1° - 3°$	54	7500	2000	15
Teledesic	576 (64 HBs)	≥ 64	396	75	"Small"	50	6400	747	10

rates in the range of 16 kbps to 9.216 Mbps (with the larger antennas and higher powers being required for the higher rates.) These terminals would interconnect with the terrestrial switched network via gateway stations employing 2.4- or 4.5-m antennas with up to 200 W of power. The latter could operate at rates of 50 to 150 Mbps.

The Astrolink system would presumably be built in stages, with only one satellite initially located in each orbital slot. Even so, each satellite will be large and will represent a considerable financial investment. To date Lockheed Martin has raised $1.325 billion toward the cost of financing its system.

19.4.2 Galaxy/Spaceway

Hughes has proposed to construct a fleet of GEO satellites for a system called Galaxy/Spaceway. The proposed constellation would consist of 21 satellites in 16 orbital locations. The Spaceway portion of this system resembles that proposed by Lockheed Martin for Astrolink in that it consists of 9 satellites placed in 5 orbital locations, with ISLs between 4 of the locations. In subsequent FCC filings, Hughes asked for and was granted permission to launch 2 Ka-band satellites into each location at 101°W, 99°W, 49°W, 25°E, and 111°E, as well as single satellites into 54°E, 101°E, and 164°E. Hughes also requested permission to launch hybrid Ku/Ka-band satellites at 36°E, 40°E, 124.5°E, 149°E, 173°E, 67°W. Permission to operate Ka-band satellites at these locations was granted. (The FCC has not yet acted on the request to operate at Ku-band.) In all, Hughes secured permission to operate 20 satellites in 15 orbital locations, making it potentially the owner of the largest fleet of geostationary Ka-band satellites.

In the Spaceway portion of the system, 2 satellites are to be deployed in each of 4 orbit locations. By operating each satellite over the allowed 500 MHz of spectrum, but with different polarizations, Hughes obtains an equivalent 1000 MHz of bandwidth. Each satellite can support 68 transponders, 64 of which occupy 125 MHz each (for the users), and 2 that occupy 250 MHz (for gateways), thus achieving further frequency reuse. These transponders operate into narrow spot beams (59-dBW EIRP) and wide spot beams (52.3-dBW EIRP) arranged to cover the land masses visible to the satellites. The ISLs operate at 60 GHz and have a data rate of 1 Gbps.

Communications services will be provided at rates of 16 kbps to 1.544 Mbps via terminals with antennas in the range of 66 to 200 cm in diameter and uplink transmitters of up to 2 W. Onboard processing of arriving packets is used to route traffic between beams and to merge the traffic in a given transponder into a single 92-Mbps stream.

Uplink power control of the terminals is planned to combat rain fade. When a terminal is operated at 0.5 W, a rain fade of 7.6 to 2.6 dB can be overcome in the narrow beam, depending on whether the user is at the center or edge of coverage. For the broad beam, the corresponding numbers are 6.7 to 1.7 dB. The rate assumed is 384 kbps. It is claimed that by using the larger (2-m) antenna, an availability exceeding 99.5% can be achieved anywhere in the world (see Figure 19.5). General Motors, which owns Hughes Communications, has pledged $1.4 billion toward the construction of the Spaceway system.

19.4.3 Teledesic

The Teledesic system was originally conceived as an LEO system because it was felt that very large data rates were incompatible with the delay (or latency) encountered with geostationary distances. Also, to mitigate rain fading, the service area of each satellite was

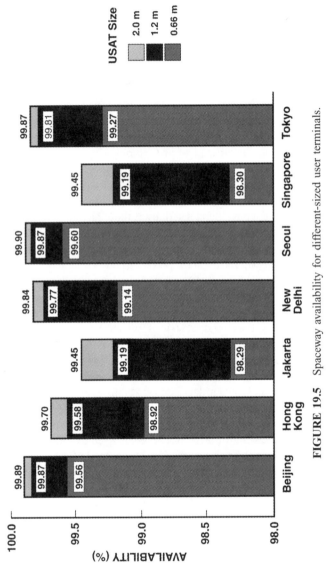

FIGURE 19.5 Spaceway availability for different-sized user terminals.

limited by the requirement that it be viewed by an earth station only when the elevation angle exceeded 40° (to minimize rain attenuation). This, together with the low-altitude (700-km) orbit chosen, then drove up the number of satellites needed for global coverage to 840. Each satellite in the system was capable of crosslinking with its 8 nearest neighbors and employed phased-array antennas to scan "cells" on the ground from which to collect and deliver traffic.

Teledesic selected Boeing as its prime contractor, who is believed to have concluded that the announced cost ($9 billion) of the system was too low. The design was then changed by raising the orbit altitude to 1400 km, thereby allowing the number of satellites required to be reduced to 288. The figures given in Table 19.5 are for the 840 satellite design.

In the second round of Ka-band filings Motorola filed for a 63-satellite LEO system called Celestri. Teledesic maintained that this would interfere with its licensed system. While not accepting that conclusion, Motorola eventually decided to join forces with Teledesic and committed $750 million in financing toward it, receiving in return the right to construct the satellites. (Together with other amounts invested, this gave Teledesic over $1 billion with which to commence building the system). Motorola's move may have forestalled efforts by Hughes to capture the satellite construction contract. However, it is understood that Motorola's engineers wished to redesign the entire Teledesic system with a view to reducing the number of satellites involved (to less than half). The resulting disagreement with Teledesic has clouded the prospects of the program at present. The date given for the start of service has now slipped out to 2004–2005, and there have been reports of Teledesic exploring other investment opportunities, and looking to other partners with which to proceed with the Teledesic system.

19.4.4 Euroskyway

The one European Ka-Band system that appears to have some prospect of being fielded is a proposal by Alenia Aerospazio of Italy to construct a fleet of five geostationary satellites known as Euroskyway. Unlike the Astrolink and Spaceway concepts in which global coverage is sought, Euroskyway's five satellites cover only Europe and the Middle East. In the first operational phase two co-located satellites would be deployed providing coverage of Europe and adjoining regions (so-called extended Europe) via 32 adjacent fixed beams. A second operational phase is envisaged in which three additional satellites are added to increase capacity and extend the coverage into Africa and Asia. The satellites are to be connected via intersatellite links operating at V-band (56 to 64 GHz).

Euroskyway intends to support asynchronous transfer mode (ATM) traffic between three classes of user terminals (including *ultra* and *very* small), two classes of service provider terminals, and gateway terminals that are to be connected to the public switched network. A downlink rate toward all these terminals of 32.768 Mbps has been selected, while the service provider terminals and gateway terminals uplink at multiples of 6.144 and 32.768 Mbps, respectively. An onboard demultiplexer/demodulator delivers cells to an ATM switch that routes them according to the packet header. The uplink rate from user terminals is at 160, 618.26 and 2048 kbps depending on the class of terminal. Traffic is to be managed through the onboard switch via user requests for connections (i.e., a call set up and reservation policy), but overload will be prevented by a means of a feedback based congestion prevention mechanism. This is implemented by periodically broadcasting the occupational status of the traffic resource pools for each uplink and downlink spot. The total satellite capacity of 7.6 Gbps (9.2 Gbps including signaling) is comparable to that claimed by Astrolink.

19.4.5 SkyBridge

The difficulties caused by rain attenuation (Section 19.4.6) at Ka-band are sufficiently severe that enormous competitive advantage would be enjoyed by a global system operating at Ku-band, if some way were to be found to avoid any interference with existing fixed Ku-band services. Alcatel Alsthom of France has proposed such a system and secured International Telecommunications Union (ITU) approval for its construction at the 1997 World Administrative Radio Conference, with provisional power flux density limits imposed to protect the existing services.

As originally proposed, the SkyBridge system would have been made up of 2 Walker constellations of 32 satellites each, displaced from one another such that satellites crossed the sky in pairs. An exclusion zone of ±10° about the equator is to be established, and no earth station would communicate with any satellite in this part of the orbit. Instead the earth station would switch to the satellite of the pair that remained outside the exclusion region. When the FCC opened its NPRM for nongeostationary Ku-band systems, SkyBridge LLC filed an amendment to its original application of February 1997. The amendment (dated January 1999) represents the culmination of further design work performed on the SkyBridge system. The principal change has been to increase the number of proposed satellites from 64 to 80, while reducing their individual capabilities. The satellites are in circular orbits at 1469-km altitude, arranged with 4 in each of 20 planes inclined at 53°. The planes are equally spaced around the equator leading to an 18° separation, and satellites in adjacent planes are phased by 67.5°. The orbital period is 115 min.

This new satellite configuration provides dual satellite visibility 100% of the time between 25° and 60° latitude and ≥ 95° of the time from the equator to 25° (Figure 19.6). At least one satellite is *available* for use at all times, up to about 70° latitude, when allowance is made for an exclusion zone of ±10° about the geostationary orbital arc and an earth station elevation requirement of ≥ 10°.

Each satellite has a forward capacity of 12 Gbps and SkyBridge claims a system capacity of ∼215 Gbps over land areas. To achieve this capacity, SkyBridge requests 2 GHz of contiguous spectrum for the downlinks (10.7 to 12.7 GHz) and 1.65 GHz of spectrum for the uplinks, arranged in four 500-MHz windows between 12.85 and 14.5 GHz with a fifth at 17.3 to 17.8 GHz.

The satellites operate as conventional "bent pipe" repeaters and have no onboard processing or intersatellite links. Each satellite forms 24 spot beams from phased-array antennas, each beam illuminates a service area 350 km in radius. The total area served by a satellite has a radius of 3,000 km. These beams are held fixed on their respective service areas as a satellite transits overhead, and then "hopped" to new service areas. This requires that the beams be continuously steered within a ±53° cone about the nadir. One option under consideration appears to be to employ steerable parabolas fed by phased arrays that could be adjusted (zoomed) to keep the cell sizes constant as the satellite transits.

The frequency reuse plan for these beams is not spelled out, nor is the design of the payload. The claimed capacity suggests that a fourfold frequency reuse plan will be employed. It appears that most of the capacity is arranged in a "loop-back" fashion, requiring that a gateway earth station be placed in each service area. To achieve the

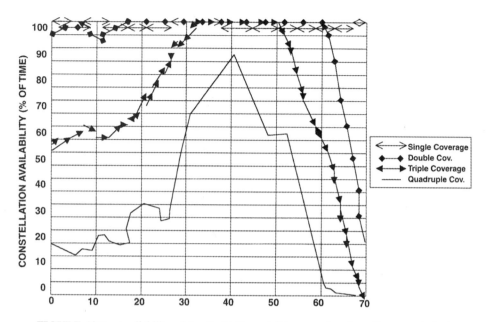

FIGURE 19.6   Availability of the SkyBridge satellites as a function of gateway latitude.

claimed global capacity of 215 Gbps then would require 430 beams to be active (i.e., 430 gateway earth stations). SkyBridge hopes to halve this number by cross-strapping some of the transponders to permit some gateways to serve the adjoining service areas. This feature will also permit gateway stations in adjoining cells to communicate with one another. Handover from satellite to satellite is to be coordinated by these gateway stations, which will send instructions to all the user terminals they support just prior to handoff.

Three types of earth station terminals are envisaged: gateways with either 2.5- or 4.5-m-diameter antennas, "residential" user terminals with antennas that appear to be 33 cm in diameter, and "professional" user terminals with 3.3-m-diameter antennas. Residential terminals would transmit a single CDMA carrier at a rate of 2.56 Mbps and peak power of 2.4 W. They would receive a single CDMA carrier with a peak forward rate of 20.48 Mbps. Professional terminals can transmit several 2.56 Mb/s carriers, each at a maximum power of 6 W, and receive several 20.48 Mbps forward links. Gateways operate at 10 W per 20.48 Mbps outbound carrier. Power control over a range of 20 dB is to be employed at each terminal (and in the satellites) to compensate for range changes and rain fading. This strategy, together with operation at reduced bit rates, allows residential terminals everywhere to achieve an availability of better than 99.90% and professional terminals to be better than 99.95%.

19.4.6   Technical Challenges Facing Multimedia Service Providers

One of the major challenges facing proponents of the new satellite systems discussed here will be ensuring an adequate supply of low-cost user terminals. To penetrate a consumer market, it is widely believed that the cost of the terminal should not exceed $1000, and it is difficult to see how the price can be lowered to this level without the production of very large quantities (hundreds of thousands). While the cost of the indoor unit can be lowered through the use of very large scale integrated (VLSI) circuits, this is not the case for the outdoor unit, which requires expensive devices to generate appreciable power (watts) at Ka-band, as well as up- and down-converters. The situation is aggravated for the LEO systems, which will require users to have tracking antennas. These are most likely to be mechanically steered small reflectors, and two may be required to minimize time lost due to handover. The prospects for providing such devices inexpensively, which will operate unattended reliably for several years, seem particularly unpromising.

A second difficulty for the Ka-band systems is the attenuation caused by rain. The effect is severest in the heavy rain associated with thunderstorms and thus is more frequent in the tropics than elsewhere. Figure 19.7 shows rain outages at the uplink frequencies for Ku (14 GHz) and Ka-bands (29 GHz) for four different rain regions. The cost of overcoming the rain attenuation in all regions becomes prohibitively expensive, thus most systems designers are attempting only to achieve an availability of 99.5%. This means that for 40 h per year the service may be unavailable or severely degraded. As such, telephone service, distance learning, tele medicine, and some other services may attract few customers.

One strategy for mitigating the rain fade problem is to employ uplink power control during periods of rain impairment. The decision to increase power can be based on the strength of the received signals, the sky noise temperature, or satellite command. Realistic changes in the power level are probably in the range of 5 to 10 times, achieving 7 to 10 dB of link enhancement. A second strategy is to switch to lower bit rates during periods of

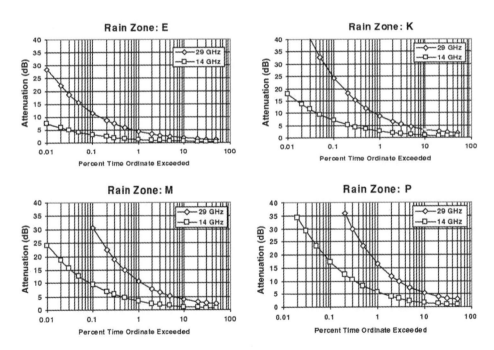

FIGURE 19.7 Amount of excess link margin required in four different rain regions to achieve various levels of availability. Washington, D.C. is in rain zone K while rain zone P represents the tropics. Curves are presented for the uplink frequencies (14 and 29 GHz) for Ku- and Ka-band, respectively.

rain. This approach would be unsuitable for many applications, but might be satisfactory for some such as Internet access.

A third likely approach to achieving good performance involves the use of convolutional coding. However, this scheme fails if more bits are in error than the scheme is capable of correcting. A burst of errors is then produced. To combat this, the bits are read into the "columns" of a trellis matrix, and then read out in "rows." This spreads out the time between the errored bits, allowing a second "outer" block coder to correct them. The signal-to-noise ratio required to achieve a bit error ratio (BER) of 10^{-10} using this strategy can be decreased from about 13 to 5 dB. Unfortunately, there is a price to be paid for this improvement—it increases the complexity of any processor onboard the satellite, since the coding must be removed to read the address of a data packet.

A third technical challenge facing most of the systems is that some form of onboard routing is necessary for a satellite that employs a large number (say, $n \geq 5$) of interconnected spot beams. This is because the number of paths through the satellite increases as n^2. The situation is aggravated in some of the concepts by the use of intersatellite links (i.e., crosslinks between satellites that allow communications around the globe). In most designs, the plan is to operate the satellite with a packet switch on board to route traffic on a packet-by-packet basis to its destination, based on an address contained in the header. However, to read the header it is necessary to sort the arriving packets by transmission frequency (i.e., demultiplex them) and then demodulate them. Next, the coding that was introduced at the earth station must be removed so that the address can be read. After routing the packet to the proper downlink, the coding must be reintroduced.

One attractive feature of onboard processing is that it permits all of the signals destined for a given transmit beam to be sent as part of a single time series. This allows the high-power amplifier in the satellite transponder to be operated at full power, which is the most efficient mode.

These onboard processors must, however, handle enormous amounts of data and are likely to consume several kilowatts of prime power introducing thermal dissipation problems. Since nothing quite so advanced has yet been flown on a commercial communications satellite, they represent additional risk.

The commercial success of these new systems is by no means guaranteed. On a cost basis they are unlikely to compete with the terrestrial alternatives for delivering high-speed data, that is, cable TV modems, or asymmetric digital subscriber (ADSL) modems. There will also be competition from terrestrial wireless systems, so the size of the addressable market is both uncertain and likely to change with time as the terrestrial infrastructure is deployed. This makes it incumbent on the designers of the geostationary satellite systems to incorporate as much flexibility as possible to permit the satellite resources (power and bandwidth) to be reassigned to different regions as traffic migrates. Unfortunately, achieving this usually imposes severe weight and cost penalty.

19.5 CONCLUDING REMARKS

Satellite projects are currently enjoying a period of great interest, spurred by a number of factors. The overall telecommunications market is growing rapidly, in part due to growth in international trade, but also due to reduced prices for many services and the introduction of new services such as cellular or PCS and access to the Internet. Deregulation and the opening of overseas markets offer new opportunities for telecommunications companies.

Satellites can provide almost "instant" infrastructure, requiring little in the way of the civil works needed to install other systems. In addition, U.S. aerospace companies are seeking new opportunities in the civil sector, given that defense orders have declined. Also, Lockheed Martin, Loral, and Motorola appear intent on vertically integrating into the service business, as Hughes has successfully done. Finally, there is a realization that the resources (frequencies and geostationary orbital slots) available for satellite projects are quite limited, and that the FCC's notices of proposed rule-making probably represent some of the last opportunities to lay claim to a limited resource.

Over the long term, communications satellites will continue to be used only when they provide a clear competitive advantage. These situations include broadcasting, distributing the same information (e.g., television) to a large number of subscribers (such as schools or cable head ends), and mobile applications (ships, planes, travelers in remote places, etc.).

The introduction of PCS sysems (by Iridium) has been inauspicious. It remains to be seen whether there is a large enough market to support one or more of the global PCS satellite systems (Globalstar and ICO) that are being deployed. Some of the regional PCS systems (e.g., ACeS and Agrani) would seem to offer better prospects for commercial success owing to their low cost and the possibility that they could be employed to provide rural telephony in some countries (with government support).

If some of the Ka-band systems described here are successful, then satellites may enjoy another role not heretofore theirs—that of providing "last mile" connections to homes and businesses for broadband data, multimedia, and related services. Success in this area depends critically on bringing down the cost of user terminals, and this in turn can be achieved only if terminals are mass-produced using specially developed chip sets for all functions. Since it will be particularly difficult to bring the price low enough for terminals that require tracking antennas, it seems that the proposed LEO systems will have greater difficulty serving the consumer market. Also, timing is critical, since in the developed countries where the largest markets are initially to be found, other technologies are being pursued to fill this need.

Satellites will also continue to be used for the foreseeable future to deliver telephony and other public switched services to countries that are not linked by fiber-optic cables and to support private (e.g., VSAT) networks that cover a wide geographic area. However, these roles appear inadequate to justify all of the current interest in new satellite systems.

PART 6

WIRELESS ATM NETWORKS

CHAPTER 20

Wireless ATM Networks

D. RAYCHAUDHURI, P. NARASIMHAN, B. RAJAGOPALAN, and D. REININGER

20.1   INTRODUCTION

Wireless asynchronous transfer mode (ATM), first proposed by Raychaudhuri and Wilson [1, 2], is a specific technical approach for implementing next-generation wireless networks capable of delivering broadband services to portable multimedia devices. During the past few years, broadband wireless networks exemplified by wireless ATM have evolved from research stage to standardization and early commercialization. A number of R&D organizations have demonstrated proof-of-concept prototypes [3–6], and early broadband wireless products based on either Internet protocol (IP) or ATM approaches have just started to reach the market. Several standards activities, including the ATM Forum's Wireless ATM (WATM) working group [7], the European Telecommunications Standards Institute's (ETSI's) Broadband Radio Access Networks (BRAN) [8], and Japan's Mobile Multimedia Access Communication (MMAC) [9] are currently in progress and should produce initial specifications for broadband wireless core network and radio air interface protocols by late 1999 or early 2000.

Wireless ATM systems discussed in this chapter are intended to provide semimobile integrated voice/data/video services in both private and public microcellular wireless network scenarios. While the service concept (shown schematically in Figure 20.1) is essentially frequency independent, initial applications of the technology tend to be targeted toward ∼25 Mbps services in the 5-Ghz unlicensed bands which have recently been allocated for high-speed data in various parts of the world [9–11]. Subsequent application to licensed frequency bands anywhere between 2 and 60 Ghz would be likely as the service concept matures. Figure 20.2 shows the application regime of wireless ATM relative to familiar wireless technologies such as cellular personal communications services (PCS), and wireless local area networks (LAN). It can be seen from the figure that wireless ATM technology which is designed to support bit rates up to 25 Mbps at pedestrian mobility

Wireless Communications in the 21ˢᵗ Century, Edited by Shafi, Ogose, and Hattori.
ISBN 0-471-155041-X © 2002 by the IEEE.

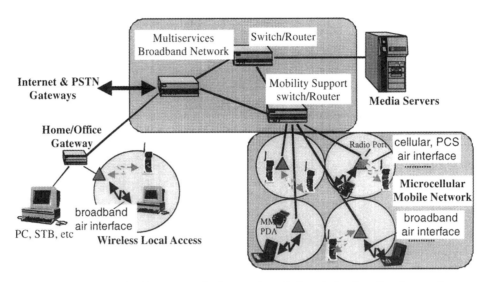

FIGURE 20.1   Service concept for future microcellular wireless broadband network.

speeds (~5–10 km/h), would typically be applied to "broadband PCS" and "high-speed wireless LAN" scenarios both private and public. Wireless ATM can thus serve as a useful building block within the framework of next-generation International Mobile Telephony (IMT)-2000 wireless service scenarios by complementing lower bit rate, higher mobility third-generation cellular service with semimobile broadband PCS and data services.

Of course, it is still premature to predict the long-term architecture of third-generation wireless networks, which will inevitably reflect the results of the ongoing Telecom/Internet convergence process. The original broadband ISDN vision of "ATM everywhere" has been replaced by the more pragmatic reality of a multiprotocol broadband network involving a mix of IP, SS7, frame-relay, and so forth with ATM. It is reasonable to

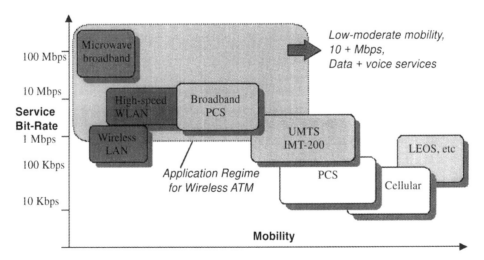

FIGURE 20.2   Application regime of wireless ATM in terms of service bit rate and mobililty.

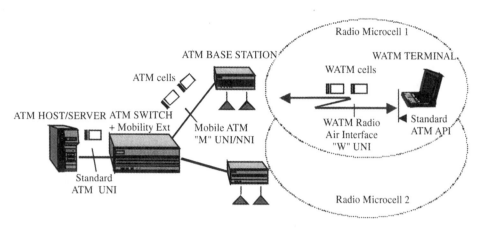

FIGURE 20.3 Wireless ATM network components.

expect that both ATM and IP protocols will play significant roles in third generation wireless networks, much as they do in wired broadband networks being deployed today. We believe that wireless ATM technology will provide a useful foundation for quality-of-service (QoS)-based multimedia services in many future systems, including those involving support of legacy cellular network protocols such as Global Systems for Mobile (GSM) communications and/or Internet (IP) data services. This important issue will be addressed in more detail later in this chapter, after first completing an overview of the basic technology.

The following sections present an overview of wireless ATM network architecture, design details for major subsystems (including radio access and mobile network infrastructure technologies), and some considerations of system level QoS. It is remarked here that since wireless ATM architectures and subsystem designs are still evolving, the contents of this chapter are intended to provide an understanding of general principles rather than firm specifications. The reader is referred to the literature [12–14] for architecture and review studies on wireless ATM that may be useful for supplementing the material presented here.

20.2 WIRELESS ATM ARCHITECTURE

20.2.1 WATM Components and Interfaces

The major hardware/software components that constitute a basic wireless ATM are shown in Figure 20.3. From the figure, it can be seen that a wireless ATM system typically consists of three major components: (1) ATM switches with standard user network interface/network network interface (UNI/NNI) capabilities together with additional mobility support software; (2) ATM "base stations" or "radio ports" also with mobility-enhanced UNI/NNI software and radio interface capabilities*; and (3) wireless ATM

---

*Note that there are three terms used in this context: ATM base station, ATM radio port, and WATM access point. In this chapter, these terms are used with the following interpretation: base station: ATM switching (UNI/NNI) + radio interface(s); radio port: ATM interface (UNI) + radio interface(s); and access point: WATM radio air interface only.

terminal with a WATM radio network interface card (NIC) and mobility and radio enhanced UNI software. Thus, there are two new hardware components, ATM base station (which can be viewed as a small mobility-enhanced switch with both radio and fiber ports) and WATM NIC, to be developed for a wireless ATM system. New software components include the mobile ATM protocol extensions for switches and base stations, as well as the WATM UNI driver needed to support mobility and radio features on the user side.

The system shown involves two new protocol interfaces: (a) the "W" UNI between mobile/wireless user terminal and ATM base station and (b) the "M" UNI/NNI interface between mobility-capable ATM network devices including switches and base stations. Both these interfaces are required to support end-to-end ATM services at a mobile terminal such as that shown in Figure 20.3. In particular, the WATM terminal sets up a connection using standard ATM signaling (UNI) capabilities to communicate with the ATM base station and network switches. All data transmitted by the WATM terminal is segmented into ATM cells with an additional radio link level header specified within the "W" interface. Mobility of the WATM terminal (i.e., handoff and location management) is handled via switch-to-switch (NNI) signaling protocol extensions specified in the "M" interface.

It is noted here that subsets of full WATM systems outlined above may also be used as building blocks for other important wireless or mobile network scenarios. For example, the "M" UNI/NNI specification can be used to construct a generic infrastructure for existing PCS, cellular, and wireless data systems in addition to end-to-end WATM services. In this configuration, ATM base stations convert wireless access protocols such as GSM, IS-136, code division multiple access (CDMA) or IEEE802.11 to a common ATM format that supports network-level mobility requirements in a generic way. Such an ATM-based mobile infrastructure provides important service integration and cost/performance advantages over existing mobile networks, while facilitating smooth migration to broadband WATM services [15]. Alternatively, the WATM radio air interface specified in the "W" UNI can be used for wireless broadband access from stationary ATM devices in offices or homes. This type of WATM access is expected to be an useful alternative to wired broadband access methods such as SONET fiber or ADSL/VDSL, offering advantages of rapid deployment and statistical multiplexing of switch port resources.

20.2.2 Protocol Architecture

The wireless ATM protocol stack is shown in Figure 20.4. It can be seen from the figure that the approach is to fully integrate new wireless link-specific functions into the standard ATM stack, while adding necessary mobility support features to existing control plane modules such as signaling and routing. The WATM protocol stack includes a new radio physical layer that interfaces to the ATM network layer through additional medium access control (MAC) and data link control (DLC) sublayers. This means that standard ATM services such as addressing/routing, virtual circuit (VC) establishment, and QoS control continue to be used in wireless ATM networks, with wireless link and mobility-specific functions added in a backward-compatible manner. As indicated in the figure, the only change in the higher layers is the provision for mobility-related extensions to the signaling/routing protocols in the ATM control plane.

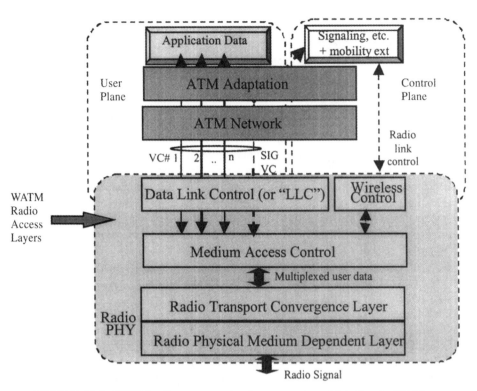

FIGURE 20.4   WATM radio access protocol components and interface to ATM layer.

20.3 WATM RADIO ACCESS LAYER

20.3.1 Radio Access Architecture

As outlined above, the standard ATM protocol stack needs to be augmented with radio access protocol layers to handle radio-link specific functions such as high-speed radio transmission/reception, sharing of available bandwidth among multiple users, error control to reduce the effects of radio channel impairments, and radio resource management. The radio access protocol layers that have been integrated into the standard ATM protocol stack are as follows:

- Radio physical layer (PHY) provides a high-speed radio link.
- Medium access control (MAC), to share available radio bandwidth among multiple users in such a way that standard ATM traffic classes are supported (along with their QoS guarantees) while maintaining high utilization of the bandwidth.
- Data link control (DLC), to provide an additional layer of error protection to ATM services to reduce the perceived error rate over the radio link by an error recovery mechanism. The DLC protocol should be applied to both real-time and non-real-time traffic, but with the recovery algorithms being adapted to accommodate the QoS requirements of each individual connection.

- Wireless control, which supports radio resource management functions (e.g., registration upon mobile terminal power-up, monitoring of the radio link, measuring signal strength on neighboring channels, transmitter power control, etc.).

20.3.2 Radio Physical Layer

The radio PHY layer is partitioned into the radio physical medium dependent (RPMD) sublayer, which defines the low-level functional details of the physical interface, and the radio transmission convergence (RTC) sublayer, which is responsible for structuring the information to the appropriate transmission format.

A high-speed radio modem that is capable of providing reasonably reliable transmission within a range of 100 to 500 m for low-speed mobile users is required for wireless ATM. The modem must be able to support burst operation with very short preambles so that the overhead incurred at the radio PHY layer is low. Typical target bit rates are $\sim$25 Mbps (with per-user bit rate requirements of around 6 Mbps sustained and 10 Mbps peak) with an expected operating frequency band of around 20 to 25 MHz. The wireless ATM devices are expected to operate, initially, in the 5-GHz U-NII band in the United States, HIPERLAN band in Europe, and MMAC band in Japan.

Some of the modulation methods that have been considered for the wireless ATM PHY layer are equalized quadrature phase shift keying (QPSK)/quadrature amplitude modulation (QAM) [16], Gaussian modulation shift keying (GMSK) [17], and multicarrier orthogonal frequency division multiplexing (OFDM) [18]. The ETSI BRAN (Hiperlan2) project committee has initially adopted a 52-carrier, 20-MHz OFDM approach with selectable bit rates of 6/9/12/18/27/36/54 Mbps. The selection of OFDM was based on its relative robustness to interference and multipath, although equalized single-carrier methods have also shown to perform well (note that this committee decision is still subject to final ratification).

The RTC sublayer specifies how the data transmitted over the wireless link is structured. Typical WATM RTC proposals are based on a TDMA/TDD frame structure with a fixed-length frame (e.g., $\sim$1 to 2 ms) and slots capable of carrying n ($= 6$, 8 etc.) bytes of data. A group of N frames (e.g., $N = 8$) is considered to constitute a superframe that is used for MAC layer scheduling and wireless control sleep-mode functions.

20.3.3 MAC Layer

The MAC layer in wireless ATM is required to share available radio bandwidth among multiple contending users in such a way that the bandwidth is utilized efficiently and standard ATM services with associated QoS requirements are supported. Some of the techniques considered in the literature include dynamic TDMA/TDD [2, 19], packet reservation multiple access (PRMA) [20] with extensions, and CDMA [21].

Most wireless ATM MAC proposals are based on a dynamic TDMA/TDD protocol with support for multiple service classes with QoS. The bandwidth allocation is controlled by the base station, with limited scheduling control for uplink bursts retained at the mobile terminals. The base station schedules all the control packets and data cells on the downlink. It allocates uplink bandwidth to each mobile terminal based on negotiated contracts and current requirements as requested by the mobile terminal. In a distributed control implementation, the MAC layer at the mobile terminal is responsible for the scheduling of control packets and/or WATM cells within the uplink burst. It is also

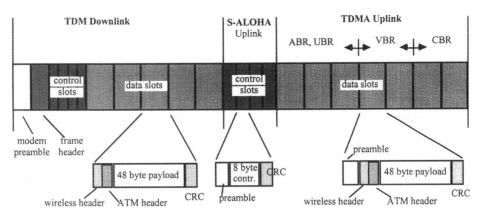

FIGURE 20.5 Outline of representative dynamic TDMA/TDD protocol used for WATM MAC.

possible to have a more centralized implementation in which slots each user virtual circuit or control message are allocated by the base station controller.

The structure of the TDMA frame suited for the above class of dynamic TDMA/TDD MAC protocol is shown in Figure 20.5. The frame consists of one downlink burst followed by one or more uplink transmission bursts. At the end of the frame is a random-access channel where short control packets can be transmitted using a random-access protocol like slotted ALOHA. All the downlink information from the base station to the mobile terminals is transmited in one burst. The downlink consists of physical layer synchronization, followed by a frame header that carries broadcast information about both the base station and the particular TDMA frame (e.g., location and size of different subframes). Downlink control and downlink data information follow to complete the downlink burst.

20.3.4 WATM Packet Formats

WATM Cell The format of a wireless ATM cell is harmonized with that of a standard ATM cell, which is augmented with a wireless header and a wireless trailer to support radio access protocols. In a representative implementation in the WATMnet system [6, 22, 23], an 8-bit cell sequence number (CSN) is added in the header, while a 16-bit cyclic redundancy check (CRC) forms the wireless trailer for a total of 56 bytes per cell on the radio link (Figure 20.6). Optionally, the generic flow control (GFC) field in the ATM cell can be used to support additional radio link features. The CRC is used to detect errors, while the CSN is used to identify cells in the DLC acknowledgement (ACK) packets to recover/retransmit lost cells. ETSI BRAN is currently considering an alternative design based on 54-byte cells, each with 10-bit sequence number and 24-bit CRC, and a 1-byte compressed ATM (or other) tag.

Control Packets Each protocol layer in the WATM protocol stack requires the use of control packets to communicate with its peer on the other side of the radio link. For example, the MAC layers at the base station and the mobile terminals exchange bandwidth request and bandwidth allocation packets. The DLC layers exchange ACK packets. Note that these control packets are used only across the WATM protocol layers and, hence, are never passed up to the network layer.

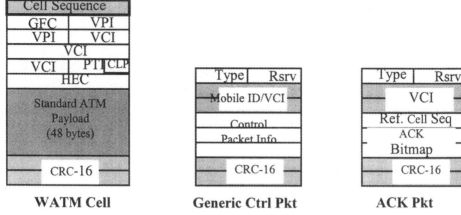

FIGURE 20.6   Representative cell and control packet formats for wireless ATM radio link.

Figure 20.6 also shows the format of a generic control packet used on the radio link in a representative WATMnet implementation. The Type field identifies the type of control packet. The mobile terminal identifier/virtual circuit identifier (MTID/VCI) field contains either a mobile terminal ID or a VCI depending on the type of control packet. For instance, in an ACK packet, this field will contain the VCI of the connection for which this ACK packet is meant, while a bandwidth request/allocation packet will contain the mobile terminal ID. All control packets are protected by a 16-bit CRC for error detection.

Control packets in the ETSI BRAN standard are composed of information elements (IE) and IE blocks. An IE is 8 bytes long and an IE block consists of 3 IEs and a 24-bit CRC (thus, adding up to 27 bytes). Control information is always transmitted in IE blocks with dummy IEs used if there are not enough information IEs to fill the block. An IE consists of an IE flag (1 bit, reserved for future use), a 4-bit IE-type field, and 59 bits of IE information.

20.3.5 MAC Algorithm

The base station is the master scheduler determining the manner in which available bandwidth is shared. Bandwidth allocation for uplink transmissions are signaled with bandwidth allocation packets sent in the downlink control subframe. Requests for uplink bandwidth are sent by the mobile temrinals, usually, as part of their uplink transmissions. In the case when a mobile terminal is in sleep mode (all of the data connections are in a temporary inactive state, so that the MT does not need to wake up in every frame to listen for control/data on the downlink), then bandwidth requests may be sent in the random-access channel.

The MAC layer supports two classes of service on the radio link: rate mode, for connections that have a notion of a rate of bandwidth usage, and burst mode, mostly used for best-effort services. All user-level traffic classes are mapped into one or a combination of the above two classes. For example, while a constant bit rate (CBR) connection can be mapped into the rate mode service, an available bit rate (ABR) connection could be mapped into a combination of the two.

The rate mode service is further split into two subclasses: guaranteed and on-demand. The former would include that part of the connection's bandwidth that is required on a periodic and guaranteed basis [the complete bandwidth of a CBR connection, e.g., or the minimum cell rate (MCR) of an ABR connection]. The on-demand subclass would include that part of the bandwidth that is not guaranteed. It is allocated on-demand (usually, for a short while to cover temporary burstiness as in the case of a VBR connection when the instantaneous rate is higher than its sustained rate, subject to UPC constraints) based on availability and other pending requests [24].

During connection setup for the rate mode services, the MAC layers at the base station and the mobile terminal record bandwidth and QoS requirements. This information is used by the base station scheduler to ensure all guaranteed bandwidth requirements are met. For the on-demand part, the mobile terminal sends bandwidth request packets (BRP) signaling either an increase or decrease in this part of the bandwidth. For the burst mode connections, the mobile terminals aggregate bandwidth requirements for all the connections and send BRPs to the base station indicating their current requirements.

For downlink traffic, the MAC scheduler at the base station has complete information on the status of the data buffers for each connection. In the uplink direction, the MAC scheduler at the mobile terminal (MT) aggregates requirements from the individual connections and sends a single BRP to the base station. The base station then uses this information and a scheduling algorithm to allocate bandwidth between connections on the downlink and MTs on the uplink.

20.3.6 Data Link Control

A DLC layer provides an extra layer of error protection to minimize the effects of the higher error rate experienced on the radio link. DLC techniques for wireless ATM that are applicable to both packet mode and stream mode services have been proposed in [25].

A cell-based selective repeat automatic repeat request (ARQ) scheme (with certain time limits for error recovery in the case of real-time services) can be used to recover cells that are not received correctly at the receiving end. To provide connection-specific QoS, the DLC layer should maintain separate transmission and reception state on a perconnection basis. This allows the protocol to be flexible to adapt to different service classes with varying QoS requirements. For example, a time limit is imposed on error recovery for real-time traffic classes, while such limits may either be much longer for best effort traffic.

Since cell error detection and retransmission requires each WATM cell to be identified uniquely, a cell sequence number (CSN) is added to the wireless ATM cell header. This CSN is valid only on the wireless hop and is stripped off before a cell is forwarded to the network convergence sublayer. To detect transmission errors, a 16-bit CRC is added to the WATM cell trailer.

The DLC layers at either end use ACK messages [Figure 20.7(a)] to indicate the reception status of WATM cells in the active transmission window. A group acknowledgment mechanism permits the ACK/NACK of up to 16 cells with a single ACK packet, thus reducing the overhead due to the DLC protocol.

Error Recovery Modes The DLC error recovery operation can be categorized into two different modes. The first one is attributed as zero loss mode where each lost cell is recovered with a relatively long recovery time-out value (compared to the duration of a TDMA frame). This mode of operation is usually performed on packet mode ABR and

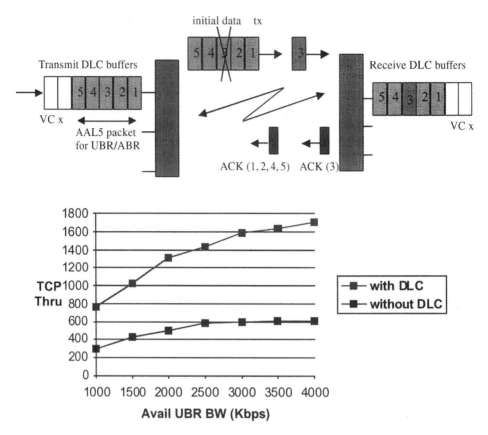

FIGURE 20.7 (a) Data link control mechanism for error recovery on WATM radio links. (b) Experimental results showing TCP throughout increase with data link control [27].

UBR connections. For example, for a UBR connection carrying TCP traffic, this mode will try to recover lost cells on the radio link rather than an end-to-end recovery by TCP, which is relatively more expensive (including additional traffic on the wired segment of the connection). The other mode is referred to as fixed recovery window mode. Attempts to recover lost cells are made within a preset and fixed time window. Connections carrying real-time traffic, which are usually more loss tolerant but delay intolerant, are most suited for this mode of operation.

Experimental results on the WATMnet prototype [6] confirm that suitable DLC provides a significant increase in end-to-end TCP throughput at the WATM terminal, as shown in Figure 20.7(b). Favorable results have also been obtained for CBR services, demonstrating that a substantial reduction in cell loss rate can be achieved with moderate buffering delay ($\sim$10 ms) within the DLC.

20.3.7 Wireless Control

A wireless control layer is used to support radio resource management functions.

Registration When a mobile terminal powers up, it registers with the strongest base station by sending a registration packet. The base station responds with a registration response that contains a 16-bit mobile terminal ID. The mobile terminal ID, which is unique as long as the MT is within the coverage area of the current base station, is used to identify mobile terminals in both uplink and downlink control packets.

Power Measurement   The mobile terminal is required to monitor received power levels in neighboring cells to determine when a handoff is required and to determine the target base station if a handoff is required. The base station can indicate to the mobile terminal (or vice versa) when mobile terminals could monitor link quality on neighboring channels. The base station, if necessary, can also request that a mobile terminal transmit the results of the power measurements to the base station.

Power Control   The base station may indicate to the mobile terminal on the status/quality of the signal received from the MT. This information can be used by the BS/MT to adjust the transmission power on uplink bursts.

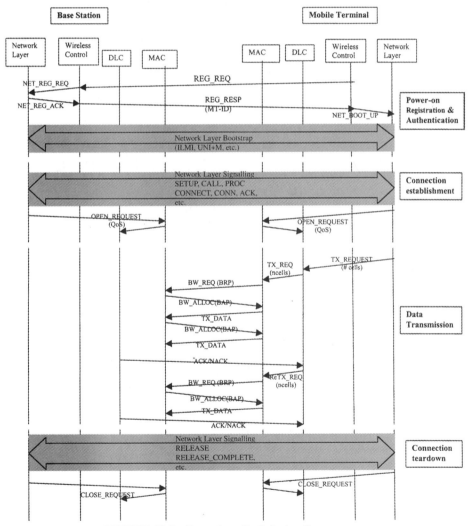

FIGURE 20.8   Example radio link signaling sequence.

20.3.8 Radio Link Signaling Sequence

Figure 20.8 shows a timing diagram of an example signaling sequence over the WATM radio link with the protocol components outlined above. As shown in the figure, the complete signaling sequence setup consists of the following phases:

- Power-on, registration, and authentication
- Network layer bootstrap, including establishment of signaling VCs, ILMI, and so forth
- Connection establishment, including ATM layer signaling as well MAC control signaling as required to set up necessary QoS resources for each VC
- Data transfer phase, including necessary MAC layer control signaling for dynamic resource allocation, data link control ACK/NACK, and the like
- Connection tear-down phase, releasing both radio-level and network-level resources

20.4 MOBILE ATM NETWORK

20.4.1 Mobility Management Protocols

Allowing end-user mobility in ATM networks gives rise to the problem of mobility management, that is, maintaining service to UNI devices regardless of their location [26]. Mobility management functions to be provided by the mobile ATM network include:

- *Location Management*   This enables unique identification of the mobile user and/or terminal, and the routing of connections to the mobile terminal regardless of its location. A key requirement here is that the correspondent systems need not be aware of the mobility or the current location of the MT.
- *Security Management*   Authentication of mobile users (or terminals) and establishing cryptographic procedures for secure communications based on the user (or terminal) profile [27].
- *Service Management*   Maintaining service features as a user (or terminal) roams among networks managed by different administrative entities. Security and service management can be incorporated as part of location management procedures.

Mobility Management Functional Entities   Location management is based on the notions of permanent and temporary ATM addresses. A permanent ATM address is a location-invariant, unique address assigned to each MT. As the MT attaches to different points in a WATM network, it may be assigned different temporary ATM addresses. Both permanent and temporary addresses are routable addresses. The permanent address is allocated from the address space of the home service provider or network. Similarly, the temporary address is assigned from the address space of the visited network. The network entities that implement the mobility management functions are:

- *Location Server (LS)*   This is a logical entity maintaining the database of associations between the user identity and permanent and temporary ATM enabled service areas (AESAs) of the corresponding MT. The LS may also keep track of service-specific information for each MT if this capability is implemented.

- *Authentication Server (AUS)* This is a logical entity maintaining a secure database of authentication and privacy-related information for each user. The AUS may physically be a part of the LS.
- *Mobile Terminal (MT)* The MT is required to execute certain functions to initiate location updates and participate in authentication and privacy protocols.
- *End-System Mobility-Supporting ATM Switch (EMAS)* EMAS's are required to identify connection set-up messages destined to MTs and invoke location resolution functions. These EMASs, for instance, could be gateway EMASs to an MATM network. Such EMASs must have the ability to redirect a connection setup message. Additionally, each edge EMAS (EMAS-E) must be able to maintain a local cache of MT address associations.

Reference Configuration The mobility management reference configuration is shown in Figure 20.9. Specifically, the figure depicts two mobility-enhanced ATM (MATM) domains, each with its own set of LS and AUS entities. A single domain may have more than one LS or AUS entity. Alternatively, a single LS or AUS entity may serve an entire domain. A mapping exists in the EMASs that allows them to seek the appropriate LS or AUS as derived from the given service identifier. The functional interfaces between various entities are shown in the figure.

Local Mobility and Roaming Scenarios The reference configuration in Figure 20.9 depicts support for user and terminal mobility within and across independently managed WATM networks. Mobility support within a single WATM network domain is referred to as local mobility support. Independent WATM networks may be interconnected via backbone fixed networks thereby enabling wide-area mobility or roaming of MTs. Support for roaming typically involves the implementation of interprovider roaming agreements, methods for remote authentication of users, accounting, and the like. These aspects are not described here. Rather, only location management for roaming users is considered.

Radio Layer Requirements for Location Management Areas To implement location management (LM), it is necessary to define location areas that are radio coverage regions with a common ATM network prefix. An MT within a location area is reachable with a temporary ATM address whose network prefix is the same as that of the location area. When the MT moves to a different location area, its temporary address changes and a location update from the MT to the network is required to change the address association. The determination of the current location area is a function integrated into the radio layer. Specifically, the radio layer utilizes a downlink control channel to broadcast system-specific information, such as network identification, MT control information (power control, access control, etc), and location area information. The location area information should be available at the MATM layer.

Paging In cellular systems, a single location area can consist of multiple radio coverage areas (or cells). Since the precise location of the MT within the location area is not known, a broadcast page message must be sent on all cells of a location area to reach the MT during call setup. There are two possibilities in wireless ATM systems:

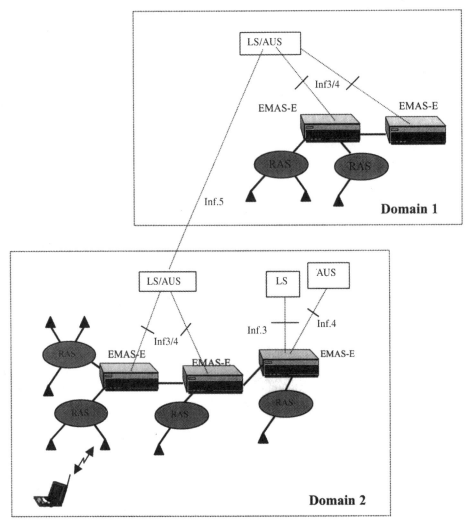

FIGURE 20.9   Mobile ATM reference configuration.

1. Each EMAS-E controls a unique location area: In this case, a location update will not be performed if an MT moves from one ATM point of attachment to another connected to the same EMAS-E. Therefore, the MT must be paged over all the EMAS-E ports during mobile-terminated call setup. The MT can be paged using its unique service identifier(s).

2. Each radio coverage region is considered a separate location area: In this case, there is a unique location area corresponding to each WATM port of the EMAS-E.

Location Management Procedures and Control Flow Identifiers   From a location management perspective, the following identifiers are of significance:

- *Service Identifier* This identifies the home service provider of the user and uniquely establishes the user's identity at the home service provider. The service identifier can be mapped to the addresses of the home LS and AUS during location update and authentication.
- *Permanent MT Address* This is the unique, location-independent address of the MT. This address is assigned from the address space of the home network, and it can be distinguished by border switches in the network as a mobile address.
- *Temporary MT Address* This is the current address of the MT. The temporary address is assigned by the EMAS-E when the MT first associates. The temporary address need not be conveyed to the MT.

Location Update The location update (LU) control flow is shown in Figure 20.10. Here, the MT sends a location update message to the network when it detects a change in location area. The location update message carries the service identifier, the permanent ATM address of the MT and the previous location area the MT visited. The EMAS-E receiving the location update message carries out a procedure to update the location server(s) with current information about the MT's location. Basically, the location update procedure consists of setting up a local table entry with the permanent AESA of the MT and current ATM address and then notifying the visited LS. At the visited LS, the MT's identity is authenticated during the first such access and the service identifier is examined to determine whether the MT is a valid roaming mobile. Depending on the results of the validation, an "LU Confirm" or "LU Reject" message is sent back to the MT, while creating a new or modified entry for that MT. If the MT is a roaming mobile, a corresponding location update is also sent back to the home LS.

Thus, location update is hierarchical. At the home LS, a new or a modified entry for the MT is created, with the service identifier, the permanent address, and the address of the visited network. An "LU Confirm" is also sent to the LS from which the location update message was received.

When the MT is switched off, it is required to send a "Location Cancel" message to the EMAS-E whose location area it is camped on. The EMAS-E then deletes the local location information and also sends a "Location Cancel" to the visited LS. The visited LS then sends a "Location Cancel" to the home LS, if the MT is a roaming mobile. Further details on the above procedure can be found in [28].

User Authentication User authentication, as shown in Figure 20.10, is carried out during location update. As shown, the authentication procedures are carried out between the MT and the EMAS-E, the EMAS-E and the visited LS, and for roaming mobiles, between the visitor location server (VLS) and the home location server (HLS).

Connection Routing, Location Query, and Rerouting The manner in which a mobile-terminated connection is routed depends on whether the MT is in its home network or roaming, and where the connection setup originates. Figure 20.11 depicts the case of an MT in its home network, with the connection setup originating from an external network. The connection setup is routed to the permanent address of the MT, which is indicated as C.2.1.1 (step 1). It is assumed that the permanent address is associated with a specific EMAS in the home network. This EMAS sends a location query to the appropriate LS (step 2). The LS returns the MT's current address (C.1.1, step 3) and the switch reroutes

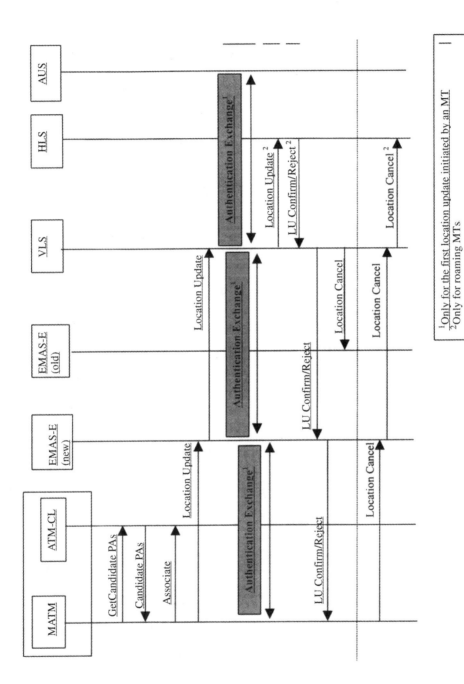

FIGURE 20.10 Location management control flow example.

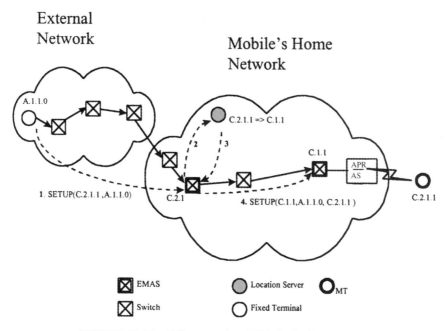

FIGURE 20.11 Call setup when MT is in the home network.

the connection setup to this address (step 4). This message carries an IE indicating the permanent MT AESA (which also indicates that the called party is mobile). In the topology shown, this redirection does not result in a nonoptimal path, but in general this may not be the case. Also, the routing of the connection set-up originating from within the home network is similar. Finally, in this example, the border switch in the home network could do the location resolution if it were an EMAS and the MT addresses were assigned from a separate address space known to all EMASs.

Figure 20.12 depicts the case where the MT is in an external network (roaming scenario). As before, the connection setup is routed to the home EMAS (step 1), which determines that the current address of the MT is B.1.1 (steps 2 and 3). The connection setup is then rerouted to the current location of the MT (step 4). This message carries an IE indicating the permanent AESA of the MT. After receiving this message, the gateway in the visited network recognizes that the called party is a roaming mobile and it queries its LS to obtain the exact location of the MT (steps 5 and 6). Finally, the setup is rerouted to the current location of the MT (step 7). Figure 20.13 summarizes the control flow for mobile-terminated call setup. The location response may indicate the inability of the LS to find the queried information (e.g., location canceled).

Connection Handover Wireless ATM implementations, as well as the standards being developed by the ATM Forum, rely on mobile-initiated handovers whereby the MT is responsible for monitoring the radio link quality and decide when to initiate a handover [28–30]. A handover process typically involves the following steps:

1. *Link Quality Monitoring* When there are active connections, the MT constantly monitors the strength of the signal it receives from each radio port (RP) within range.

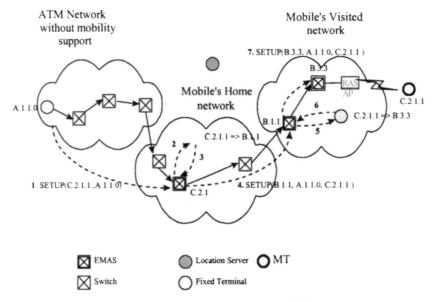

FIGURE 20.12 Call setup when MT is in Visited network.

2. *Handover Trigger* At a given instance, all the connections from/to the MT are routed through the same RP, but deterioration in the quality of the radio link triggers the handover procedure.

3. *Handover Initiation* Once a handover is triggered, the MT initiates the procedure by sending a signal to the edge EMAS with which it is in direct contact. This signal indicates to the EMAS the list of candidate RPs to which active connections can be handed over.

4. *Target RP Selection* The edge EMAS selects one RP as the handover target from the list of candidates sent by the MT. This step may make use of network-specific criteria for spreading the traffic load among various RPs and interaction between the edge EMAS and other EMASs housing the candidate RPs.

5. *Connection Rerouting* After the target RP is selected, the edge EMAS initiates the rerouting of all connections from/to the MT within the MATM network to the target RP. The complexity of this step depends on the specific procedures chosen for rerouting connections, as described next. Due to constraints on the network or radio resources, it is possible that not all connections are successfully rerouted at the end of this step.

6. *Handover Completion* The MT is notified of the completion of handover for one or more active connnections. The MT may then associate with the new RP and begin sending/receiving data over these connections.

Specific implementations may differ in the precise sequence of events during handover. Furthermore, the handover complexity and capabilities may be different. For instance, some systems may implement *lossless* handover whereby cell loss and missequencing of cells are avoided during handover by buffering cells inside the network [26]. Two types of handovers may be distinguished:

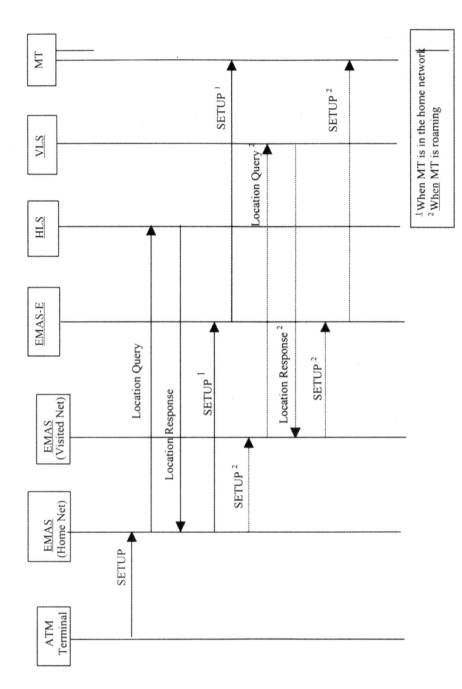

FIGURE 20.13   Control flow for mobile terminated call setup.

425

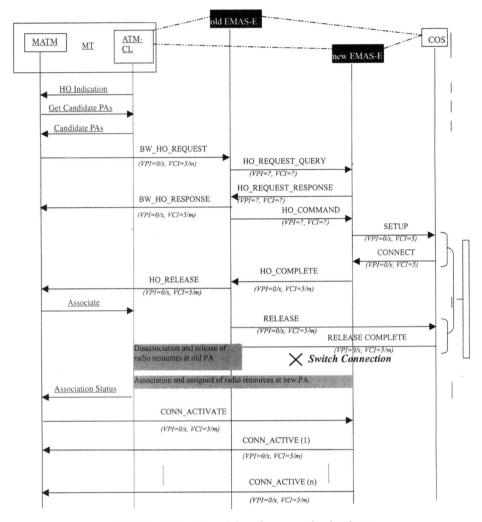

FIGURE 20.14   Control flow for connection handover.

- *Backward Handover*   The MT initiates handover through the current RP it is connected to, before loss of radio contact. This is the normal scenario.
- *Forward Handover*   The MT loses connectivity to the current RP due to a sudden degeneration of the radio link. It then chooses a new RP and initiates handover of active connections.

In the following, the control flow for backward handover is described briefly. Further details on this procedure, as well as the alternate procedure for forward handover can be found in [28].

Backward Handover Description   Figure 20.14 illustrates the control flow during backward handover. The following events take place:

1. The radio layer at the MT indicates to the MATM layer that backward handover is required. The MATM layer, in response, begins the backward handover procedure.

2. The MATM layer requests the radio layer for the identity of the candidate points of attachements (PAs). Each of these is identified by a unique ATM address.

3. The MATM layer then initiates backward handover by sending a BW_HO_ REQUEST message to the EMAS-E (referred to as the "old" EMAS-E). This message may include the prioritized list of call references and a prioritized list of candidate Pas.

4. Upon receiving the BW_HO_REQUEST message, the old EMAS-E considers the list of candidate PAs, either selects a target PA (if locally managed by that EMAS-E) or sends an HO_REQUEST_QUERY message to switch supporting the candidate PAs. This "new" EMAS-E then invokes local procedures to select a target PA.

5. The EMAS-E then sends an HO_REQUEST_RESPONSE message to the old EMAS-E over the handoff control channel (HCC), indicating the target PA chosen and the call references of connections that can be handed over.

6. After receiving HO_REQUEST_RESPONSE messages from all the queried EMAS-Es, the old EMAS-E selects a target PA (perhaps under a new EMAS-E). The EMAS-E then sends a BW_HO_RESPONSE message to the MT, either indicating the target PA (or the failure to select a target PA).

7. The old EMAS-E sends an HO_COMMAND to the EMAS-E managing the target PA (referred to as the "new" EMAS-E). This message contains the address of the target PA, the identities of all connections to be handed over, and the ATM traffic and QoS parameters of the connections. The source EMAS-E also indicates the address of the cross-over switch (COS) and the COS selection procedure code. The COS is the EMAS in the current connection path to which the new EMAS-E establishes a partial connection segment. The COS can be selected in different ways depending on how rerouting is performed during handover [31]:

   (a) *VC Extension*   In this case, a connection segment is established between the old and new EMAS-Es. The COS therefore is the old EMAS-E.

   (b) *Anchor-Based Rerouting*   In this case, an EMAS in the connection path is permanently chosen as the COS and used in all handovers involving the connection. Such a COS can be established via configuration or selected when the connection is initially set up. During handover, a new connection segment is established from the new EMAS-E to the COS and the existing segment between the COS and the old EMAS-E is released.

   (c) *Dynamic COS Discovery*   In this case, the COS is selected dynamically as described in [28].

8. Upon receiving the HO_COMMAND message, the new EMAS-E interacts with the appropriate radio resource manager to allocate radio resources for the connections being handed over. It then utilizes the COS address in the message as the destination, determines a connection route, and generates a PNNI SETUP message for each connection being handed over. Each SETUP message also contains the COS selection procedure code.

9. If the setup was successful, the new EMAS-E sends an HO_COMPLETE message to the old EMAS-E with the identity of the concerned connection.

10. The old EMAS-E waits to receive an HO_COMPLETE (or an HO_FAILURE) message for each connection being handed over. When this condition is satisfied or if a timer expires, the old EMAS-E sends an HO_RELEASE message to the MT. This message indicates to the MT that it can dissociate itself from current PA and associate with the new PA.

11. The old EMAS-E sends a partial RELEASE for each connection. This releases the connection segment between the old EMAS-E and the COS for each connection.

12. Upon receiving the RELEASE message, the COS responds with a RELEASE COMPLETE message and switches the connection to the new connection segment.

13. Meanwhile, the MT dissociates from old PA and associates with the new PA. It then sends a CONN_ACTIVATE message to the new EMAS-E.

14. Upon receiving the CONN_ACTIVATE message from the MT, new EMAS-E responds with a CONN_ACTIVE message. The CONN_ACTIVE message indicates the identities of connections that have been handed over so far along with their new VPI/VCI.

The description above has left open some questions. Among them: What mechanisms are used to reliably exchange control messages between the various entities that take part in handover? What actions are taken when network or radio link failures occur during handover? How can lossless handover be included in the control flow? What effect does transient disruptions in service during handover have on application behavior, and what are the performance impacts of signaling for handover? The short answers to these questions are: reliability can be incorporated by implementing a reliable transfer protocol for those control messages that do not already use such a transport (SETUP, RELEASE, etc., do, but HO_REQUEST_QUERY e.g., requires attention). Actions taken during network failures require further analysis, but forward handover can be used to recover from radio link failures during handover. Lossless handover requires inband signaling within each connection and buffering in the network. Details on this can be found in [26]. The effect of transient disruptions on applications can be minimal, depending on how rerouting is implemented during handover. This is described in detail in [32]. Finally, some of the performance issues related to mobility management are investigated in [33].

20.5 QoS CONTROL iN WIRELESS ATM

Quality-of-service (QoS) support in wireless ATM involves several considerations beyond those addressed in earlier work on conventional ATM networks. In particular, QoS control in a WATM scenario must deal with the following additional issues:

- Potentially limited bandwidth on the shared-medium radio access link
- Varying radio access link and core network traffic conditions during handoff/path rerouting associated with terminal mobility
- Increased heterogeneity in terms of display, throughput, and computational limitations of mobile terminals [laptops, personal digital assistants (PDA), telephones, etc.]

The above factors imply the need for a robust and scalable QoS framework that permits some degree of variation in delivered service quality during the course of a connection [34,

35]. Such a "soft QoS" control framework has been proposed in [36] in context of nonmobile ATM and IP networks, as a practical means for achieving high statistical multiplexing gains with nonstationary multimedia and video traffic. This approach is based on the concept of a "VBR+" service class that permits dynamic renegotiation of usage parameter control (UPC) values during the course of a connection. The resulting end-to-end QoS framework for a WATM network is schematically illustrated in Figure 20.14.

20.5.1 The Soft QoS Model

It is observed that the QoS model in Figure 20.15 requires some degree of application scalability in order to operate robustly under varying conditions. This can be achieved if applications are designed to exhibit "soft" user satisfaction versus bit-rate profiles typical of the "S-curves" shown on the right of Figure 20.14. Many real-time media sources (including compressed audio and video) have been shown to exhibit such soft degradation [37, 38], so that the principle can be used in most practical cases. Once the S-curve for an application component has been found, it is possible to design a QoS controller at network entities that takes into account specified parameters such as a "minimum satisfaction index" to allocate bandwidth between active connections, and to block new connection requests when congestion is anticipated.

Within the network elements, a soft-QoS control mechanism utilizes the softness profiles to allocate bandwidth while maintaining uniform satisfaction among contending applications [39]. Two key components of the soft-QoS control framework are: dynamic bandwidth allocation and call admission control (CAC). Dynamic bandwidth allocation is used to match the variable bit rate requirements of multimedia applications to available network resources while the connection is in progress. CAC checks the availability of resources within a coverage area at the time of connection establishment. A new connection is accepted if sufficient resources are available for the connection to operate within the specified service contract parameters, and are otherwise blocked.

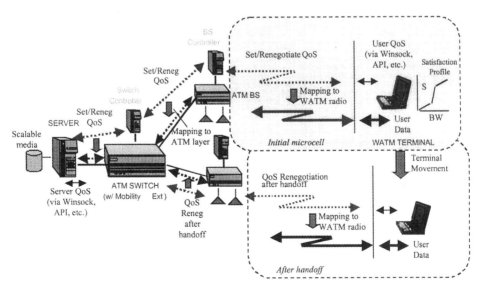

FIGURE 20.15 Overview of soft QoS framework for wireless ATM systems.

20.5.2   Dynamic Bandwidth Allocation

Dynamic bandwidth allocation is used within each wireless ATM microcell to maintain high utilization while providing soft QoS guarantees. MTs can signal the base station to request a bandwidth allocation appropriate for their traffic needs. In addition, base stations can use a bandwidth reallocation algorithm to reassign bandwidth among active connections when congestion occurs; connections that have the highest satisfaction index are selected as primary candidates to get bandwidth from in order to maximize the minimum satisfaction index of all active connections. An iterative algorithm is used, based on the softness profile of active connections to determine how much bandwidth to reallocate from each connection to maintain the satisfaction of active connections above minimum. Details of the algorithm can be found in [39].

The message overhead and processing requirement of dynamic bandwidth allocation has been studied via simulation. For an example with 10 active connections at a base station, each with bandwidth between 0.5 and 2 Mbps, resulting in 0.6 utilization on a 25-Mbps WATM microcell, the bandwidth negotiation rate is about 4 signals/min/ connection, or 40 requests/min/cell. The impact of dynamic bandwidth allocation on QoS is statistically captured by the low satisfaction rate (LSR) and the low satisfaction mean duration (LSMD). LSR indicates the probability that bandwidth allocation becomes insufficient to meet the QoS contract (i.e., the satisfaction falls below the minimum defined in the softness profile). LSMD gives the expected duration of quality outage. For 0.6 utilization, LSR = 5% and LSMD = 3 s. Such parameters can be used to specify the service-level agreement (SLA) for multimedia in a WATM system. At about 60% network utilization, the SLA will specify 30% blocking probability with 95% success rate for bandwidth renegotiation; bandwidth renegotiation failure results on average of 3 s of below minimum QoS.

20.5.3   Mobile Call Admission Control

In broadband wireless networks, the impact of establishing a new connection on other ongoing connections depends not only on the connection's traffic descriptor and QoS expectations, but also on the terminals' mobility. In this scenario, QoS guarantees may be violated after a handoff into another wireless cell even if an appropriate CAC mechanism was used locally within the cell where the connection originated. Thus, if bandwidth is reserved only at the cell where the connection originated, session QoS cannot be guaranteed. Advance reservation of bandwidth in all adjacent radio cells has been proposed as a possible solution, but this may not be practical due to a corresponding drop in efficiency by a factor equal to the frequency reuse number (typically 5 to 10 for microcellular systems).

A number of proposals for bandwidth allocation mechanisms based on terminal mobility profiling have been discussed in the literature [40]. Most of the proposals are adequate in limited coverage areas for short-term, slow-moving connections, but fail to provide session-level QoS for long-term multimedia connections. For example, some require prediction of mobile trajectory and/or regular recomputing of the mobile's bandwidth requirements as it roams from one cell to another. In [41], a scalable CAC for mobile networks is presented that borrows from the temporal statistical multiplexing principles that allow efficient bandwidth allocation for variable bit rate connections. The proposal models the spatial mobility of a terminal from one cell to another as an ON–OFF traffic source. In the ON state (the mobile is within the cell), with a bandwidth requirement R, and in the OFF state (the mobile is outside the cell), the bandwidth requirement is 0. This model allows decomposition of the global CAC problem into a series of stationary CAC algorithms distributed over all cells within the coverage area, each using the bandwidth requirement derived from the ON–OFF model.

To apply stationary methods, it is necessary to compute the time a mobile is expected to spend in the ON and OFF states. If a coverage area comprises N cells, $C_1, \ldots, C_N$ (see Figure 20.16), in every interval T, there is a probability p_{ij} for a mobile to leave C_i for C_j. The probabilities p_{ij} are easily observable and collectable from the network. Within the WATM system they are called normalized handoff rate (NHR), and they represent the number of handoffs from C_i to C_j divided by the number of active connections in C_i. NHRs can be exchanged between cells on a regular basis over preestablished VCs to facilitate the calculation of the model's stationary probabilities. NHRs can also be calculated off-line, for several temporal periods (time of day, weekend, weekday, etc). The model can take into account special traffic patterns that could develop to spatial (roads, buildings, tracks) and temporal (rush hour) effects. The NHRs are the basis to define a Markov model with N states corresponding to the cells in the coverage region. The stationary probabilities p_k are the solutions of the Markov chain driven by the NHRs, and say that during a long session, the mobile terminal spends on average p_k of its session time within the cell C_k and $1 - p_k$ in other cells.

Since the mobile is expected to spend p_k of its time within the cell C_k, the average bandwidth requirement from that cell is Rp_k. Any cell C_k within the covered area is expected to need, on average, Rp_k units of bandwidth to accommodate the connection at the desired QoS. Conceptually, we say that the connection with a bandwidth requirement R, "casts a shadow" on other cells of the covered region. The "shadow" cast over the coverage region is represented by the vector of coordinates Rp_k, where $p_1 + \cdots + p_N = 1$. Thus, the shadow approach is based on aggregated statistical information measured at the base stations (handoff rates between cells, arrival and departure rates) and does not require additional knowledge on terminal's mobility patterns.

When a base station receives a request for a connection setup, it first checks if sufficient resources are available within that cell based on the connection's traffic descriptor and QoS requirements. The request is rejected (the connection is blocked) if there are insufficient

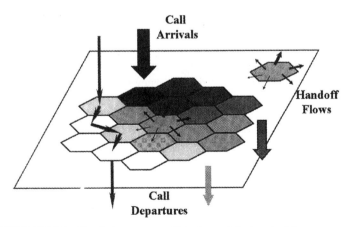

FIGURE 20.16   Simulation events within the mobile terminals' coverage area.

resources. Otherwise we use the model to estimate the overall handoff dropping probability over the mobile terminal's coverage area. The connection is accepted if the resulting probability is below the threshold value, H.

The performance of this type of mobile CAC has been evaluated using simulations. The coverage area simulated consisted of 19 cells (see Figure 20.16), and the traffic had the following characteristics:

- Connection's bandwidth uniformly distributed between 0.5 and 1.5 Mbps; 25 active connections on average for a WATM microcell.
- Average session duration is 15 min.
- Handoff rate of $1/150$ s; 5 handoffs per session.

The performance of this mobile CAC was studied for various maximum targeted handoff drop probabilities (H), and measured blocking probability and average utilization (defined as the ratio of total bandwidth requirements of active connections to the total radio capacity of the coverage area). The M-CAC performance is benchmarked against a reference CAC that accepts all connections, provided there is sufficient bandwidth in the originating cell. Thus, some connections will be terminated when the terminal moves to a congested cell with insufficient bandwidth to support the connections' QoS requirements. The reference CAC has the largest handoff drop probability but the highest bandwidth utilization. Table 20.1 shows some simulation results comparing M-CAC with the reference local CAC. It is observed that the local CAC tends to accept more new calls (hence the lower blocking probability and utilization) at the expense of higher forced termination probability. Since forced termination probability is a key user performance measure, mechanisms such as the global mobile CAC discussed here are important for maintaining service QoS.

20.6   CONCLUDING REMARKS

This chapter has presented a review of wireless ATM technology, covering protocol architecture, WATM radio access, mobile ATM core network, and QoS control. Based on current trends toward multimedia and portability in the computing industry, it is reasonable to expect that broadband wireless technologies such as wireless ATM will play an increasingly important role in the future network infrastructure. Obviously, it is too early to predict the precise role of wireless/mobile ATM relative to competing IP-based alternatives. In any case, we believe that wireless ATM networks (or variations thereof) will prove an excellent foundation for building networks that support QoS and dynamic

TABLE 20.1   Simulation Results Comparing M-CAC with Reference Local CAC

	Local CAC (%)	M-CAC ($H = 1\%$) (%)	M-CAC ($H = 2\%$) (%)
Blocking	8	35	30
Utilization	90	50	67
5-Hop forced termination prob.	40	5	10

mobility, even in IP-oriented service scenarios. Availability of wireless ATM and other broadband wireless products at the right cost/performance point should stimulate future development of a new generation of mobile multimedia terminals and applications.

REFERENCES

1. D. Raychaudhuri and N. Wilson, "Multimedia Personal Communication Networks: System Design Issues," 3rd WINLAB Workshop on 3rd Gen. Wireless Information Networks, April 1992, pp. 259–288. (also in *Wireless Communications*, J. M. Holtzman and D. J. Goodman, Eds., Kluwer Academic, 1993, pp. 289–304).

2. D. Raychaudhuri and N. Wilson, "ATM Based Transport Architecture for Multiservices Wireless Personal Communication Network," *IEEE J. Selected Areas in Comm.*, pp. 1401–1414, Oct. 1994.

3. J. Porter and A. Hopper, "An Overview of the ORL Wireless ATM System," IEEE ATM Workshop, Wash. D.C., Sept. 30–Oct. 1, 1995.

4. K. Y. Eng et. al., "BAHAMA: A Broadband Ad-Hoc Wireless ATM Local Area Network," *Proc. ICC' 95*, pp. 1216–1123, 1995.

5. M. Umehira et. al., "An ATM Wireless Access System for Tetherless Multimedia Services," *Proc. ICUPC' 95*, Tokyo, Nov. 1995.

6. D. Raychaudhuri, et al., "WATMnet: A Prototype Wireless ATM System for Multimedia Personal Communication," *IEEE J. Selected Areas in Comm.*, pp. 83–95, Jan. 97.

7. "Charter, Scope and Work Plan for Proposed Wireless ATM Working Group," ATM Forum, Anchorage, April 1996, ATM Forum/96-0530/PLEN.

8. ETSI-BRAN, "BRAN: Requirements and Architectures for BRAN," WG1 TD1 Technical Draft Report, 1997.

9. "Multimedia Mobile Access Communication," MMAC Systems Promotion Council activity brochure, Association of Radio Industries and Businesses (ARIB), Tokyo, Japan, 1997.

10. U.S. Federal Communications Communication, "Operation of Unlicensed NII Devices in the 5 Ghz Range," ET Docket 96-102, Jan. 1997.

11. ETSI-RES10, "High Performance Radio Local Area Network (HIPERLAN)," Draft Standard, Sophia Antipolis, France, 1995.

12. D. Raychaudhuri, "Wireless ATM Networks: Architecture, System Design & Prototyping," *IEEE Personal Comm. Mag.*, 1996, pp. 42–49, Aug. 1996.

13. Special Issue on Wireless ATM, *IEEE J. Selected Areas in Comm.*, Jan. 1997.

14. D. Raychaudhuri, "Wireless ATM: Technology Status and Future Directions," *IEEE Proc.*, Vol. 87, No. 10, Oct. 1999.

15. D. Raychaudhuri, "Current Topics in Wireless and Mobile ATM Networks: QoS Control, IP Support and Legacy Service Integration," *Proc. PIMRC'98*, Boston, MA, Sept. 1998.

16. R. Valenzuela, "Performance of Quadrature Amplitude Modulation for Indoor Radio Communications," *IEEE Trans. on Commun.*, Vol. COM-35, No. 11, pp. 1236–1238, Nov. 1987.

17. J. Tellado-Mouerelo, E. Wesel, and J. Cioffi, "Adaptive DFE for GMSK in Indoor Radio Channels," *IEEE J. Selected Areas in Comm.*, pp. 492–501, April 1996.

18. L. Cimini, "Analysis and Simulation of a Digital Mobile Channel Using Orthogonal Frequency Division Multiplexing," *IEEE Trans. on Comm.*, pp. 665–675, July 1985.

19. G. Falk, et. al., "Integration of Voice and Data in the Wideband Packet Satellite Network," *IEEE J. Selected Areas in Comm.*, Vol. SAC-1, No. 6, pp. 1076–1083, Dec. 1983.

20. S. Nanda, D. J. Goodman, and U. Timor, "Performance of PRMA: A Packet Voice Protocol for Cellular Systems," *IEEE Trans. Veh. Tech*, Vol. VT-40, pp. 584–598, 1991.

21. N. Wilson, R. Ganesh, K. Joseph, and D. Raychaudhuri, "Packet CDMA vs. Dynamic TDMA for Access Control in an Integrated Voice/Data PCN," *IEEE J. Selected Areas in Comm.*, pp. 870–884, Aug. 1993.

22. P. Narasimhan et al., "Design and Performance of Radio Access Protocols in WATMnet, a Prototype Wireless ATM Network," Proc. Winlab. Workshop, April 1997.

23. C. A. Johnston et al., " Architecture and Implementation of Radio Access Protocols Wireless ATM Networks," *Proc. ICC'98*, Atlanta, GA, June 1998.

24. S. K. Biswas, D. J. Reininger, and D. Raychaudhuri, "Bandwidth Allocation for VBR Video in Wireless ATM Networks," Proc. ICC'97, Montreal, CA, June 1997.

25. H. Xie, P. Narasimhan, R. Yuan, and D. Raychaudhuri, "Data Link Control Protocols for Wireless ATM Access Channels," Proc. ICUPC'95, Tokyo, Nov. 1995.

26. A. Acharya, B. Rajagopalan, and D. Raychaudhuri, "Mobility Management in Wireless ATM Networks," *IEEE Commun. Mag.*, pp. 100–109, Nov. 1997.

27. D. Brown, "Techniques for Privacy and Authentication in Personal Communication Systems," *IEEE Personal Comm.*, Aug., 1985.

28. B. Rajagopalan, Ed., ATM Forum BTD-WATM-01.12, Draft Wireless ATM Capability Set 1 Specification, Sept. 1999.

29. H. Mitts et al., "Microcellular Handover for WATM Release 1.0: Proposal for Scope and Terms of Reference," ATM Forum 97-0226, April 1997.

30. J. Ala-Laurila and G. Awater, "The Magic WAND: Wireless ATM Network Demonstrator," *Proc. ACTS Mobile Summit '97*, Denmark, Oct., 1997.

31. C. K. Toh, "Crossover Switch Discovery in Wireless ATM LAN's," *ACM Mobile Networks and Nomadic Appl. J.*, Vol. 1, No. 4, 1(4), 1996.

32. P. Mishra and M. Srivastava, "Effect of Connection Rerouting on Application Performance in Mobile Networks," *Proc. IEEE Conf. on Distributed Computing Syst.*, May, 1997.

33. G. P. Pollini, K. S. Meier-Hellstern, and D. J. Goodman, "Signaling Traffic Volume Generated by Mobile and Personal Communications," *IEEE Commun. Mag.*, June, 1995.

34. S. Singh, "Quality of Service Guarantees in Mobile Computing," *Comp. Commun.*, Vol. 19, pp. 359–371, 1996.

35. A. Campbell, C. Aurrecoechea, and L. Hauw, "A Review of QoS Architectures, *ACM Multimedia Syst. J.*, 1996.

36. D. Reininger, D. Raychaudhuri, and J. Hui, "Dynamic Bandwidth Allocation for VBR Video over ATM Networks," *IEEE JSAC*, Vol. 14, No. 6, pp. 1076–1086, Aug. 1996.

37. J. G. Lourens, H. H. Malleson, and C. C Theron, "Optimization of Bit-Rates for Digitally Compressed Television Services as a Function of Acceptable Picture Quality and Picture Complexity," *Proc. IEE Colloq. Digitally Compressed TV by Satellite*, 1995.

38. E. Nakasu et al., "A Statistical Analysis of MPEG-2 Picture Quality for Television Broadcasting," *SMPTE J.*, pp. 702–11.

39. D. Reininger, R. Izmailov, B. Rajagopalan, M. Ott, and D. Raychaudhuri, "Soft QoS control in the WATMnet Broadband Wireless System," *IEEE Personal Commun.*, pp. 34–43, Feb. 1999.

40. R. Jain and E. Knightly, "A Framework for Design and Evaluation of Admission Control Algorithms in Multi-Service Mobile Networks," *Proc. IEEE INFOCOM'99*, 1999.

41. D. Reininger and R. Izmailov, "Admission and Bandwidth Allocation for Soft-QoS Guarantees in Mobile Multimedia Networks," *Proc. IEEE WCNC'99*, New Orleans, September, 1999.

ABOUT THE EDITORS

Mansoor Shafi, *IEEE Fellow*, is Principal Advisor of Wireless Systems at Telecom New Zealand. He has been employed by Telecom NZ for more than 20 years and has extensively published on many subjects relating to the physical layer of communication systems. He serves as an NZ delegate to the meetings of ITU-R TG 8/1 and WP 8F, the specialist groups responsible for the preparation of IMT 2000 standards. He was awarded the IEEE Comsoc Public Service Award in 1992.

Shigeaki Ogose, *IEEE Senior Member*, joined NTT in 1977. Since then, he has been engaged in the research and development of the digital mobile communication systems including PHS. He has been with the Faculty of Engineering, Kagawa University, since 1998. He received his B.S.E.E. and M.S.E.E. degrees from Hiroshima University, Japan, in 1975 and 1977, respectively. He received Ph.D. degree from Kyoto University, Japan, in 1986. Professor Ogose is a member of the IEICE.

Takeshi Hattori, *IEEE Member*, joined NTT in 1974. Since then, he has been engaged in research and development of cellular systems, paging systems, maritime systems and advanced cordless systems. He has proposed the concept of Personal Handy Phone System (PHS). In 1981, Dr. Hattori was awarded the IEEE Vehicular Technology Society Paper of the Year. He has been with the Faculty of Science and Technology, Sophia University, since 1997. He received his B.S.E.E., M.S.E.E. and Ph.D. degrees from the University of Tokyo, Japan, in 1969, 1971 and 1974, respectively. Professor Hattori is a member of the IEICE.